Distributional Reinforcement Learning

Adaptive Computation and Machine Learning
Francis Bach, editor

A complete list of books published in the Adaptive Computation and Machine Learning series appears at the back of this book.

Distributional Reinforcement Learning

Marc G. Bellemare, Will Dabney, and Mark Rowland

The MIT Press
Cambridge, Massachusetts
London, England

The MIT Press would like to thank the anonymous peer reviewers who provided comments on drafts of this book. The generous work of academic experts is essential for establishing the authority and quality of our publications. We acknowledge with gratitude the contributions of these otherwise uncredited readers.

This book was set in LaTeX by the authors. Printed and bound in the United States of America.

Library of Congress Cataloging-in-Publication Data

Names: Bellemare, Marc G., author. | Dabney, Will, author. | Rowland, Mark (Research scientist), author.
Title: Distributional reinforcement learning / Marc G. Bellemare, Will Dabney, Mark Rowland.
Description: Cambridge, Massachusetts : The MIT Press, [2023] | Series: Adaptive computation and machine learning | Includes bibliographical references and index.
Identifiers: LCCN 2022033240 (print) | LCCN 2022033241 (ebook) | ISBN 9780262048019 (hardcover) | ISBN 9780262374019 (epub) | ISBN 9780262374026 (pdf)
Subjects: LCSH: Reinforcement learning. | Reinforcement learning–Statistical methods.
Classification: LCC Q325.6 .B45 2023 (print) | LCC Q325.6 (ebook) | DDC 006.3/1–dc23/eng20221102
LC record available at https://lccn.loc.gov/2022033240
LC ebook record available at https://lccn.loc.gov/2022033241

10 9 8 7 6 5 4 3 2 1

Contents

Preface

The history of this book begins one evening in November 2016, when after an especially unfruitful day of research, Will and Marc decided to try a different approach to reinforcement learning. The idea took inspiration from the earlier "Compress and Control" algorithm (Veness et al. 2015) and recent successes in using classification algorithms to perform regression (van den Oord et al. 2016), yet was unfamiliar, confusing, exhilarating. Working from one of the many whiteboards in DeepMind offices at King's Cross, there were many false starts and much reinventing the wheel. But eventually C51, a distributional reinforcement learning algorithm, came to be. The analysis of the distributional Bellman operator proceeded in parallel with algorithmic development, and by the ICML 2017 deadline, there was a theorem regarding the contraction of this operator in the Wasserstein distance and state-of-the-art performance at playing Atari 2600 video games. These results were swiftly followed by a second paper that aimed to explain the fairly large gap between the contraction result and the actual C51 algorithm. The trio was completed when Mark joined for a summer internship, and at that point the first real theoretical results came regarding distributional reinforcement learning algorithms. The QR-DQN, Implicit Quantile Networks (IQN), and expectile temporal-difference learning algorithms then followed. In parallel, we also began studying how one could theoretically explained just *why* distributional reinforcement learning led to better performance in large-scale settings; the first results suggested said that it should not, only deepening a mystery that we continue to work to solve today.

One of the great pleasures of working on a book together has been to be able to take the time to produce a more complete picture of the scientific ancestry of distributional reinforcement learning. Bellman (1957b) himself expressed in passing that quantities other than the expected return should be of interest; Howard and Matheson (1972) considered the question explicitly. Earlier studies focused on a single characteristic of the return distribution, often a criterion to be optimized: for example, the variance of the return (Sobel 1982). Similarly, many results in risk-sensitive reinforcement learning have focused on optimizing a specific measure of risk, such as variance-penalized expectation (Mannor and Tsitsiklis 2011) or conditional-value-at-risk (Chow and Ghavamzadeh 2014). Our contribution to this vast body of work is perhaps to treat these criteria and characteristics in a more unified manner, focusing squarely on the return distribution as the main object of interest, from which everything can be derived. We see signs of this unified treatment paying off in answering related questions (Chandak et al. 2021). Of course, we have only been able to get there because of relatively recent advances in the study of probability metrics (Székely 2002; Rachev et al. 2013), better tools with which to study recursive distributional relationships (Rösler 1992; Rachev and Rüschendorf 1995), and key results from stochastic approximation theory.

Our hope is that, by providing a more comprehensive treatment of distributional reinforcement, we may pave the way for further developments in sequential decision-making and reinforcement learning. The most immediate effects should be seen in deep reinforcement learning, which has since that first ICML paper used distributional predictions to improve performance across a wide variety of problems, real and simulated. In particular, we are quite excited to see how risk-sensitive reinforcement learning may improve the reliability and effectiveness of reinforcement learning for robotics (Vecerik et al. 2019; Bodnar et al. 2020; Cabi et al. 2020). Research in computational neuroscience has already demonstrated the value of taking a distributional perspective, even to explain biological phenomena (Dabney et al. 2020b). Eventually, we hope that our work can generally help further our understanding of what it means for an agent to interact with its environment.

In developing the material for this book, we have been immensely lucky to work with a few esteemed mentors, collaborators, and students who were willing to indulge us in the first steps of this journey. Rémi Munos was instrumental in shaping this first project and helping us articulate its value to DeepMind and the scientific community. Yee Whye Teh provided invaluable advice, pointers to the statistics literature, and lodging and eventually brought the three of us together. Pablo Samuel Castro and Georg Ostrovski built, distilled, removed technical hurdles, and served as the voice of reason. Clare Lyle, Philip Amortila, Robert Dadashi, Saurabh Kumar, Nicolas Le Roux, John Martin, and Rosie Zhao helped answer a fresh set of questions that we had until then lacked the formal language to describe, eventually creating more problems than answers – such is the way of science. Yunhao Tang and Harley Wiltzer gracefully accepted to be the first consumers of this book, and their feedback on all parts of the notation, ideas, and manuscript has been invaluable.

We are very grateful for the excellent feedback – narrative and technical – provided to us by Adam White and our anonymous reviewers, which allowed us to make substantial improvements on the original draft. We thank Rich Sutton, Andy Barto, Csaba Szepesvári, Kevin Murphy, Aaron Courville, Doina Precup, Prakash Panangaden, David Silver, Joelle Pineau, and Dale Schuurmans, for discussions on book writing and serving as role models on taking an effort larger than anything else we had previously done. We appreciate the technical and conceptual input of many of our colleagues at Google, DeepMind, Mila, and beyond: Bernardo Avila Pires, Jason Baldridge, Pierre-Luc Bacon, Yoshua Bengio, Michael Bowling, Sal Candido, Peter Dayan, Thomas Degris, Audrunas Gruslys, Hado van Hasselt, Shie Mannor, Volodymyr Mnih, Derek Nowrouzezahrai, Adam Oberman, Bilal Piot, Tom Schaul, Danny Tarlow, and Olivier Pietquin. We further thank the many people who reviewed parts of this book and helped fill in some of the gaps in our knowledge: Yinlam Chow, Erick Delage, Pierluca D'Oro, Doug Eck, Amir-massoud Farahmand, Jesse Farebrother, Chris Finlay, Tadashi Kozuno, Hugo Larochelle, Elliot Ludvig, Andrea Michi, Blake Richards, Daniel Slater, and Simone Totaro. We further thank Vektor Dewanto, Tyler Kastner, Karolis Ramanauskas, Rylan Schaeffer, Eugene Tarassov, and Jun Tian for their feedback on the online draft and the COMP-579 students at McGill University for beta-testing our presentation of the material. We were lucky to perform this research within DeepMind and Google Brain, which provided support both moral and material and inspiration to take on ever larger challenges. Finally, we thank Francis Bach, Elizabeth Swayze, Matt Valades, and the team at MIT Press for championing this work and making it a possibility.

Marc gives further thanks to Judy Loewen, Frédéric Lavoie, Jacqueline Smith, Madeleine Fugère, Samantha Work, Damon MacLeod, and Andreas Fidjeland, for support along the scientific journey, and to Lauren Busheikin, for being an incredibly supportive partner. Further thanks go to CIFAR and the Mila academic community for providing the fertile scientific ground from which the writing of this book began.

Will wishes to additionally thank Zeb Kurth-Nelson and Matt Botvinick for their patience and scientific rigor as we explored distributional reinforcement learning in neuroscience; Koray Kavukcuoglu and Demis Hassabis for their enthusiasm and encouragement surrounding the project; Rémi Munos for supporting our pursuit of random, risky research ideas; and Blair Lyonev for being a supportive partner, providing both encouragement and advice surrounding the challenges of writing a book.

Mark would like to thank Maciej Dunajski, Andrew Thomason, Adrian Weller, Krzysztof Choromanski, Rich Turner, and John Aston for their supervision and mentorship, and his family and Kristin Goffe for all their support.

1 Introduction

A hallmark of intelligence is the ability to adapt behavior to reflect external feedback. In reinforcement learning, this feedback is provided as a real-valued quantity called the *reward*. Stubbing one's toe on the dining table or forgetting soup on the stove are situations associated with negative reward, while (for some of us) the first cup of coffee of the day is associated with positive reward.

We are interested in agents that seek to maximize their cumulative reward – or *return* – obtained from interactions with an *environment*. An agent maximizes its return by making decisions that either have immediate positive consequences or steer it into a desirable state. A particular assignment of reward to states and decisions determines the agent's objective. For example, in the game of Go, the objective is represented by a positive reward for winning. Meanwhile, the objective of keeping a helicopter in flight is represented by a per-step negative reward (typically expressed as a cost) proportional to how much the aircraft deviates from a desired flight path. In this case, the agent's return is the total cost accrued over the duration of the flight.

Often, a decision will have uncertain consequences. Travelers know that it is almost impossible to guarantee that a trip will go as planned, even though a three-hour layover is usually more than enough to catch a connecting flight. Nor are all decisions equal: transiting through Chicago O'Hare may be a riskier choice than transiting through Toronto Pearson. To model this uncertainty, reinforcement learning introduces an element of chance to the rewards and to the effects of the agent's decisions on its environment. Because the return is the sum of rewards received along the way, it too is random.

Historically, most of the field's efforts have gone toward modeling the mean of the random return. Doing so is useful, as it allows us to make the right decisions: when we talk of "maximizing the cumulative reward," we typically mean "maximizing the expected return." The idea has deep roots in probability, the law of large numbers, and subjective utility theory. In fact, most reinforcement learning textbooks axiomatize the maximization of expectation. Quoting Richard Bellman, for example:

> The general idea, and this is fairly unanimously accepted, is to use some average of the possible outcomes as a measure of the value of a policy.

This book takes the perspective that modeling the expected return alone fails to account for many complex, interesting phenomena that arise from interactions with one's environment. This is evident in many of the decisions that we make: the first rule of investment states that expected profits should be weighed against volatility. Similarly, lottery tickets offer negative expected returns but attract buyers with the promise of a high payoff. During a snowstorm, relying on the average frequency at which buses arrive at a stop is likely to lead to disappointment. More generally, hazards big and small result in a wide range of possible returns, each with its own probability of occurrence. These returns and their probabilities can be collectively described by a *return distribution*, our main object of study.

1.1 Why Distributional Reinforcement Learning?

Just as a color photograph conveys more information about a scene than a black and white photograph, the return distribution contains more information about the consequences of the agent's decisions than the expected return. The expected return is a scalar, while the return distribution is infinite-dimensional; it is possible to compute the expectation of the return from its distribution, but not the other way around (to continue the analogy, one cannot recover hue from luminance).

By considering the return distribution, rather than just the expected return, we gain a fresh perspective on the fundamental problems of reinforcement learning. This includes understanding of how optimal decisions should be made, methods for creating effective representations of an agent's state, and the consequences of interacting with other learning agents. In fact, many of the tools we develop here are useful beyond reinforcement learning and decision-making. We call the process of computing return distributions *distributional dynamic programming*. Incremental algorithms for learning return distributions, such as quantile temporal-difference learning (Chapter 6), are likely to find uses wherever probabilities need to be estimated.

Throughout this book, we will encounter many examples in which we use the tools of distributional reinforcement learning to characterize the consequences of the agent's choices. This is a modeling decision, rather than a reflection of some underlying truth about these examples. From a theoretical perspective,

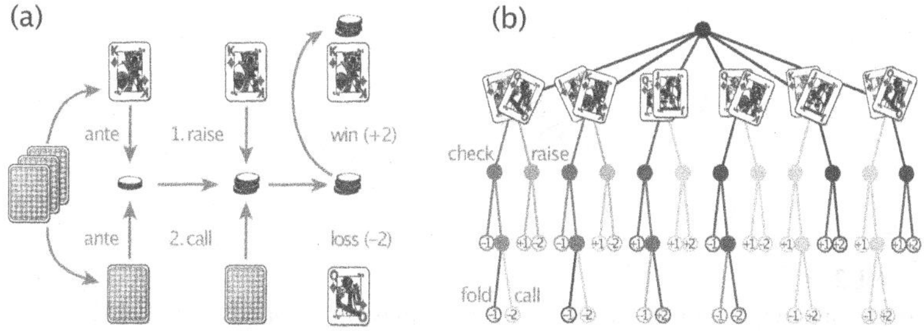

Figure 1.1
(**a**) In Kuhn poker, each player is dealt one card and then bets on whether they hold the highest card. The diagram depicts one particular play through; the house's card (bottom) is hidden until betting is over. (**b**) A game tree with all possible states shaded according to their frequency of occurrence in our example. Leaf nodes depict immediate gains and losses, which we equate with different values of the received reward.

justifying our use of the distributional model requires us to make a number of probabilistic assumptions. These include the notion that the random nature of the interactions is intrinsically irreducible (what is sometimes called *aleatoric uncertainty*) and unchanging. As we encounter these examples, the reader is invited to reflect on these assumptions and their effect on the learning process.

Furthermore, there are many situations in which such assumptions do not completely hold but where distributional reinforcement learning still provides a rich picture of how the environment operates. For example, an environment may appear random because some parts of it are not described to the agent (it is said to be *partially observable*) or because the environment changes over the course of the agent's interactions with it (multiagent learning, the topic of Section 11.1, is an example of this). Changes in the agent itself, such as a change in behavior, also introduce nonstationarity in the observed data. In practice, we have found that the return distributions are a valuable reflection of the underlying phenomena, even when there is no aleatoric uncertainty at play. Put another way, the usefulness of distributional reinforcement learning methods does not end with the theorems that characterize them.

1.2 An Example: Kuhn Poker

Kuhn poker is a simplified variant of the well-known card game. It is played with three cards of a single suit (jack, queen, and king) and over a single round of betting, as depicted in Figure 1.1a. Each player is dealt one card and must first bet a fixed ante, which for the purpose of this example we will take to be £1. After looking at their card, the first player decides whether to *raise*, which

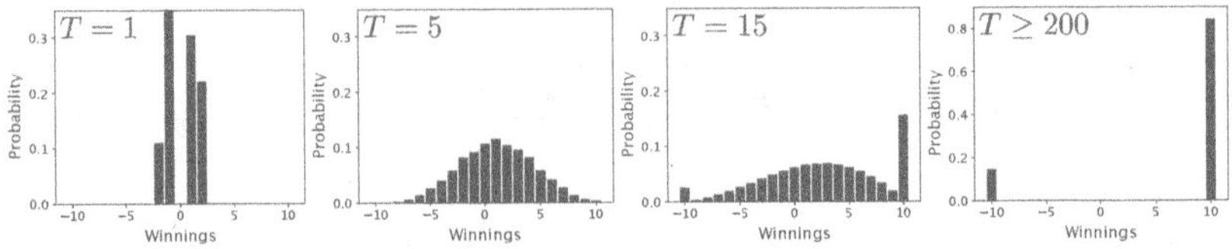

Figure 1.2
Distribution over winnings for the player after playing T rounds. For $T = 1$, this corresponds to the distribution of immediate gains and losses. For $T = 5$, we see a single mode appear roughly centered on the expected winnings. For larger T, two additional modes appear, one in which the agent goes bankrupt and one where the player has successfully doubled their stake. As $T \to \infty$, only these two outcomes have nonzero probability.

doubles their bet, or *check*. In response to a raise, the second player can *call* and match the new bet or *fold* and lose their £1 ante. If the first player chooses to *check* instead (keep the bet as-is), the option to raise is given to the second player, symmetrically. If neither player folded, the player with the higher card wins the *pot* (£1 or £2, depending on whether the ante was raised). Figure 1.1b visualizes a single play of the game as a fifty-five-state game tree.

Consider a player who begins with £10 and plays a total of up to T hands of Kuhn poker, stopping early if they go bankrupt or double their initial stake. To keep things simple, we assume that this player always goes first and that their opponent, *the house*, makes decisions uniformly at random. The player's strategy depends on the card they are dealt and also incorporates an element of randomness. There are two situations in which a choice must be made: whether to raise or check at first and whether to call or fold when the other player raises. The following table of probabilities describes a concrete strategy as a function of the player's dealt card:

Holding a ...	Jack	Queen	King
Probability of raising	⅓	0	1
Probability of calling	0	⅔	1

If we associate a positive and negative reward with each round's gains or losses, then the agent's random return corresponds to their total winnings at the end of the T rounds and ranges from -10 to 10.

How likely is the player to go bankrupt? How long does it take before the player is more likely to be ahead than not? What is the mean and variance of the player's winnings after $T = 15$ hands have been played? These three questions (and more) can be answered by using distributional dynamic programming to determine the distribution of returns obtained after T rounds. Figure 1.2

shows the distribution of winnings (change in money held by the player) as a function of T. After the first round ($T = 1$), the most likely outcome is to have lost £1, but the expected reward is positive. Consequently, over time, the player is likely to be able to achieve their objective. By the fifteenth round, the player is much more likely to have doubled their money than to have gone broke, with a bell-shaped distribution of values in between. If the game is allowed to continue until the end, the player has either gone bankrupt or doubled their stake. In our example, the probability that the player comes out a winner is approximately 85 percent.

1.3 How Is Distributional Reinforcement Learning Different?

In reinforcement learning, the value function describes the expected return that one would counterfactually obtain from beginning in any given state. It is reasonable to say that its fundamental object of interest – the expected return – is a *scalar* and that algorithms that operate on value functions operate on *collections of scalars* (one per state). On the other hand, the fundamental object of distributional reinforcement learning is a probability distribution over returns: the return distribution. The return distribution characterizes the probability of different returns that can be obtained as an agent interacts with its environment from a given state. Distributional reinforcement learning algorithms operate on collections of probability distributions that we call *return-distribution functions* (or simply *return functions*).

More than a simple type substitution, going from scalars to probability distributions results in changes across the spectrum of reinforcement learning topics. In distributional reinforcement learning, equations relating scalars become equations relating random variables. For example, the Bellman equation states that the expected return at a state x, denoted $V^\pi(x)$, equals the expectation of the immediate reward R, plus the discounted expected return at the next state X':

$$V^\pi(x) = \mathbb{E}_\pi \left[R + \gamma V^\pi(X') \mid X = x\right].$$

Here π is the agent's policy – a description of how it chooses actions in different states. By contrast, the distributional Bellman equation states that the *random return* at a state x, denoted $G^\pi(x)$, is itself related to the random immediate reward and the random next-state return according to a distributional equation:[1]

$$G^\pi(x) \stackrel{\mathcal{D}}{=} R + \gamma G^\pi(X'), \quad X = x.$$

1. Later we will consider a form that equates probability distributions directly.

In this case, $G^\pi(x)$, R, X', and $G^\pi(X')$ are random variables, and the superscript $\mathcal{D}$ indicates equality between their distributions. Correctly interpreting the distributional Bellman equation requires identifying the dependency between random variables, in particular between R and X'. It also requires understanding how discounting affects the probability distribution of $G^\pi(x)$ and how to manipulate the collection of random variables G^π implied by the definition.

Another change concerns how we quantify the behavior of learning algorithms and how we measure the quality of an agent's predictions. Because value functions are real-valued vectors, the distance between a value function estimate and the desired expected return is measured as the absolute difference between those two quantities. On the other hand, when analyzing a distributional reinforcement learning algorithm, we must instead measure the distance between probability distributions using a *probability metric*. As we will see, some probability metrics are better suited to distributional reinforcement learning than others, but no single metric can be identified as the "natural" metric for comparing return distributions.

Implementing distributional reinforcement learning algorithms also poses some concrete computational challenges. In general, the return distribution is supported on a range of possible returns, and its shape can be quite complex. To represent this distribution with a finite number of parameters, some approximation is necessary; the practitioner is faced with a variety of choices and trade-offs. One approach is to discretize the support of the distribution uniformly and assign a variable probability to each interval, what we call the *categorical representation*. Another is to represent the distribution using a finite number of uniformly weighted particles whose locations are parameterized, called the *quantile representation*. In practice and in theory, we find that the choice of distribution representation impacts the quality of the return function approximation and also the ease with which it can be computed.

Learning return distributions from sampled experience is also more challenging than learning to predict expected returns. The issue is particularly acute when learning proceeds by *bootstrapping*: that is, when the return function estimate at one state is learned on the basis of the estimate at successor states. When the return function estimates are defined by a deep neural network, as is common in practice, one must also take care in choosing a loss function that is compatible with a stochastic gradient descent scheme.

For an agent that only knows about expected returns, it is natural (almost necessary) to define optimal behavior in terms of maximizing this quantity. The Q-learning algorithm, which performs credit assignment by maximizing over state-action values, learns a policy with exactly this objective in mind. Knowledge of the return function, however, allows us to define behaviors that depend on the full distributions of returns – what is called *risk-sensitive reinforcement learning*. For example, it may be desirable to act so as to avoid states that carry a high probability of failure or penalize decisions that have high variance. In many circumstances, distributional reinforcement learning enables behavior that is more robust to variations and, perhaps, better suited to real-world applications.

1.4 Intended Audience and Organization

This book is intended for advanced undergraduates, graduate students, and researchers who have some exposure to reinforcement learning and are interested in understanding its distributional counterpart. We present core ideas from classical reinforcement learning as they are needed to contextualize distributional topics but often omit longer discussions and a presentation of specialized methods in order to keep the exposition concise. The reader wishing a more in-depth review of classical reinforcement learning is invited to consult one of the literature's many excellent books on the topic, including Bertsekas and Tsitsiklis (1996), Szepesvári (2010), Bertsekas (2012), Puterman (2014), Sutton and Barto (2018), and Meyn (2022).

Already, an exhaustive treatment of distributional reinforcement learning would require a substantially larger book. Instead, here we emphasize key concepts and challenges of working with return distributions, in a mathematical language that aims to be both technically correct but also easily applied. Our choice of topics is driven by practical considerations (such as scalability in terms of available computational resources), a topic's relative maturity, and our own domains of expertise. In particular, this book contains only one chapter about what is commonly called the *control problem* and focuses on dynamic programming and temporal-difference algorithms over Monte Carlo methods. Where appropriate, in the bibliographical remarks, we provide references on these omitted topics. In general, we chose to include proofs when they pertain to major results in the chapter or are instructive in their own right. We defer the proof of a number of smaller results to exercises.

Each chapter of this book is structured like a hiking trail.[2] The first sections (the "foothills") introduce a concept from classical reinforcement learning and extend it to the distributional setting. Here, a knowledge of undergraduate-level probability theory and computer science usually suffices. Later sections (the "incline") dive into more technical points: for example, a proof of convergence or more complex algorithms. These may be skipped without affecting the

2. Based on one of the author's experience hiking around Banff, Canada.

reader's understanding of the fundamentals of distributional reinforcement learning. Finally, most chapters end on a few additional results or remarks that are interesting yet easily omitted (the "side trail"). These are indicated by an asterisk (*). For the latter part of the chapter's journey, the reader may wish to come equipped with tools from advanced probability theory; our own references are Billingsley (2012) and Williams (1991).

The book is divided into three parts. The first part introduces the building blocks of distributional reinforcement learning. We begin by introducing our fundamental objects of study, the return distribution and the distributional Bellman equation (Chapter 2). Chapter 3 then introduces categorical temporal-difference learning, a simple algorithm for learning return distributions. By the end of Chapter 3, the reader should understand the basic principles of distributional reinforcement learning and be able to use them in simple practical settings.

The second part develops the theory of distributional reinforcement learning. Chapter 4 introduces a language for measuring distances between return distributions and operators for transforming with these distributions. Chapter 5 introduces the notion of a probability representation, needed to implement distributional reinforcement learning; it subsequently considers the problem of computing and approximating return distributions using such representations, introducing the framework of distributional dynamic programming. Chapter 6 studies how return distributions can be learned from samples and in a incremental fashion, giving a formal construction of categorical temporal-difference learning as well as other algorithms such as quantile temporal-difference learning. Chapter 7 extends these ideas to the setting of optimal decision-making (also called the control setting). Finally, Chapter 8 introduces a different perspective on distributional reinforcement learning based on the notion of statistical functionals. By the end of the second part, the reader should understand the challenges that arise when designing distributional reinforcement learning algorithms and the available tools to address these challenges.

The third and final part develops distributional reinforcement learning for practical scenarios. Chapter 9 reviews the principles of linear value function approximation and extends these ideas to the distributional setting. Chapter 10 discusses how to combine distributional methods with deep neural networks to obtain algorithms for deep reinforcement learning. Chapter 11 discusses the emerging use of distributional reinforcement learning in two further domains of research (multiagent learning and neuroscience) and concludes.

Code for examples and exercises, as well as standard implementations of the algorithms presented here, can be found at http://distributional-rl.org.

1.5 Bibliographical Remarks

1.0. The quote is due to Bellman (1957b).
1.1. Kuhn poker is due to Kuhn (1950), who gave an exhaustive characterization of the game's Nash equilibria. The player's strategy used in the main text forms part of such a Nash equilibrium.

2 The Distribution of Returns

Training for a marathon. Growing a vegetable garden. Working toward a piano recital. Many of life's activities involve making decisions whose benefits are realized only later in the future (whether to run on a particular Saturday morning; whether to add fertilizer to the soil). In reinforcement learning, these benefits are summarized by the return received following these decisions. The return is a random quantity that describes the sum total of the consequences of a particular activity – measured in dollars, points, bits, kilograms, kilometers, or praise.

Distributional reinforcement learning studies the random return. It asks questions such as: How should it be described, or approximated? How can it be predicted on the basis of past observations? The overarching aim of this book is to establish a language with which such questions can be answered. By virtue of its subject matter, this language is somewhere at the intersection of probability theory, statistics, operations research, and of course reinforcement learning itself. In this chapter, we begin by studying how the random return arises from sequential interactions and immediate rewards. From this, we establish the fundamental relationship of random returns: the distributional Bellman equation.

2.1 Random Variables and Their Probability Distributions

A quantity that we wish to model as random can be represented via a *random variable*. For example, the outcome of a coin toss can be represented with a random variable, which may take on either the value "heads" or "tails". We can reason about a random variable through its *probability distribution*, which specifies the probability of its possible realizations.

Example 2.1. Consider driving along a country road toward a railway crossing. There are two possible states that the crossing may be in. The crossing may be open, in which case you can drive straight through, or it may be closed, in which case you must wait for the train to pass and for the barriers to lift before driving on. We can model the state of the crossing as a random variable C with two outcomes, "open" and "closed." The distribution of C is specified by a *probability mass function*, which provides the probability of each possible outcome:

$$\mathbb{P}(C = \text{“open”}) = p\,, \quad \mathbb{P}(C = \text{“closed”}) = 1 - p\,,$$

for some $p \in [0, 1]$. △

Example 2.2. Suppose we arrive at the crossing described above and the barriers are down. We may model the waiting time T (in minutes) until the barriers are open again as a uniform distribution on the interval $[0, 10]$; informally, any real value between 0 and 10 is equally likely. In this case, the probability distribution can be specified through a *probability density function*, a function $f : \mathbb{R} \to [0, \infty)$. In the case of the uniform distribution above, this function is given by

$$f(t) = \begin{cases} \frac{1}{10} & \text{if } t \in [0, 10] \\ 0 & \text{otherwise} \end{cases}.$$

The density then provides the probability of T lying in any interval $[a, b]$ according to

$$\mathbb{P}(T \in [a, b]) = \int_a^b f(t)\mathrm{d}t\,. \quad △$$

In this book, we will encounter instances of random variables – such as rewards and returns – that are discrete, have densities, or in some cases fall in neither category. To deal with this heterogeneity, one solution is to describe probability distributions over $\mathbb{R}$ using their *cumulative distribution function* (CDF), which always exists. The cumulative distribution function associated with a random variable Z is the function $F_Z : \mathbb{R} \to [0, 1]$ defined by

$$F_Z(z) = \mathbb{P}(Z \leq z)\,.$$

In distributional reinforcement learning, common operations on random variables include summation, multiplication by a scalar, and indexing into collections of random variables. Later in the chapter, we will see how to describe these operations in terms of cumulative distribution functions.

Example 2.3. Suppose that now we consider the random variable T' describing the total waiting time experienced at the railroad crossing. If we arrive at the crossing and it is open ($C =$ "open"), there is no need to wait and $T' = 0$. If, however, the barrier is closed ($C =$ "closed"), the waiting time is distributed uniformly on [0, 10]. The cumulative distribution function of T' is

$$F_{T'}(t) = \begin{cases} 0 & t < 0 \\ p & t = 0 \\ p + \frac{(1-p)t}{10} & 0 < t < 10 \\ 1 & t \geq 10 . \end{cases}$$

Observe that there is a nonzero probability of T' taking the value 0 (which occurs when the crossing is open), so the distribution cannot have a density. Nor can it have a probability mass function, as there are a continuum of possible waiting times from 0 to 10 minutes. △

Another solution is to treat probability distributions atomically, as elements of the space $\mathscr{P}(\mathbb{R})$ of probability distributions. In this book we favor this approach over the use of cumulative distribution functions, as it lets us concisely express operations on probability distributions. The probability distribution that puts all of its mass on $z \in \mathbb{R}$, for example, is the Dirac delta denoted δ_z; the uniform distribution on the interval from a to b is $\mathcal{U}([a, b])$. Both distributions belong to $\mathscr{P}(\mathbb{R})$. The distribution of the random variable T' in Example 2.3 also belongs to $\mathscr{P}(\mathbb{R})$. It is a *mixture* of distributions and can be written in terms of its constituent parts:

$$p\delta_0 + (1-p)\mathcal{U}([0, 10]) .$$

More formally, these objects are *probability measures*: functions that associate different subsets of outcomes with their respective probabilities. If Z is a real-valued random variable and ν is its distribution, then for a subset $S \subseteq \mathbb{R}$,[3] we write

$$\nu(S) = \mathbb{P}(Z \in S) .$$

In particular, the probability assigned to S by a mixture distribution is the weighted sum of probabilities assigned by its constituent parts: for $\nu_1, \nu_2 \in \mathscr{P}(\mathbb{R})$ and $p \in [0, 1]$, we have

$$(p\nu_1 + (1-p)\nu_2)(S) = p\nu_1(S) + (1-p)\nu_2(S) .$$

With this language, the cumulative distribution function of $\nu \in \mathscr{P}(\mathbb{R})$ can be expressed as

$$F_\nu(y) = \nu((-\infty, y]) .$$

This notation also extends to distributions over outcomes that are not real-valued. For instance, $\mathscr{P}(\{\text{"open"}, \text{"closed"}\})$ is the set of probability distributions over the state of the railroad crossing in Example 2.1. Probability distributions (as

3. Readers expecting to see the qualifier "measurable" here should consult Remark 2.1.

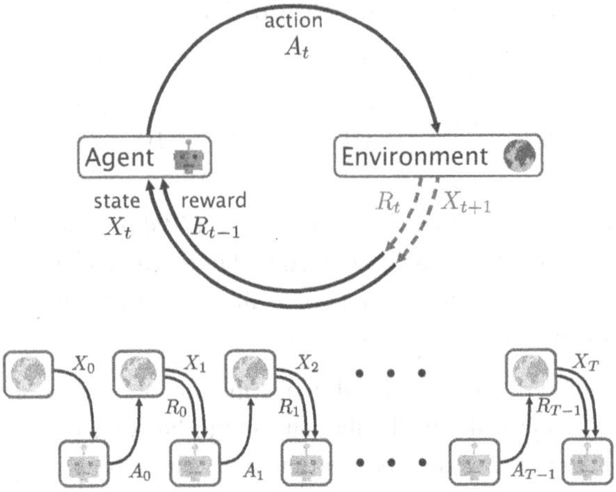

Figure 2.1
Top: The Markov decision process model of the agent's interactions with its environment. **Bottom:** The same interactions, unrolled to show the sequence of random variables $(X_t, A_t, R_t)_{t\geq 0}$, beginning at time $t = 0$ and up to the state X_T.

elements of $\mathscr{P}(\mathbb{R})$) make it possible to express some operations on distributions that would be unwieldy to describe in terms of random variables.

2.2 Markov Decision Processes

In reinforcement learning, an *environment* is any of a wide variety of systems that emit observations, can be influenced, and persist in one form or another over time. A data center cooling system, a remote-controlled helicopter, a stock market, and a video game console can all be thought of as environments. An *agent* interacts with its environment by making choices that have consequences in this environment. These choices may be implemented simply as an IF statement in a simulator or they may require a human to perform some task in our physical world.

We assume that interactions take place or are recorded at discrete time intervals. These give rise to a sequential process in which at any given time $t \in \mathbb{N} = \{0, 1, 2, \dots\}$, the current situation is described by a *state* X_t from a finite set $\mathcal{X}$.[4] The initial state is a random variable X_0 with probability distribution $\xi_0 \in \mathscr{P}(\mathcal{X})$.

4. Things tend to become more complicated when one considers infinite state spaces; see Remark 2.3.

The agent influences its future by choosing an *action* A_t from a finite set of actions $\mathcal{A}$. In response to this choice, the agent is provided with a real-valued *reward* R_t.[5] This reward indicates to the agent the usefulness or worth of its choice. The action also affects the state of the system; the new state is denoted X_{t+1}. An illustration of this sequential interaction is given in Figure 2.1. The reward and next state are modeled by the *transition dynamics* $P : \mathcal{X} \times \mathcal{A} \to \mathscr{P}(\mathbb{R} \times \mathcal{X})$ of the environment, which provides the joint probability distribution of R_t and X_{t+1} in terms of the state X_t and action A_t. We say that R_t and X_{t+1} are *drawn* from this distribution:

$$R_t, X_{t+1} \sim P(\cdot, \cdot \mid X_t, A_t) . \qquad (2.1)$$

In particular, when R_t is discrete, Equation 2.1 can be directly interpreted as

$$\mathbb{P}(R_t = r, X_{t+1} = x' \mid X_t = x, A_t = a) = P(r, x' \mid x, a).$$

Modeling the two quantities jointly is useful in problems where the reward depends on the next state (common in board games, where the reward is associated with reaching a certain state) or when the state depends on the reward (common in domains where the state keeps track of past rewards). In this book, however, unless otherwise noted, we make the simplifying assumption that the reward and next state are independent given X_t and A_t, and separate the transition dynamics into a *reward distribution* and *transition kernel*:

$$R_t \sim P_{\mathcal{R}}(\cdot \mid X_t, A_t)$$
$$X_{t+1} \sim P_{\mathcal{X}}(\cdot \mid X_t, A_t) .$$

A *Markov decision process* (MDP) is a tuple $(\mathcal{X}, \mathcal{A}, \xi_0, P_{\mathcal{X}}, P_{\mathcal{R}})$ that contains all the information needed to describe how the agent's decisions influence its environment. These decisions are not themselves part of the model but instead arise from a *policy*. A policy is a mapping $\pi : \mathcal{X} \to \mathscr{P}(\mathcal{A})$ from states to probability distributions over actions such that

$$A_t \sim \pi(\cdot \mid X_t) .$$

Such policies choose the action A_t solely on the basis of the immediately preceding state X_t and possibly a random draw. Technically, these are a special subset of decision-making rules that are both *stationary* (they do not depend on the time t at which the decision is to be taken, except through the state X_t) and *Markov* (they do not depend on events prior to time t). Stationary Markov policies will be enough for us for most of this book, but we will study more general policies in Chapter 7.

5. An alternative convention is to denote the reward that follows action A_t by R_{t+1}. Here we prefer R_t to emphasize the association between action and reward.

Finally, a state $x_\varnothing$ from which no other states can be reached is called absorbing or *terminal*. For all actions $a \in \mathcal{A}$, its next-state distribution is

$$P_\mathcal{X}(x_\varnothing \mid x_\varnothing, a) = 1.$$

Terminal states correspond to situations in which further interactions are irrelevant: once a game of chess is won by one of the players, for example, or once a robot has successfully accomplished a desired task.

2.3 The Pinball Model

The game of American pinball provides a useful metaphor for how the various pieces of a Markov decision process come together to describe real systems. A classic American pinball machine consists of a slanted, glass-enclosed play area filled with bumpers of various shapes and sizes. The player initiates the game by using a retractable spring to launch a metal ball into the play area, a process that can be likened to sampling from the initial state distribution. The metal ball progresses through the play area by bouncing off the various bumpers (the transition function), which reward the player with a variable number of points (the reward function). The game ends when the ball escapes through a gap at the bottom of the play area, to which it is drawn by gravity (the terminal state). The player can prevent this fate by controlling the ball's course with a pair of flippers on either side of the gap (the action space). Good players also use the flippers to aim the ball toward the most valuable bumpers or other, special high-scoring zones and may even physically shake the pinball cabinet (called *nudging*) to exert additional control. The game's state space describes possible arrangements of the machine's different moving parts, including the ball's location.

Turning things around, we may think of any Markov decision process as an abstract pinball machine. Initiated by the equivalent of inserting the traditional quarter into the machine, we call a single play through the Markov decision process a *trajectory*, beginning from the random initial state and lasting until a terminal state is reached. This trajectory is the sequence $X_0, A_0, R_0, X_1, A_1, \ldots$ of random interleaved states, actions, and rewards. We use the notation $(X_t, A_t, R_t)_{t \geq 0}$ to express this sequence compactly.

The various elements of the trajectory depend on each other according to the rules set by the Markov decision process. These rules can be summarized by a collection of *generative equations*, which tell us how we might write a program for sampling a trajectory one variable at a time. For a time step $t \in \mathbb{N}$, let us denote by $X_{0:t}$, $A_{0:t}$, and $R_{0:t}$ the subsequences $(X_0, X_1, \ldots, X_t)$, $(A_0, A_1, \ldots, A_t)$, and $(R_0, R_1, \ldots, R_t)$, respectively. The generative equations are

$$X_0 \sim \xi_0 ;$$
$$A_t \mid (X_{0:t}, A_{0:t-1}, R_{0:t-1}) \sim \pi(\cdot \mid X_t), \text{ for all } t \geq 0 ;$$
$$R_t \mid (X_{0:t}, A_{0:t}, R_{0:t-1}) \sim P_{\mathcal{R}}(\cdot \mid X_t, A_t), \text{ for all } t \geq 0 ;$$
$$X_{t+1} \mid (X_{0:t}, A_{0:t}, R_{0:t}) \sim P_{\mathcal{X}}(\cdot \mid X_t, A_t), \text{ for all } t \geq 0 .$$

We use the notation $Y \mid (Z_0, Z_1, \ldots)$ to indicate the basic dependency structure between these variables. The equation for A_t, for example, is to be interpreted as

$$\mathbb{P}(A_t = a \mid X_0, A_0, R_0, \cdots, X_{t-1}, A_{t-1}, R_{t-1}, X_t) = \pi(a \mid X_t).$$

For $t = 0$, the notation $A_{0:t-1}$ and $R_{0:t-1}$ denotes the empty sequence. Because the policy fixes the "decision" part of the Markov decision process formalism, the trajectory can be viewed as a Markov chain over the space $\mathcal{X} \times \mathcal{A} \times \mathbb{R}$. This model is sometimes called a *Markov reward process*.

By convention, terminal states yield no reward. In these situations, it is sensible to end the sequence at the time $T \in \mathbb{N}$ at which a terminal state is first encountered. It is also common to notify the agent that a terminal state has been reached. In other cases (such as Example 2.4 below), the sequence might (theoretically) go on forever.

We use the notion of a joint distribution to formally ask questions (and give answers) about the random trajectory. This is the joint distribution over all random variables involved, denoted $\mathbb{P}_\pi(\cdot)$. For example, the probability that the agent begins in state x and finds itself in that state again after t time steps when acting according to a policy π can be mathematically expressed as $\mathbb{P}_\pi(X_0 = x, X_t = x)$. Similarly, the probability that a positive reward will be received at some point in time is

$$\mathbb{P}_\pi(\text{there exists } t \geq 0 \text{ such that } R_t > 0) .$$

The explicit policy subscript is a convention to emphasize that the agent's choices affect the distribution of outcomes. It also lets us distinguish statements about random variables derived from the random trajectory from statements about other, arbitrary random variables.

The joint distribution $\mathbb{P}_\pi$ gives rise to the expectation $\mathbb{E}_\pi$ over real-valued random variables. This allows us to write statements such as

$$\mathbb{E}_\pi[2 \times R_0 + \mathbb{1}_{\{X_1 = X_0\}} R_1] .$$

Remark 2.1 provides additional technical details on how this expectation can be constructed from $\mathbb{P}_\pi$.

Example 2.4. The *martingale* is a betting strategy popular in the eighteenth century and based on the principle of doubling one's ante until a profit is made. This strategy is formalized as a Markov decision process where the (infinite) state space $\mathcal{X} = \mathbb{Z}$ is the gambler's loss thus far, with negative losses denoting gains. The action space $\mathcal{A} = \mathbb{N}$ is the gambler's bet.[6] If the game is fair, then for each state x and each action a,

$$P_{\mathcal{X}}(x + a \mid x, a) = P_{\mathcal{X}}(x - a \mid x, a) = 1/2.$$

Placing no bet corresponds to $a = 0$. With $X_0 = 0$, the martingale policy is $A_0 = 1$ and for $t > 0$,

$$A_t = \begin{cases} X_t + 1 & X_t > 0 \\ 0 & \text{otherwise.} \end{cases}$$

Formally speaking, the policy maps each loss $x > 0$ to δ_{x+1}, the Dirac distribution at $x + 1$, and all negative states (gains) to δ_0. Simple algebra shows that for $t > 0$,

$$X_t = \begin{cases} 2^t - 1 & \text{with probability } 2^{-t}, \\ -1 & \text{with probability } 1 - 2^{-t}. \end{cases}$$

That is, the gambler is assured to eventually make a profit (since $2^{-t} \to 0$), but arbitrary losses may be incurred before a positive gain is made. Calculations show that the martingale strategy has nil expected gain ($\mathbb{E}_\pi[X_t] = 0$ for $t \geq 0$), while the variance of the loss grows unboundedly with the number of rounds played ($\mathbb{E}_\pi[X_t^2] \to \infty$); as a result, it is frowned upon in many gambling establishments. △

The notation $\mathbb{P}_\pi$ makes clear the dependence of the distribution of the random trajectory $(X_t, A_t, R_t)_{t \geq 0}$ on the agent's policy π. Often, we find it also useful to view the initial state distribution ξ_0 as a parameter to be reasoned explicitly about. In this case, it makes sense to more explicitly denote the joint distribution by $\mathbb{P}_{\xi_0,\pi}$. The most common situation is when we consider the same Markov decision process but initialized at a specific starting state x. In this case, the distribution becomes $\xi_0 = \delta_x$, meaning that $\mathbb{P}_{\xi_0,\pi}(X_0 = x) = 1$. We then use the fairly standard shorthand

$$\mathbb{P}_\pi(\,\cdot \mid X_0 = x) = \mathbb{P}_{\delta_x,\pi}(\,\cdot\,), \qquad \mathbb{E}_\pi[\,\cdot \mid X_0 = x] = \mathbb{E}_{\delta_x,\pi}[\,\cdot\,].$$

Technically, this use of the conditioning bar is an abuse of notation. We are directly modifying the probability distributions of these random variables, rather than conditioning on an event as the notation normally signifies.[7] However, this notation is convenient and common throughout the reinforcement learning

6. We are ignoring cash flow issues here.

7. If we attempt to use the actual conditional probability distribution instead, we find that it is ill-defined when $\xi_0(x) = 0$. See Exercise 2.2.

literature, and we therefore adopt it as well. It is also convenient to modify the distribution of the first action A_0, so that rather than this random variable being sampled from $\pi(\cdot \mid X_0)$, it is fixed at some action $a \in \mathcal{A}$. We use similar notation as above to signify the resulting distribution over trajectories and corresponding expectations:

$$\mathbb{P}_\pi(\cdot \mid X_0 = x, A_0 = a), \qquad \mathbb{E}_\pi[\cdot \mid X_0 = x, A_0 = a].$$

2.4 The Return

Given a *discount factor* $\gamma \in [0, 1)$, the *discounted return*[8] (or simply the *return*) is the sum of rewards received by the agent from the initial state onward, discounted according to their time of occurrence:

$$G = \sum_{t=0}^{\infty} \gamma^t R_t. \tag{2.2}$$

The return is a sum of scaled, real-valued random variables and is therefore itself a random variable. In reinforcement learning, the success of an agent's decisions is measured in terms of the return that it achieves: greater returns are better. Because it is the measure of success, the return is the fundamental object of reinforcement learning.

The discount factor encodes a preference to receive rewards sooner than later. In settings that lack a terminal state, the discount factor is also used to guarantee that the return G exists and is finite. This is easily seen when rewards are bounded on the interval $[R_{\text{MIN}}, R_{\text{MAX}}]$, in which case we have

$$G \in \left[\frac{R_{\text{MIN}}}{1-\gamma}, \frac{R_{\text{MAX}}}{1-\gamma}\right]. \tag{2.3}$$

Throughout this book, we will often write V_{MIN} and V_{MAX} for the endpoints of the interval above, denoting the smallest and largest return obtainable when rewards are bounded on $[R_{\text{MIN}}, R_{\text{MAX}}]$. When rewards are not bounded, the existence of G is still guaranteed under the mild assumption that rewards have finite first moment; we therefore adopt this assumption throughout the book.

Assumption 2.5. For each state $x \in \mathcal{X}$ and action $a \in \mathcal{A}$, the reward distribution $P_{\mathcal{R}}(\cdot \mid x, a)$ has finite first moment. That is, if $R \sim P_{\mathcal{R}}(\cdot \mid x, a)$, then

$$\mathbb{E}[|R|] < \infty. \qquad \triangle$$

8. In parts of this book, we also consider the *undiscounted return* $\sum_{t=0}^{\infty} R_t$. We sometimes write $\gamma = 1$ to denote this return, indicating a change from the usual setting. One should be mindful that we then need to make sure that the sum of rewards converges.

Proposition 2.6. Under Assumption 2.5, the random return G exists and is finite with probability 1, in the sense that

$$\mathbb{P}_\pi\big(G \in (-\infty, \infty)\big) = 1 . \qquad \triangle$$

In the usual treatment of reinforcement learning, Assumption 2.5 is taken for granted (there is little to be predicted otherwise). We still find Proposition 2.6 useful as it establishes that the assumption is *sufficient* to guarantee the finiteness of G and hence the existence of its probability distribution. The phrase "with probability 1" allows for the fact that in some realizations of the random trajectory, G is infinite (for example, if the rewards are normally distributed) – but the probability of observing such a realization is nil. Remark 2.2 gives more details on this topic, along with a proof of Proposition 2.6; see also Exercise 2.17.

We call the probability distribution of G the *return distribution*. The return distribution determines quantities such as the expected return

$$\mathbb{E}_\pi[G] = \mathbb{E}_\pi\Big[\sum_{t=0}^{\infty} \gamma^t R_t\Big],$$

which plays a central role in reinforcement learning, the variance of the return

$$\mathrm{Var}_\pi(G) = \mathbb{E}_\pi\big[(G - \mathbb{E}_\pi[G])^2\big],$$

and tail probabilities such as

$$\mathbb{P}_\pi\Big(\sum_{t=0}^{\infty} \gamma^t R_t \geq 0\Big),$$

which arise in risk-sensitive control problems (discussed in Chapter 7). As the following examples illustrate, the probability distributions of random returns vary from simple to intricate, according to the complexity of interactions between the agent and its environment.

Example 2.7 (Blackjack). The card game of blackjack is won by drawing cards whose total value is greater than that of a dealer, who plays according to a known, fixed set of rules. Face cards count for 10, while aces may be counted as either 1 or 11, to the player's preference. The game begins with the player receiving two cards, which they can complement by *hitting* (i.e., requesting another card). The game is lost immediately if the player's total goes over 21, called *going bust*. A player satisfied with their cards may instead *stick*, at which point the dealer adds cards to their own hand, until a total of 17 or more is reached.

If the player's total is greater than the dealer's or the dealer goes bust, they are paid out their ante; otherwise, the ante is lost. In the event of a tie, the player keeps the ante. We formalize this as receiving a reward of 1, −1, or 0 when play concludes. A *blackjack* occurs when the player's initial cards are an ace and a 10-valued card, which most casinos will pay 3 to 2 (a reward of $3/2$) provided the dealer's cards sum to less than 21. After seeing their initial two cards, the player may also choose to *double down* and receive exactly one additional card. In this case, wins and losses are doubled (2 or −2 reward).

Since the game terminates in a finite number of steps T and the objective is to maximize one's profit, let us equate the return G with the payoff from playing one hand. The return takes on values from the set

$$\{-2, -1, 0, 1, 3/2, 2\}.$$

Let us denote the cards dealt to the player by $C_0, C_1, \ldots, C_T$, the sum of these cards by Y_p, and the dealer's card sum by Y_d. With this notation, we can handle the ace's two possible values by adding 10 to Y_p or Y_d when at least one ace was drawn, and the sum is 11 or less. We define $Y_\text{p} = 0$ and $Y_\text{d} = 0$ when the player or dealer goes bust, respectively. Consider a player who doubles when dealt cards whose total is 11. The probability distribution of the return is

$$\begin{aligned}
\mathbb{P}_\pi(G = 3/2) &\overset{(a)}{=} 2\mathbb{P}_\pi(C_0 = 1, C_1 = 10, Y_\text{d} \neq 21) \\
\mathbb{P}_\pi(G = -2) &\overset{(b)}{=} \mathbb{P}_\pi(C_0 + C_1 = 11, C_0 \neq 1, C_1 \neq 1, C_0 + C_1 + C_2 < Y_\text{d}) \\
\mathbb{P}_\pi(G = 2) &\overset{(c)}{=} \mathbb{P}_\pi(C_0 + C_1 = 11, C_0 \neq 1, C_1 \neq 1, C_0 + C_1 + C_2 > Y_\text{d}) \\
\mathbb{P}_\pi(G = -1) &= \mathbb{P}_\pi(Y_\text{p} < Y_\text{d}) - \mathbb{P}_\pi(G = -2) \\
\mathbb{P}_\pi(G = 0) &= \mathbb{P}_\pi(Y_\text{p} = Y_\text{d}) \\
\mathbb{P}_\pi(G = 1) &= \mathbb{P}_\pi(Y_\text{p} > Y_\text{d}) - \mathbb{P}_\pi(G = 3/2) - \mathbb{P}_\pi(G = 2);
\end{aligned}$$

(a) implements the blackjack rule (noting that either C_0 or C_1 can be an ace) while (b) and (c) handle doubling (when there is no blackjack). For example, if we assume that cards are drawn with replacement, then

$$\mathbb{P}(Y_\text{d} = 21) \approx 0.12$$

and

$$\mathbb{P}_\pi(G = 3/2) = 2 \times (1 - \mathbb{P}_\pi(Y_\text{d} = 21)) \times \tfrac{1}{13}\tfrac{4}{13} \approx 0.042,$$

since there are thirteen card types to drawn from, one of which is an ace and four of which have value 10.

Computing the probabilities for other outcomes requires specifying the player's policy in full (when do they hit?). Were we to do this, we could then estimate these probabilities from simulation or, more tediously, calculate them by hand. △

Example 2.8. Consider a simple solitaire game that involves repeatedly throwing a single six-sided die. If a 1 is rolled, the game ends immediately. Otherwise, the player receives a point and continues to play. Consider the undiscounted return

$$\sum_{t=0}^{\infty} R_t = \underbrace{1 + 1 + \cdots + 1}_{T \text{times}},$$

where T is the time at which the terminating 1 is rolled. This is an integer-valued return ranging from 0 to ∞. It has the geometric distribution

$$\mathbb{P}_\pi\Big(\sum_{t=0}^{\infty} R_t = k\Big) = \tfrac{1}{6}\big(\tfrac{5}{6}\big)^k, \quad k \in \mathbb{N}$$

corresponding to the probability of seeing k successes before the first failure (rolling a 1). Choosing a discount factor less than 1 (perhaps modeling the player's increasing boredom) changes the support of this distribution but not the associated probabilities. The partial sums of the geometric series correspond to

$$\sum_{t=0}^{k} \gamma^t = \frac{1 - \gamma^{k+1}}{1 - \gamma}$$

and it follows that for $k \geq 0$,

$$\mathbb{P}_\pi\Big(\sum_{t=0}^{\infty} \gamma^t R_t = \tfrac{1-\gamma^k}{1-\gamma}\Big) = \tfrac{1}{6}\big(\tfrac{5}{6}\big)^k. \quad △$$

When a closed-form solution is tedious or difficult to obtain, we can sometimes still estimate the return distribution from simulations. This is illustrated in the next example.

Example 2.9. The Cliffs of Moher, located in County Clare, Ireland, are famous for both their stunning views and the strong winds blowing from the Atlantic Ocean. Inspired by the scenic walk from the nearby village of Doolin to the Cliffs' tourist center, we consider an abstract cliff environment (Figure 2.2) based on a classic domain from the reinforcement learning literature (Sutton and Barto 2018).[9]

The walk begins in the cell "S" (Doolin village) and ends with a positive reward (+1) when the "G" cell (tourist center) is reached. Four actions are available, corresponding to each of the cardinal directions – however, at each step, there is a probability $p = 0.25$ that the strong winds take the agent in an

9. There are a few differences, which we invite the reader to discover.

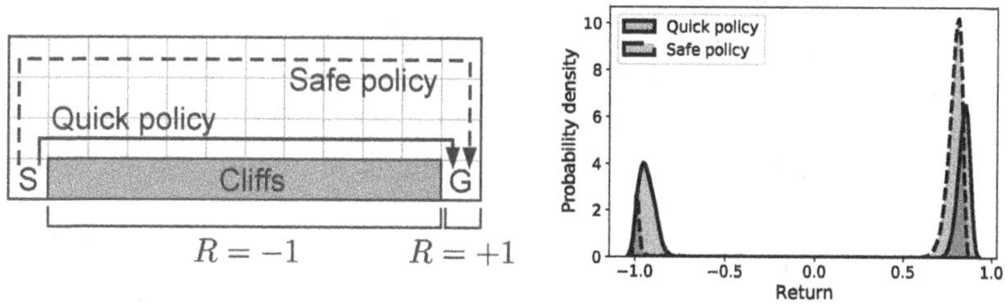

Figure 2.2
Left: The Cliffs environment, along with the path preferred by the quick and safe policies. **Right**: The return distribution for these policies, estimated by sampling 100,000 trajectories from the environment. The figure reports the distribution as a probability density function computed using kernel density estimation (with bandwidth 0.02).

unintended (random) direction. For simplicity, the edges of the grid act as walls. We model falling off the cliff using a negative reward of -1, at which point the episode also terminates. Reaching the tourist center yields a reward of $+1$ and also ends the episode. The reward is zero elsewhere. A discount factor of $\gamma = 0.99$ incentivizes the agent to not dally too much along the way.

Figure 2.2 depicts the return distribution for two policies: a quick policy that walks along the cliff's edge and a safe policy that walks two cells away from the edge.[10] The corresponding returns are bimodal, reflecting the two possible classes of outcomes. The return distribution of the faster policy, in particular, is sharply peaked around -1 and 1: the goal may be reached quickly, but the agent is more likely to fall. △

The next two examples show how even simple dynamics that one might reasonably encounter in real scenarios can result in return distributions that are markedly different from the reward distributions.

Example 2.10. Consider a single-state, single-action Markov decision process, so that $\mathcal{X} = \{x\}$ and $\mathcal{A} = \{a\}$. The initial distribution is $\xi_0 = \delta_x$ and the transition kernel is $P_{\mathcal{X}}(x \mid x, a) = 1$. The reward has a Bernoulli distribution, with

$$P_{\mathcal{R}}(0 \mid x, a) = P_{\mathcal{R}}(1 \mid x, a) = 1/2\,.$$

Suppose we take the discount factor to be $\gamma = 1/2$. The return is

$$G = R_0 + \tfrac{1}{2}R_1 + \tfrac{1}{4}R_2 + \cdots\,. \tag{2.4}$$

10. More precisely, the safe policy always attempts to move up when it is possible to do so, unless it is in one of the rightmost cells. The quick policy simply moves right until one of these cells is reached; afterward, it goes down toward the "G" cell.

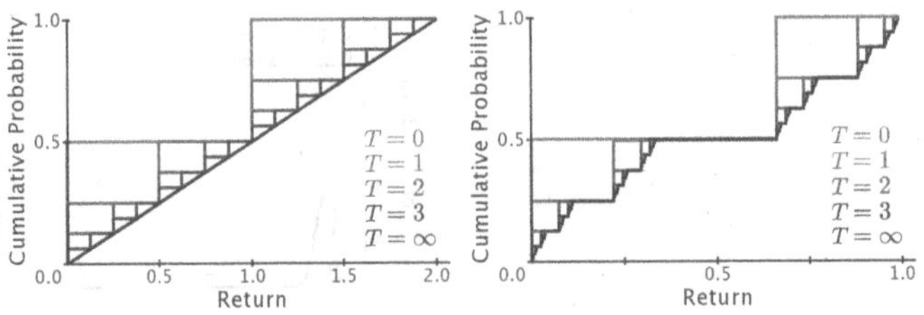

Figure 2.3
Left: Illustration of Example 2.10, showing cumulative distribution functions (CDFs) for the truncated return $G_{0:T} = \sum_{t=0}^{T} \left(\frac{1}{2}\right)^t R_t$ as T increases. The number of steps in this function doubles with each increment of T, reaching the uniform distribution in the limit as $T \to \infty$. **Right**: The same illustration, now for Example 2.11.

We will show that G has a uniform probability distribution over the interval $[0, 2]$. Observe that the right-hand side of Equation 2.4 is equivalent to the binary number

$$R_0.R_1R_2\ldots \tag{2.5}$$

Since $R_t \in \{0, 1\}$ for all t, this implies that the support of G is the set of numbers that have a (possibly infinite) binary expansion as in Equation 2.5. The smallest such number is clearly 0, while the largest number is 2 (the infinite sequence of 1s). Now

$$\mathbb{P}_\pi(G \in [0, 1]) = \mathbb{P}_\pi(R_0 = 0) \quad \mathbb{P}_\pi(G \in [1, 2]) = \mathbb{P}_\pi(R_0 = 1),$$

so that it is equally likely that G falls either in the interval $[0, 1]$ or $[1, 2]$.[11] If we now subdivide the lower interval, we find that

$$\begin{aligned}\mathbb{P}_\pi(G \in [0.5, 1]) &= \mathbb{P}_\pi(G \in [0, 1])\mathbb{P}_\pi(G \geq 0.5 \mid G \in [0, 1]) \\ &= \mathbb{P}_\pi(R_0 = 0)\mathbb{P}_\pi(R_1 = 1 \mid R_0 = 0) \\ &= \mathbb{P}_\pi(R_0 = 0)\mathbb{P}_\pi(R_1 = 1),\end{aligned}$$

and analogously for the intervals $[0, 0.5]$, $[1, 1.5]$, and $[1.5, 2]$. We can repeat this subdivision recursively to find that any *dyadic interval* $[a, b]$ whose endpoints belong to the set

$$\mathcal{Y} = \Big\{ \sum_{j=0}^{n} (\tfrac{1}{2})^j a_j : n \in \mathbb{N}, a_j \in \{0, 1\} \text{ for } 0 \leq j \leq n \Big\}$$

11. The probability that $G = 1$ is zero.

has probability $\frac{b-a}{2}$: that is,

$$\mathbb{P}_\pi(G \in [a,b]) = \frac{b-a}{2} \text{ for } a, b \in \mathcal{Y}, a < b\,. \tag{2.6}$$

Because the uniform distribution over the interval $[0, 2]$ satisfies Equation 2.6, we conclude that G is distributed uniformly on $[0, 2]$. A formal argument requires us to demonstrate that Equation 2.6 uniquely determines the cumulative distribution function of G (Exercise 2.6); this argument is illustrated (informally) in Figure 2.3. △

Example 2.11 (*). If we substitute the reward distribution of the previous example by

$$P_{\mathcal{R}}(0 \mid x, a) = P_{\mathcal{R}}(2/3 \mid x, a) = 1/2$$

and take $\gamma = 1/3$, the return distribution becomes the *Cantor distribution* (see Exercise 2.7). The Cantor distribution has no atoms (values with probability greater than zero) or a probability density. Its cumulative distribution function is the Cantor function (see Figure 2.3), famous for violating many of our intuitions about mathematical analysis. △

2.5 The Bellman Equation

The cliff-walking scenario (Example 2.9) shows how different policies can lead to qualitatively different return distributions: one where a positive reward for reaching the goal is likely and one where high rewards are likely, but where there is also a substantial chance of a low reward (due to a fall). Which should be preferred? In reinforcement learning, the canonical way to answer this question is to reduce return distributions to scalar values, which can be directly compared. More precisely, we measure the quality of a policy by the expected value of its random return,[12] or simply *expected return*:

$$\mathbb{E}_\pi\Big[\sum_{t=0}^{\infty} \gamma^t R_t\Big]. \tag{2.7}$$

Being able to determine the expected return of a given policy is thus central to most reinforcement learning algorithms. A straightforward approach is to enumerate all possible realizations of the random trajectory $(X_t, R_t, A_t)_{t\geq 0}$ up to length $T \in \mathbb{N}$. By weighting them according to their probability, we obtain the approximation

$$\mathbb{E}_\pi\Big[\sum_{t=0}^{T-1} \gamma^t R_t\Big]. \tag{2.8}$$

12. Our assumption that the rewards have finite first moment (Assumption 2.5) guarantees that the expectation in Expression 2.7 is finite.

However, even for reasonably small T, this is problematic, because the number of partial trajectories of length $T \in \mathbb{N}$ may grow exponentially with T.[13] Even for problems as small as cliff-walking, enumeration quickly becomes impractical. The solution lies in the *Bellman equation*, which provides a concise characterization of the expected return under a given policy.

To begin, consider the expected return for a policy π from an initial state x. This is called the *value* of x, written

$$V^\pi(x) = \mathbb{E}_\pi\Big[\sum_{t=0}^{\infty} \gamma^t R_t \,\Big|\, X_0 = x\Big]. \tag{2.9}$$

The *value function* V^π describes the value at all states. As the name implies, it is formally a mapping from a state to its expected return under policy π. The value function lets us answer counterfactual questions ("how well would the agent do from this state?") and also allows us to determine the expected return in Equation 2.7, since (by the generative equations)

$$\mathbb{E}_\pi\big[V^\pi(X_0)\big] = \mathbb{E}_\pi\Big[\sum_{t=0}^{\infty} \gamma^t R_t\Big].$$

By linearity of expectations, the expected return can be decomposed into an immediate reward R_0 (which depends on the initial state X_0) and the sum of future rewards:

$$\mathbb{E}_\pi\Big[\sum_{t=0}^{\infty} \gamma^t R_t\Big] = \mathbb{E}_\pi\Big[R_0 + \underbrace{\sum_{t=1}^{\infty} \gamma^t R_t}_{\text{future rewards}}\Big].$$

The Bellman equation expresses this relationship in terms of the value of different states; we give its proof in the next section.

Proposition 2.12 (The Bellman equation). Let V^π be the value function of policy π. Then for any state $x \in \mathcal{X}$, it holds that

$$V^\pi(x) = \mathbb{E}_\pi\big[R_0 + \gamma V^\pi(X_1) \mid X_0 = x\big]. \tag{2.10}$$

△

The Bellman equation transforms an infinite sum (Equation 2.7) into a recursive relationship between the value of a state x, its expected reward, and the value of its successor states. This makes it possible to devise efficient algorithms for determining the value function V^π, as we shall see later.

13. If rewards are drawn from a continuous distribution, there is in fact an infinite number of possible sequences of length T.

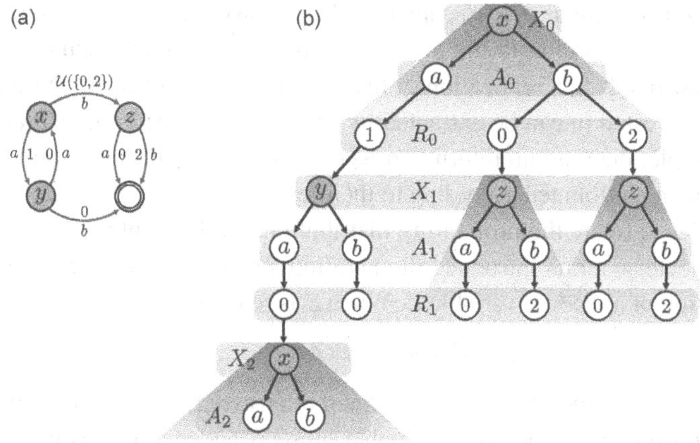

Figure 2.4
(a) An example Markov decision process with three states x, y, z and a terminal state denoted by a double circle. Action a gives a reward of 0 or 1 depending on the state, while action b gives a reward of 0 in state y and 2 in state z. In state x, action b gives a random reward of either 0 or 2 with equal probability. Arrows indicate the (deterministic) transition kernel from each state. **(b)** An illustration of the Markov property: the subsequences rooted in state $X_1 = z$ have the same probability distribution; and the time-homogeneity property: the subsequence rooted at $X_2 = x$ has the same probability distribution as the full sequence beginning in $X_0 = x$.

An alternative to the value function is the *state-action value function* Q^π. The state-action value function describes the expected return when the action A_0 is fixed to a particular choice a:

$$Q^\pi(x, a) = \mathbb{E}_\pi\Big[\sum_{t=0}^{\infty} \gamma^t R_t \mid X_0 = x, A_0 = a\Big].$$

Throughout this book, we will see a few situations for which this form is preferred, either because it simplifies the exposition or makes certain algorithms possible. Unless otherwise noted, the results we will present hold for both value functions and state-action value functions.

2.6 Properties of the Random Trajectory

The Bellman equation is made possible by two fundamental properties of the Markov decision process, *time-homogeneity* and the *Markov property*. More precisely, these are properties of the random trajectory $(X_t, R_t, A_t)_{t\geq 0}$, whose probability distribution is described by the generative equations of Section 2.3.

To understand the distribution of the random trajectory $(X_t, A_t, R_t)_{t\geq 0}$, it is helpful to depict its possible outcomes as an infinite tree (Figure 2.4). The root of the tree is the initial state X_0, and each level consists of a state-action-reward triple, drawn according to the generative equations. Each branch of the tree then corresponds to a realization of the trajectory. Most of the quantities that we are interested in can be extracted by "slicing" this tree in particular ways. For example, the random return corresponds to the sum of the rewards along each branch, discounted according to their level.

In order to relate the probability distributions of different parts of the random trajectory, let us introduce the notation $\mathcal{D}(Z)$ to denote the probability distribution of a random variable Z. When Z is real-valued we have, for $S \subseteq \mathbb{R}$,

$$\mathcal{D}(Z)(S) = \mathbb{P}(Z \in S).$$

One advantage of the $\mathcal{D}(\cdot)$ notation is that it often avoids unnecessarily introducing (and formally characterizing) such a subset S and is more easily extended to other kinds of random variables. Importantly, we write $\mathcal{D}_\pi$ to refer to the distribution of random variables derived from the joint distribution $\mathbb{P}_\pi$.[14] For example, $\mathcal{D}_\pi(R_0 + R_1)$ and $\mathcal{D}_\pi(X_2)$ are the probability distribution of the sum $R_0 + R_1$ and of the third state in the random trajectory, respectively.

The *Markov property* states that the trajectory from a given state x is independent of the states, actions, and rewards that were encountered prior to x. Graphically, this means that for a given level of the tree k, the identity of the state X_k suffices to determine the probability distribution of the subtree rooted at that state.

Lemma 2.13 (Markov property). The trajectory $(X_t, A_t, R_t)_{t \geq 0}$ has the Markov property. That is, for any $k \in \mathbb{N}$, we have

$$\mathcal{D}_\pi\Big((X_t, A_t, R_t)_{t=k}^{\infty} \mid (X_t, A_t, R_t)_{t=0}^{k-1} = (x_t, a_t, r_t)_{t=0}^{k-1}, X_k = x\Big) = \mathcal{D}_\pi\Big((X_t, A_t, R_t)_{t=k}^{\infty} \mid X_k = x\Big)$$

whenever the conditional distribution of the left-hand side is defined. △

As a concrete example, Figure 2.4 shows that there are two realizations of the trajectory for which $X_1 = z$. By the Markov property, the expected return from this point on is independent of whether the immediately preceding reward (R_0) was 0 or 2.

One consequence of the Markov property is that the expectation of the discounted return from time k is independent of the trajectory prior to time k, given X_k:

$$\mathbb{E}_\pi\Big[\sum_{t=k}^{\infty} \gamma^t R_t \mid (X_t, A_t, R_t)_{t=0}^{k-1} = (x_t, a_t, r_t)_{t=0}^{k-1}, X_k = x\Big] = \mathbb{E}_\pi\Big[\sum_{t=k}^{\infty} \gamma^t R_t \mid X_k = x\Big].$$

14. Because the random trajectory is infinite, the technical definition of $\mathcal{D}_\pi$ requires some care and is given in Remark 2.4.

Time-homogeneity states that the distribution of the trajectory from a state x does not depend on the time k at which this state is visited. Graphically, this means that for a given state x, the probability distribution of the subtree rooted in that state is the same irrespective of the level k at which it occurs.

Lemma 2.14 (Time-homogeneity). The trajectory $(X_t, A_t, R_t)_{t\geq 0}$ is time-homogeneous, in the sense that for all $k \in \mathbb{N}$,

$$\mathcal{D}_\pi\Big((X_t, A_t, R_t)_{t=k}^{\infty} \mid X_k = x\Big) = \mathcal{D}_{\delta_x,\pi}\Big((X_t, A_t, R_t)_{t=0}^{\infty}\Big), \tag{2.11}$$

whenever the conditional distribution on the left-hand side is defined. △

While the left-hand side of Equation 2.11 is a proper conditional distribution, the right-hand side is derived from $\mathbb{P}_\pi(\cdot \mid X_0 = x)$ and by convention is not a conditional distribution. With this is mind, this is why we write $\mathcal{D}_{\delta_x,\pi}$ rather than the shorthand $\mathcal{D}_\pi(\cdot \mid X_0 = x)$.

Lemmas 2.13 and 2.14 follow as consequences of the definition of the trajectory distribution. A formal proof requires some measure-theoretic treatment; we provide a discussion of some of the considerations in Remark 2.4.

Proof of Proposition 2.12 (the Bellman equation). The result follows straightforwardly from Lemmas 2.13 and 2.14. We have

$$\begin{aligned}
V^\pi(x) &= \mathbb{E}_\pi\Big[\sum_{t=0}^{\infty} \gamma^t R_t \mid X_0 = x\Big] \\
&\overset{(a)}{=} \mathbb{E}_\pi\big[R_0 \mid X_0 = x\big] + \gamma\, \mathbb{E}_\pi\Big[\sum_{t=1}^{\infty} \gamma^{t-1} R_t \mid X_0 = x\Big] \\
&\overset{(b)}{=} \mathbb{E}_\pi\big[R_0 \mid X_0 = x\big] + \gamma \mathbb{E}_\pi\Bigg[\mathbb{E}_\pi\Big[\sum_{t=1}^{\infty} \gamma^{t-1} R_t \mid X_0 = x, A_0, X_1\Big] \mid X_0 = x\Bigg] \\
&\overset{(c)}{=} \mathbb{E}_\pi\big[R_0 \mid X_0 = x\big] + \gamma \mathbb{E}_\pi\Bigg[\mathbb{E}_\pi\Big[\sum_{t=1}^{\infty} \gamma^{t-1} R_t \mid X_1\Big] \mid X_0 = x\Bigg] \\
&\overset{(d)}{=} \mathbb{E}_\pi\big[R_0 \mid X_0 = x\big] + \gamma \mathbb{E}_\pi\big[V^\pi(X_1) \mid X_0 = x\big] \\
&= \mathbb{E}_\pi\left[R_0 + \gamma V^\pi(X_1) \mid X_0 = x\right],
\end{aligned}$$

where (a) is due to the linearity of expectations, (b) follows by the law of total expectation, (c) follows by the Markov property, and (d) is due to time-homogeneity and the definition of V^π. □

The Bellman equation states that the expected value of a state can be expressed in terms of the immediate action A_0, reward R_0, and the successor state X_1, omitting the rest of the trajectory. It is therefore convenient to define a generative model that only considers these three random variables along with the initial state X_0. Let $\xi \in \mathscr{P}(\mathcal{X})$ be a distribution over states. The *sample transition model* assigns a probability distribution to the tuple (X, A, R, X') taking values in $\mathcal{X} \times \mathcal{A} \times \mathbb{R} \times \mathcal{X}$ according to

$$\begin{aligned} &X \sim \xi\,; \\ &A \,|\, X \sim \pi(\cdot \,|\, X)\,; \\ &R \,|\, (X, A) \sim P_{\mathcal{R}}(\cdot \,|\, X, A)\,; \\ &X' \,|\, (X, A, R) \sim P_{\mathcal{X}}(\cdot \,|\, X, A)\,. \end{aligned} \tag{2.12}$$

We also write $\mathbb{P}_\pi$ for the joint distribution of these random variables. We will often find it useful to consider a single *source state* x, such that as before $\xi = \delta_x$. We write $(X = x, A, R, X')$ for the resulting random tuple, with probability distribution and expectation

$$\mathbb{P}_\pi(\cdot \,|\, X = x) \text{ and } \mathbb{E}_\pi[\cdot \,|\, X = x]\,.$$

The sample transition model allows us to omit time indices in the Bellman equation, which simplifies to

$$V^\pi(x) = \mathbb{E}_\pi\left[R + \gamma V^\pi(X') \,|\, X = x\right].$$

In the sample transition model, we call ξ the *state distribution*. It is generally different from the initial state distribution ξ_0, which describes a property of the environment. In Chapters 3 and 6, we will use the state distribution to model part of a learning algorithm's behavior.

2.7 The Random-Variable Bellman Equation

The Bellman equation characterizes the expected value of the random return from any state x compactly, allowing us to reduce the generative equations (an infinite sequence) to the sample transition model. In fact, we can leverage time-homogeneity and the Markov property to characterize all aspects of the random return in this manner. Consider again the definition of this return as a discounted sum of random rewards:

$$G = \sum_{t=0}^{\infty} \gamma^t R_t\,.$$

As with value functions, we would like to relate the return from the initial state to the random returns that occur downstream in the trajectory. To this end, let us define the *return function*

$$G^\pi(x) = \sum_{t=0}^{\infty} \gamma^t R_t, \quad X_0 = x\,, \tag{2.13}$$

which describes the return obtained when acting according to π starting from a given state x. Note that in this definition, the notation $X_0 = x$ again modifies the initial state distribution ξ_0. Equation 2.13 is thus understood as "the discounted sum of random rewards described by the generative equations with $\xi_0 = \delta_x$." Formally, G^π is a collection of random variables indexed by an initial state x, each generated by a random trajectory $(X_t, A_t, R_t)_{t\geq 0}$ under the distribution $\mathbb{P}_\pi(\cdot \mid X_0 = x)$. Because Equation 2.13 is concerned with random variables, we will sometimes find it convenient to be more precise and call it the *return-variable function.*[15]

The infinite tree of Figure 2.4 illustrates the abstract process by which one might generate realizations from the random variable $G^\pi(x)$. We begin at the root, whose value is fixed to $X_0 = x$. Each level is drawn by sampling, in succession, the action A_t, the reward R_t, and finally the successor state X_{t+1}, according to the generative equations. The return $G^\pi(x)$ accumulates the discounted rewards $(\gamma^t R_t)_{t\geq 0}$ along the way.

The nature of random variables poses a challenge to converting this generative process into a recursive formulation like the Bellman equation. It is tempting to try and formulate a distributional version of the Bellman equation by defining a collection of random variables $\tilde{G}^\pi$ according to the relationship

$$\tilde{G}^\pi(x) = R + \gamma \tilde{G}^\pi(X'), \quad X = x \tag{2.14}$$

for each $x \in \mathcal{X}$, in direct analogy with the Bellman equation for expected returns. However, these random variables $\tilde{G}^\pi$ do not have the same distribution as the random returns G^π. The following example illustrates this point.

Example 2.15. Consider the single-state Markov decision process of Example 2.10, for which the reward R has a Bernoulli distribution with parameter $1/2$. Equation 2.14 becomes

$$\tilde{G}^\pi(x) = R + \gamma \tilde{G}^\pi(x), \quad X = x,$$

which can be rearranged to give

$$\tilde{G}^\pi(x) = \sum_{t=0}^{\infty} \gamma^t R = \frac{1}{1-\gamma} R.$$

Since R is either 0 or 1, we deduce that

$$\tilde{G}^\pi(x) = \begin{cases} 0 & \text{with probability } 1/2 \\ \frac{1}{1-\gamma} & \text{with probability } 1/2. \end{cases}$$

15. One might wonder about the joint distribution of the random returns $(G^\pi(x) : x \in \mathcal{X})$. In this book, we will (perhaps surprisingly) not need to specify this joint distribution; however, it is valid to conceptualize these random variables as independent for concreteness.

This is different from the uniform distribution identified in Example 2.10, in the case $\gamma = 1/2$. △

The issue more generally with Equation 2.14 is that it effectively reuses the reward R and successor state X' across multiple visits to the initial state $X = x$, which in general is not what we intend and violates the structure of the MDP in question.[16] Put another way, Equation 2.14 fails because it does not correctly handle the joint distribution of the sequence of rewards $(R_t)_{t\geq 0}$ and states $(X_t)_{t\geq 0}$ encountered along a trajectory. This phenomenon is one difference between distributional and classical reinforcement learning; in the latter case, the issue is avoided thanks to the linearity of expectations.

The solution is to appeal to the notion of *equality in distribution*. We say that two random variables Z_1, Z_2 are equal in distribution, denoted

$$Z_1 \stackrel{\mathcal{D}}{=} Z_2\,,$$

if their probability distributions are equal. This is effectively shorthand for

$$\mathcal{D}(Z_1) = \mathcal{D}(Z_2)\,.$$

Equality in distribution can be thought of as breaking up the dependency of the two random variables on their sample spaces to compare them solely on the basis of their probability distributions. This avoids the problem posed by directly equating random variables.

Proposition 2.16 (The random-variable Bellman equation). Let G^π be the return-variable function of policy π. For a sample transition $(X = x, A, R, X')$ independent of G^π, it holds that for any state $x \in \mathcal{X}$,

$$G^\pi(x) \stackrel{\mathcal{D}}{=} R + \gamma G^\pi(X'), \quad X = x\,. \tag{2.15}$$

△

From a generative perspective, the random-variable Bellman equation states that we can draw a sample return by sampling an immediate reward R and successor state X' and then recursively generating a sample return from X'.

Proof of Proposition 2.16. Fix $x \in \mathcal{X}$ and let $\xi_0 = \delta_x$. Consider the (partial) random return

$$G_{k:\infty} = \sum_{t=k}^{\infty} \gamma^{t-k} R_t, \qquad k \in \mathbb{N}\,.$$

16. Computer scientists may find it useful to view Equation 2.14 as simulating draws from R and X' with a pseudo-random number generator that is reinitialized to the same state after each use.

In particular, $G^\pi(x) \overset{\mathcal{D}}{=} G_{0:\infty}$ under the distribution $\mathbb{P}_\pi(\cdot \mid X_0 = x)$. Following an analogous chain of reasoning as in the proof of Proposition 2.12, we decompose the return $G_{0:\infty}$ into the immediate reward and the rewards obtained later in the trajectory:

$$G_{0:\infty} = R_0 + \gamma G_{1:\infty} .$$

We can decompose this further based on the state occupied at time 1:

$$G_{0:\infty} = R_0 + \gamma \sum_{x' \in \mathcal{X}} \mathbb{1}\{X_1 = x'\} G_{1:\infty} .$$

Now, by the Markov property, on the event that $X_1 = x'$, the return obtained from that point on is independent of the reward R_0. Further, by the time-homogeneity property, on the same event, the return $G_{1:\infty}$ is equal in distribution to the return obtained when the episode begins at state x'. Thus, we have

$$R_0 + \gamma \sum_{x' \in \mathcal{X}} \mathbb{1}\{X_1 = x'\} G_{1:\infty} \overset{\mathcal{D}}{=} R_0 + \gamma \sum_{x' \in \mathcal{X}} \mathbb{1}\{X_1 = x'\} G^\pi(x') = R_0 + \gamma G^\pi(X_1) ,$$

and hence

$$G^\pi(x) \overset{\mathcal{D}}{=} R_0 + \gamma G^\pi(X_1), \quad X_0 = x.$$

The result follows by equality of distribution between (X_0, A_0, R_0, X_1) with the sample transition model $(X = x, A, R, X')$ when $X_0 = x$. □

Note that Proposition 2.16 implies the standard Bellman equation, in the sense that the latter is obtained by taking expectations on both sides of the distributional equation:

$$\begin{aligned} & G^\pi(x) \overset{\mathcal{D}}{=} R + \gamma G^\pi(X'), \quad X = x \\ \implies & \mathcal{D}_\pi(G^\pi(x)) = \mathcal{D}_\pi(R + \gamma G^\pi(X') \mid X = x) \\ \implies & \mathbb{E}_\pi[G^\pi(x)] = \mathbb{E}_\pi\left[R + \gamma G^\pi(X') \mid X = x\right] \\ \implies & V^\pi(x) = \mathbb{E}_\pi\left[R + \gamma V^\pi(X') \mid X = x\right] \end{aligned}$$

where we made use of the linearity of expectations as well as the independence of X' from the random variables $(G^\pi(x) : x \in \mathcal{X})$.

2.8 From Random Variables to Probability Distributions

Random variables provide an intuitive language with which to express how rewards are combined to form the random return. However, a proper random-variable Bellman equation requires the notion of equality in distribution to avoid incorrectly reusing realizations of the random reward R and next-state X'.

As a consequence, Equation 2.15 is somewhat incomplete: it characterizes the distribution of the random return from x, $G^\pi(x)$, but not the random variable itself.

A natural alternative is to do away with the return-variable function G^π and directly relate the distribution of the random return at different states. The result is what we may properly call the *distributional Bellman equation*. Working with probability distributions requires mathematical notation that is somewhat unintuitive compared to the simple addition and multiplication of random variables but is free of technical snags. With precision in mind, this is why we call Equation 2.15 the random-variable Bellman equation.

For a real-valued random variable Z with probability distribution $\nu \in \mathscr{P}(\mathbb{R})$, recall the notation

$$\nu(S) = \mathbb{P}(Z \in S), \quad S \subseteq \mathbb{R}.$$

This allows us to consider the probability assigned to S by ν more directly, without referring to Z. For each state $x \in \mathcal{X}$, let us denote the distribution of the random variable $G^\pi(x)$ by $\eta^\pi(x)$. Using this notation, we have

$$\eta^\pi(x)(S) = \mathbb{P}(G^\pi(x) \in S), \quad S \subseteq \mathbb{R}.$$

We call the collection of these per-state distributions the *return-distribution function*. Each $\eta^\pi(x)$ is a member of the space $\mathscr{P}(\mathbb{R})$ of probability distributions over the reals. Accordingly, the space of return-distribution functions is denoted $\mathscr{P}(\mathbb{R})^{\mathcal{X}}$.

To understand how the random-variable Bellman equation translates to the language of probability distributions, consider that the right-hand side of Equation 2.15 involves three operations on random variables: indexing into G^π, scaling, and addition. We use analogous operations over the space of probability distributions to construct the Bellman equation for return-distribution functions; these distributional operations are depicted in Figure 2.5.

Mixing. Consider the return-variable and return-distribution functions G^π and η^π, respectively, as well as a source state $x \in \mathcal{X}$. In the random-variable equation, the term $G^\pi(X')$ describes the random return received at the successor state X', when $X = x$ and A is drawn from $\pi(\cdot \mid X)$ – hence the idea of indexing the collection G^π with the random variable X'.

Generatively, this describes the process of first sampling a state x' from the distribution of X' and then sampling a realized return from $G^\pi(x')$. If we denote the result by $G^\pi(X')$, we see that for a subset $S \subseteq \mathbb{R}$, we have

$$\begin{aligned}\mathbb{P}_\pi(G^\pi(X') \in S \mid X = x) &= \sum_{x' \in \mathcal{X}} \mathbb{P}_\pi(X' = x' \mid X = x)\mathbb{P}_\pi(G^\pi(X') \in S \mid X' = x', X = x) \\ &= \sum_{x' \in \mathcal{X}} \mathbb{P}_\pi(X' = x' \mid X = x)\mathbb{P}_\pi(G^\pi(x') \in S)\end{aligned}$$

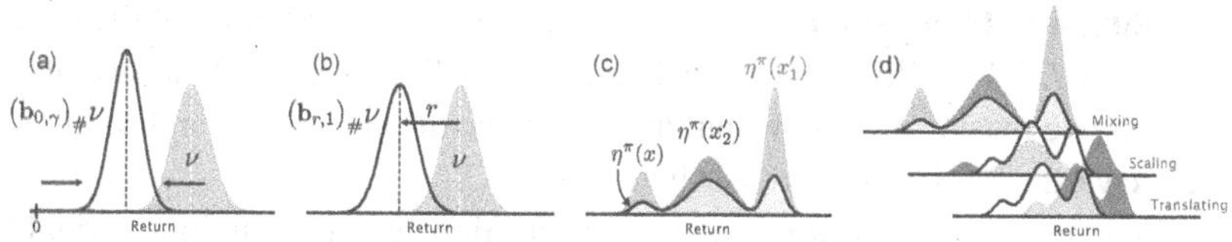

Figure 2.5
Illustration of the effects of the transformations of the random return in distributional terms, for a given source state. The discount factor is applied to the individual realizations of the return distribution, resulting in a **scaling (a)** of the support of this distribution. The arrows illustrate that this scaling results in a "narrower" probability distribution. The reward r **translates (b)** the return distribution. We write these two operations as a pushforward distribution constructed from the bootstrap function $\mathrm{b}_{r,\gamma}$. Finally, the return distributions of successor states are combined to form a **mixture distribution (c)** whose mixture weights are the transition probabilities to these successor states. The combined transformations are given in **(d)**, which depicts the return distribution at the source state (dark outline).

$$= \Big(\sum_{x' \in \mathcal{X}} \mathbb{P}_\pi(X' = x' \mid X = x) \eta^\pi(x') \Big)(S) . \tag{2.16}$$

This shows that the probability distribution of $G^\pi(X')$ is a weighted combination, or *mixture* of probability distributions from η^π. More compactly, we have

$$\begin{aligned} \mathcal{D}_\pi(G^\pi(X') \mid X = x) &= \sum_{x' \in \mathcal{X}} \mathbb{P}_\pi(X' = x' \mid X = x) \eta^\pi(x') \\ &= \mathbb{E}_\pi \left[\eta^\pi(X') \mid X = x \right] . \end{aligned}$$

Consequently, the distributional analogue of indexing into the collection G^π of random returns is the mixing of their probability distributions.

Although we prefer working directly with probability distributions, the indexing step also has a simple expression in terms of cumulative distribution functions. In Equation 2.16, taking the set S to be the half-open interval $(-\infty, z]$, we obtain

$$\begin{aligned} \mathbb{P}_\pi(G^\pi(X') \leq z \mid X = x) &= \mathbb{P}_\pi(G^\pi(X') \in (-\infty, z] \mid X = x) \\ &= \sum_{x' \in \mathcal{X}} \mathbb{P}_\pi(X' = x' \mid X = x) \mathbb{P}_\pi(G^\pi(x') \leq z) . \end{aligned}$$

Thus, the mixture of next-state return distributions can be described by the mixture of their cumulative distribution functions:

$$F_{G^\pi(X')}(z) = \sum_{x' \in \mathcal{X}} \mathbb{P}_\pi(X' = x' \mid X = x) F_{G^\pi(x')}(z) .$$

Scaling and translation. Suppose we are given the distribution of the next-state return $G^\pi(X')$. What is then the distribution of $R + \gamma G^\pi(X')$? To answer this question, we must express how multiplying by the discount factor and adding a fixed reward r transforms the distribution of $G^\pi(X')$.

This is an instance of a more general question: given a random variable $Z \sim \nu$ and a transformation $f : \mathbb{R} \to \mathbb{R}$, how should we express the distribution of $f(Z)$ in terms of f and ν? Our approach is to use the notion of a *pushforward distribution*. The pushforward distribution $f_\# \nu$ is defined as the distribution of $f(Z)$:

$$f_\# \nu = \mathcal{D}(f(Z)) .$$

One can think of it as applying the function f to the individual realizations of this distribution – "pushing" the mass of the distribution around. The pushforward notation allows us to reason about the effects of these transformations on distributions themselves, without having to involve random variables.

Now, for two scalars $r \in \mathbb{R}$ and $\gamma \in [0, 1)$, let us define the *bootstrap function*

$$\mathrm{b}_{r,\gamma} : z \mapsto r + \gamma z .$$

The pushforward operation applied with the bootstrap function scales each realization by γ and then adds r to it. That is,

$$(\mathrm{b}_{r,\gamma})_\# \nu = \mathcal{D}(r + \gamma Z) . \tag{2.17}$$

Expressed in terms of cumulative distribution functions, this is

$$F_{r+\gamma Z}(z) = F_Z\Big(\frac{z - r}{\gamma}\Big) .$$

We use the pushforward operation and the bootstrap function to describe the transformation of the next-state return distribution by the reward and the discount factor. If x' is a state with return distribution $\eta^\pi(x')$, then

$$(\mathrm{b}_{r,\gamma})_\# \eta^\pi(x') = \mathcal{D}(r + \gamma G^\pi(x')) .$$

We finally combine the pushforward and mixing operations to produce a probability distribution

$$(\mathrm{b}_{r,\gamma})_\# \, \mathbb{E}_\pi \left[\eta^\pi(X') \mid X = x\right] = \mathbb{E}_\pi \Big[(\mathrm{b}_{r,\gamma})_\# \eta^\pi(X') \mid X = x\Big] , \tag{2.18}$$

by linearity of the pushforward (see Exercises 2.11–2.13).

Equation 2.18 gives the distribution of the random return for a specific realization of the random reward R. By taking the expectation over R and X', we obtain the distributional Bellman equation.

Proposition 2.17 (The distributional Bellman equation). Let η^π be the return-distribution function of policy π. For any state $x \in \mathcal{X}$, we have

$$\eta^\pi(x) = \mathbb{E}_\pi[(\mathrm{b}_{R,\gamma})_\# \eta^\pi(X') \mid X = x] \,. \tag{2.19}$$

△

Figure 2.6 illustrates the recursive relationship between return distributions described by the distributional Bellman equation.

Example 2.18. In Example 2.10, it was determined that the random return at state x is uniformly distributed on the interval $[0, 2]$. Recall that there are two possible rewards (0 and 1) with equal probability, x transitions back to itself, and $\gamma = 1/2$. As $X' = x$, when $r = 0$, we have

$$\begin{aligned}\mathbb{E}_\pi\Big[(\mathrm{b}_{r,\gamma})_\# \eta^\pi(X') \mid X = x\Big] &= (\mathrm{b}_{0,\gamma})_\# \eta^\pi(x) \\ &= \mathcal{D}(\gamma G^\pi(x)) \\ &= \mathcal{U}([0, 2\gamma]) \\ &= \mathcal{U}([0, 1])\,.\end{aligned}$$

Similarly, when $r = 1$, we have

$$\begin{aligned}(\mathrm{b}_{1,\gamma})_\# \eta^\pi(x) &= \mathcal{D}(1 + \gamma G^\pi(x)) \\ &= \mathcal{U}([1, 2])\,.\end{aligned}$$

Consequently,

$$\begin{aligned}\mathcal{D}_\pi(R + \gamma G^\pi(X') \mid X = x) &= \frac{1}{2}(\mathrm{b}_{0,\gamma})_\# \eta^\pi(x) + \frac{1}{2}(\mathrm{b}_{1,\gamma})_\# \eta^\pi(x) \\ &= \frac{1}{2}\mathcal{U}([0, 1]) + \frac{1}{2}\mathcal{U}([1, 2]) \\ &= \mathcal{U}([0, 2]) \\ &= \eta^\pi(x)\,.\end{aligned}$$

This illustrates how the pushforward operation can be used to express, in distributional terms, the transformations at the heart of the random-variable Bellman equation. △

Proposition 2.17 can also be stated in terms of cumulative distribution functions. Halfway between Equation 2.19 and the random-variable equation, what one might call the cumulative distribution function Bellman equation is

$$F_{G^\pi(x)}(z) = \mathbb{E}_\pi\big[F_{R+\gamma G^\pi(X')}(z) \mid X = x\big]$$

$$= \mathbb{E}_\pi\Big[F_{G^\pi(X')}\Big(\frac{z-R}{\gamma}\Big) \mid X = x\Big].$$

The first form highlights the relationship between the random-variable equation and the distributional equation, whereas the second form more directly expresses the relationship between the different cumulative distribution functions; the odd indexing reflects the process of transferring probability mass "backward," from $G^\pi(X')$ to $G^\pi(x)$. Although the cumulative distribution function Bellman equation is at first glance simpler than Equation 2.19, in later chapters, we will see that working with probability distributions is usually more natural and direct.

Proof of Proposition 2.17. Let $S \subseteq \mathbb{R}$ and fix $x \in \mathcal{X}$. By the law of total expectation, we may write

$$\begin{aligned}
&\mathbb{P}_\pi(R + \gamma G^\pi(X') \in S \mid X = x) \\
&\quad = \sum_{a\in\mathcal{A}} \pi(a \mid x) \sum_{x'\in\mathcal{X}} P_{\mathcal{X}}(x' \mid x, a)\mathbb{P}_\pi(R + \gamma G^\pi(X') \in S \mid X = x, A = a, X' = x') \\
&\quad \overset{(a)}{=} \sum_{a\in\mathcal{A}} \pi(a \mid x) \sum_{x'\in\mathcal{X}} P_{\mathcal{X}}(x' \mid x, a) \times \\
&\qquad\qquad \mathbb{E}_R\Big[\mathbb{P}_\pi(R + \gamma G^\pi(x') \in S \mid X = x, A = a, R) \mid X = x, A = a\Big] \\
&\quad = \sum_{a\in\mathcal{A}} \pi(a \mid x) \sum_{x'\in\mathcal{X}} P_{\mathcal{X}}(x' \mid x, a)\,\mathbb{E}_R\Big[((\mathrm{b}_{R,\gamma})_\# \eta^\pi(x'))(S) \mid X = x, A = a\Big] \\
&\quad = \mathbb{E}_\pi\Big[((\mathrm{b}_{R,\gamma})_\# \eta^\pi(X'))(S) \mid X = x\Big] \\
&\quad = \mathbb{E}_\pi\Big[(\mathrm{b}_{R,\gamma})_\# \eta^\pi(X') \mid X = x\Big](S).
\end{aligned}$$

The notation $\mathbb{E}_R$ in (a) denotes the expectation with respect to $R \sim P_{\mathcal{R}}(\cdot \mid x, a)$; the independence of R and X' given A allows us to remove the conditional $X' = x'$ from the inner probability term. Now, by Proposition 2.16, we know that

$$G^\pi(x) \overset{\mathcal{D}}{=} R + \gamma G^\pi(X'), \quad X = x.$$

By definition, $\eta^\pi(x)$ is the distribution of $G^\pi(x)$ and hence the distribution of $R + \gamma G^\pi(X')$ when $X = x$. But the derivation above shows that that distribution is

$$\mathbb{E}_\pi\big[(\mathrm{b}_{R,\gamma})_\# \eta^\pi(X') \mid X = x\big],$$

which completes the proof. □

Equation 2.19 directly relates the return distributions at different states and does away with the return-variable function G^π. This makes it particularly useful when mathematical rigor is required, such as to prove the more formal results

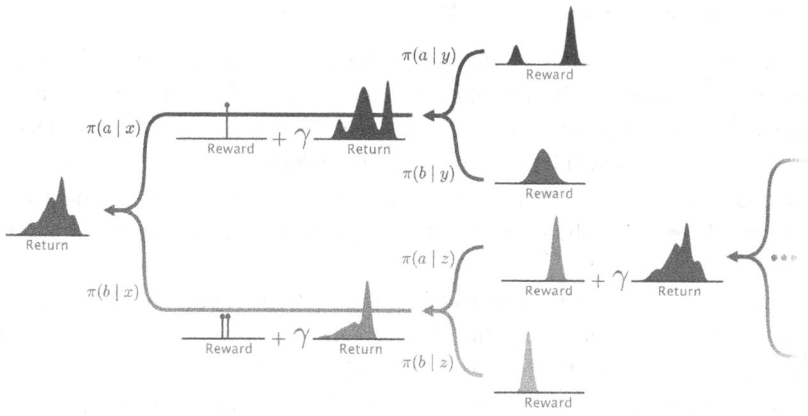

Figure 2.6
The distributional Bellman equation states that the return distribution at one state (far left) is formed by scaling the return distributions at successor states, shifting them by the realized rewards, and mixing over possible realizations.

in later chapters. The random variables X, R, and X' remain, however, inside the expectation on the right-hand side. Their role is to concisely express the three operations of mixing, scaling, and translation. It is also possible to omit these random variables and write Equation 2.19 purely in terms of probability distributions, by making the expectation explicit:

$$\eta^\pi(x) = \sum_{a \in \mathcal{A}} \pi(a \mid x) \sum_{x' \in \mathcal{X}} P_\mathcal{X}(x' \mid x, a) \int_\mathbb{R} P_\mathcal{R}(\mathrm{d}r \mid x, a)(\mathrm{b}_{r,\gamma})_\# \eta^\pi(x') \,, \qquad (2.20)$$

as was partially done in the proof above. We will revisit this form of the distributional Bellman equation in Chapter 5 when we consider algorithms for approximating the return distribution. In general, however, Equation 2.19 is more compact and applies equally well to discrete and continuous distributions.

Observe that the random action A is implicit in the random-variable and distributional Bellman equations but explicit in Equation 2.20. This is because A is needed to determine the probability distributions of R and X', which are only independent conditional on the action. This is a useful reminder that we need to treat the addition of R to $\gamma G^\pi(X')$ with care – these are not in general independent random variables, and their addition does not generally correspond to a simple convolution of distributions. We will consider this issue in more detail in Chapter 4.

2.9 Alternative Notions of the Return Distribution*

In reinforcement learning, the discount factor γ is typically presented as a scalar multiplier that shrinks the magnitude of future rewards. This is the interpretation we have used to present the return in this chapter. However, the discount factor can be given a different interpretation as a *probability of continuing*. Under this alternate interpretation, the rewards received by the agent are undiscounted, and instead there is a probability $1-\gamma$ that the sequence terminates at each time step.

Under this model, the agent's interactions with its environment end after a random *termination time* T, which is geometrically distributed with parameter γ, independently of all other random variables in the sequence. Specifically, we have $\mathbb{P}_\pi(T=t) = \gamma^t(1-\gamma)$ for all $t \geq 0$. In this case, the agent's return is now given by

$$\sum_{t=0}^{T} R_t \, .$$

We call this the *random-horizon return*.

Straightforward calculation allows us to show that the expected value of the return is unchanged (see Exercise 2.14):

$$\mathbb{E}_\pi\left[\sum_{t=0}^{T} R_t\right] = \mathbb{E}_\pi\left[\sum_{t=0}^{\infty} \gamma^t R_t\right] .$$

Because of this equivalence, from an expected-return perspective, it does not matter whether γ is interpreted as a scaling factor or probability. By contrast, this change in interpretation of γ has a significant impact on the structure of the return distribution. For example, the random variable $\sum_{t=0}^{T} R_t$ is unbounded even when the rewards are bounded, unlike the discounted return.

Just as there exists a distributional Bellman equation for the discounted return, there is also a recursive characterization of the return distribution for this variant of the return. By introducing an auxiliary random variable $I = \mathbb{1}_{\{T>0\}}$, we can write down a distributional Bellman equation for this alternative interpretation of the return. Expressing this relationship in terms of random variables, we have

$$G^\pi(x) \stackrel{\mathcal{D}}{=} R + I G^\pi(X'), \quad X = x \, .$$

In terms of distributions themselves, we have

$$\eta^\pi(x) = (1-\gamma)\mathcal{D}_\pi(R \mid X = x) + \gamma \mathbb{E}_\pi\left[(\mathrm{b}_{R,1})_\# \eta^\pi(X') \mid X = x\right],$$

where $\mathcal{D}_\pi(R \mid X = x)$ denotes the distribution of the reward R at x under π. Notice that γ in the equation above, which previously scaled the support of the distribution, now mixes two probability distributions. These results are proven in the course of Exercise 2.15 along with similar results for a more general class of return definitions. The fact that there exists a distributional Bellman equation for these variants of the return means that much of the theory of distributional reinforcement learning still applies. We encourage the reader to consider how the algorithms and analysis presented in later chapters may be modified to accommodate alternative notions of the return.

We see that there are multiple ways in which we can define distributional Bellman equations while preserving the relationship between expectations. Another approach, commonly known as *n-step bootstrapping* in reinforcement learning, relates the return distribution at state x with the return distribution n steps into the future. Exercise 2.16 describes another alternative based on first return times.

2.10 Technical Remarks

Remark 2.1. Our account of random variables in this chapter has not required a measure-theoretic formalism, and in general this will be the case for the remainder of the book. For readers with a background in measure theory, the existence of the joint distribution of the random variables $(X_t, A_t, R_t)_{t\geq 0}$ in the Markov decision process model described in this chapter can be deduced from constructing a consistent sequence of distributions for the random variables $(X_t, A_t, R_t)_{t=0}^{H}$ for each horizon $H \in \mathbb{N}$ using the conditional distributions given in the definition and then appealing to the Ionescu Tulcea theorem (Tulcea 1949). See Lattimore and Szepesvári (2020) for further discussion in the context of Markov decision processes.

We also remark that a more formal definition of probability measures restricts the subsets of outcomes under consideration to be measurable with an underlying σ-algebra. Here, and elsewhere in the book, we will generally avoid repeated mention of the qualifier "measurable" in cases such as these. The issue can be safely ignored by readers without a background in measure theory. For readers with such a background, measurability is always implicitly with respect to the power set σ-algebra on finite sets, the Borel σ-algebra on the reals, and the corresponding product σ-algebra on products of such spaces. △

Remark 2.2. The existence of the random return $\sum_{t=0}^{\infty} \gamma^t R_t$ is often taken for granted in reinforcement learning. In part, this is because assuming that rewards are deterministic or have distributions with bounded support makes its existence straightforward; see Exercise 2.17.

The situation is more complicated when the support of the reward distributions is unbounded, as there might be realizations of the random trajectory for which the sequence $(G_{0:T})_{T\geq 0}$ does not converge. If, for example, the rewards are normally distributed, then it is possible that $R_0 = 1, R_1 = -\gamma^{-1}, R_2 = \gamma^{-2}, R_3 = -\gamma^{-3}, \dots$, in which case $G_{0:T}$ oscillates between 0 and 1 indefinitely. When the truncated returns do not converge, it is not meaningful to talk about the return obtained in this realization of the trajectory.

Under the assumption that all reward distributions have finite mean, we can demonstrate convergence of the truncated return with probability 1:

$$\mathbb{P}\Big(\lim_{T\to\infty} \sum_{t=0}^{T} \gamma^t R_t \text{ exists}\Big) = 1\,, \tag{2.21}$$

and hence guarantee that we can meaningfully study the distributional properties of the random return. Intuitively, problematic sequences of rewards such as the sequence given above require reward samples that are increasingly "unlikely". Equation 2.21 is established by appealing to the martingale convergence theorem, and the details of this derivation are explored in Exercise 2.19. △

Remark 2.3. The assumption of finite state and action spaces makes several technical aspects of the presentation in this book more straightforward. As a simple example, consider a Markov decision process with countably infinite state space given by the nonnegative integers $\mathbb{N}$, for which state x deterministically transitions to state $x + 1$. Suppose in addition that the trajectory begins in state 0. If the reward at state x is $(-\gamma)^{-x}$, then the truncated return up to time T is given by

$$\sum_{t=0}^{T-1} \gamma^t R_t = \sum_{t=0}^{T-1} \gamma^t (-\gamma)^{-t} = \sum_{t=0}^{T-1} (-1)^{-t}\,.$$

In the spirit of Remark 2.2, when T is odd, this sum is 1, while for T even, it is 0. Hence, the limit $\sum_{t=0}^{\infty} \gamma^t R_t$ does not exist, despite the fact that all reward distributions have finite mean.

Dealing with continuous state spaces requires more technical sophistication still. Constructing the random variables describing the trajectory requires taking into account the topology of the state space in question, and careful consideration of technical conditions on policies (such as measurability) is necessary too. For details on how these issues may be addressed, see Puterman (2014) and Meyn and Tweedie (2012). △

Remark 2.4 (Proof of time-homogeneity and the Markov property)**.** The generative equations defining the distribution of the random trajectory $(X_t, A_t, R_t)_{t\geq 0}$ immediately give statements such as

$$\mathbb{P}_\pi(X_{k+1} = x' \mid X_{0:k}, A_{0:k}, R_{0:k}, X_k) = \mathbb{P}_\pi(X_{k+1} = x' \mid X_k)\,,$$

which shows that the next state in the trajectory enjoys the Markov property. Lemmas 2.13 and 2.14 state something stronger: that the entire future trajectory has this property. To prove these lemmas from the generative equations, the strategy is to note that the distribution of $(X_t, A_t, R_t)_{t=k}^{\infty}$ conditional on $(X_t, A_t, R_t)_{t=0}^{k-1}$ and X_k is determined by its finite-dimensional marginal distributions: that is, conditional probabilities of the form

$$\mathbb{P}_\pi(X_{t_i} = x_i, A_{t_i} = a_i, R_{t_i} \in S_i \text{ for } i = 1, \ldots, n \mid (X_t, A_t, R_t)_{t=0}^{k-1}, X_k)$$

for all $n \in \mathbb{N}^+$ and sequences $k \leq t_1 < \cdots < t_n$ of time steps (see, e.g., Billingsley 2012, Section 36). Thus, if these conditional probabilities are equal to the corresponding quantities

$$\mathbb{P}_\pi(X_{t_i} = x_i, A_{t_i} = a_i, R_{t_i} \in S_i \text{ for } i = 1, \ldots, n \mid X_k),$$

then the distributions are equal. Finally, the equality of these finite-dimensional marginals can be shown by induction from the generative equations. △

2.11 Bibliographical Remarks

An introduction to random variables and probability distributions may be found in any undergraduate textbook on the subject. For a more technical presentation, see Williams (1991) and Billingsley (2012).

2.2. Markov decision processes are generally attributed to Bellman (1957b). A deeper treatment than we give here can be found in most introductory textbooks, including those by Bertsekas and Tsitsiklis (1996), Szepesvári (2010), and Puterman (2014). Our notation is most aligned with that of Sutton and Barto (2018). Interestingly, while it is by now standard to use reward as an indicator of success, Bellman's own treatment does not make it an integral part of the formalism.

2.3. The formulation of a trajectory as a sequence of random variables is central to control as inference (Toussaint 2009; Levine 2018), which uses tools from probabilistic reasoning to derive optimal policies. Howard (1960) used the analogy of a frog hopping around a lily pond to convey the dynamics of a Markov decision process; we find our own analogy more vivid. The special consideration owed to infinite sequences is studied at length by Hutter (2005).

2.4. The work of Veness et al. (2015) makes the return (as a random variable) the central object of interest and is the starting point of our own investigations into distributional reinforcement learning. Issues regarding the existence of the random return and a proper probabilistic formulation can be found in that paper. An early formulation of the return-variable function can be found in Jaquette (1973), who used it to study alternative optimality criteria. Chapman and Kaelbling (1991) consider the related problem of predicting the discounted cumulative probability of observing a particular level of reward in order to mitigate partial observability in a simple video game. The blackjack and cliff-walking examples are adapted from Sutton and Barto (2018) and, in the latter case, inspired by one of the authors' trip to Ireland. In both cases, we put a special emphasis on the probability distribution of the random return. The uniform distribution example is taken from Bellemare et al. (2017a); such discounted sums of Bernoulli random variables also have a long history in probability theory (Jessen and Wintner 1935; see Solomyak 1995; Diaconis and Freedman 1999; Peres et al. 2000 and references therein).

2.5–2.6. The Bellman equation is standard to most textbooks on the topic. A particularly thorough treatment can be found in the work of Puterman (2014) and Bertsekas (2012). The former also provides a good discussion on the implications of the Markov property and time-homogeneity.

2.7–2.8. Bellman equations relating quantities other than expected returns were originally introduced in the context of risk-sensitive control, at varying levels of generality. The formulation of the distributional Bellman equation in terms of cumulative distribution functions was first given by Sobel (1982), under the assumption of deterministic rewards and policies. Chung and Sobel (1987) later gave a version for random, bounded rewards. See also the work of White (1988) for a review of some of these approaches and Morimura et al. (2010a) for a more recent presentation of the CDF Bellman equation.

Other versions of the distributional Bellman equation have been phrased in terms of moments (Sobel 1982), characteristic functions (Mandl 1971; Farahmand 2019), and the Laplace transform (Howard and Matheson 1972; Jaquette 1973, 1976; Denardo and Rothblum 1979), again at varying levels of generality, and in some cases using the undiscounted return. Morimura et al. (2010b) also present a version of the equation in terms of probability densities. The formulation of the distributional Bellman equation in terms of pushforward distributions is due to Rowland et al. (2018); the pushforward notation is broadly used in measure-theoretic probability, and our use of it is influenced by optimal transport theory (Villani 2008).

Distributional formulations have also been used to design Bayesian methods for reasoning about uncertainty regarding the value function (Ghavamzadeh et al. 2015). Dearden et al. (1998) propose an algorithm that maintains a posterior on the return distribution under the assumption that rewards are normally distributed. Engel et al. (2003, 2007) use a random-variable equation to derive an algorithm based on Gaussian processes.

Our own work in the field is rooted in the theory of stochastic fixed point equations (Rösler 1991, 1992; Rachev and Rüschendorf 1995), also known as recursive distributional equations (Aldous and Bandyopadhyay 2005), from which we draw the notion of equality in distribution. Recursive distributional equations have been used, among other applications, for complexity analysis of randomized algorithms (Rösler and Rüschendorf 2001), as well as the study of branching processes and objects in random geometry (Liu 1998). See Alsmeyer (2012) for a book-length survey of the field. In effect, the random-variable Bellman equation for a given state x is a recursive distributional equation in the sense of Aldous and Bandyopadhyay (2005); however, in distributional reinforcement learning, we typically emphasize the collection of random variables (the return-variable function) over individual equations.

2.9. The idea of treating the discount factor as a probability recurs in the literature, often as a technical device within a proof. Our earliest source is Derman (1970, Chap. 3). The idea is used to handle discounted, infinite horizon problems in the reinforcement learning as inference setting (Toussaint and Storkey 2006) and is remarked on by Sutton et al. (2011). The probabilistic interpretation of the discount factor has also found application in model-based reinforcement learning (Sutton 1995; Janner et al. 2020), and is closely related to the notion of termination probabilities in options (Sutton et al. 1999). The extension to the distributional setting was worked out with Tom Schaul but not previously published. See also White (2017) for a broader discussion on unifying various uses of discount factors.

2.12 Exercises

Exercise 2.1. We defined a state x as being terminal if

$$P_{\mathcal{X}}(x \mid x, a) = 1 \qquad \text{for all } a \in \mathcal{A}$$

and $P_{\mathcal{R}}(0 \mid x, a) = 1$.

(i) Explain why multiple terminal states are redundant from a return function perspective.

(ii) Suppose you are given an MDP with k terminal states. Describe a procedure that creates a new MDP that behaves identically but has a single terminal state. △

Exercise 2.2. Consider the generative equations of Section 2.3. Using Bayes's rule, explain why

$$\mathbb{P}_\pi(X_1 = x' \mid X_0 = x, A_0 = a)$$

needs to be introduced as a notational convention, rather than derived from the definition of the joint distribution $\mathbb{P}_\pi$. △

Exercise 2.3. Suppose that $\gamma < 1$ and $R_t \in [R_{\text{MIN}}, R_{\text{MAX}}]$ for all $t \in \mathbb{N}$. Show that

$$\sum_{t=0}^{\infty} \gamma^t R_t \in \left[\frac{R_{\text{MIN}}}{1-\gamma}, \frac{R_{\text{MAX}}}{1-\gamma}\right].$$ △

Exercise 2.4. Find a probability distribution ν that has unbounded support and such that

$$\mathbb{P}_\pi\Big(\Big|\sum_{t=0}^{\infty} \gamma^t Z_t\Big| < \infty\Big) = 1, \quad Z_t \overset{\text{i.i.d.}}{\sim} \nu \text{ for } t \in \mathbb{N}.$$ △

Exercise 2.5. Using an open-source implementation of the game of blackjack or your own, use simulations to estimate the probability distribution of the random return under

(i) The uniformly random policy.
(ii) The policy that doubles when starting with 10 or 11, hits on 16 and lower, and otherwise sticks.
(iii) The policy that hits on 16 and lower and never doubles. Is it a better policy? △

In all cases, aces should count as 11 whenever possible.

Exercise 2.6. Show that any random variable Z satisfying the condition in Equation 2.6 must have CDF given by

$$\mathbb{P}_\pi(Z \le z) = \tfrac{z}{2}, \; 0 \le z \le 2,$$

and hence that Z has distribution $\mathcal{U}([0, 2])$. △

Exercise 2.7 (*). The Cantor distribution has a cumulative distribution function F, defined for $z \in [0, 1]$ by expressing z in base 3:

$$z = \sum_{n=1}^{\infty} z_n 3^{-n}, \qquad z_n \in \{0, 1, 2\},$$

such that

$$F(z) = \sum_{n=1}^{\infty} \mathbb{1}\{z_n > 0, z_i \ne 1 \text{ for all } 1 \le i < n\} 2^{-n}.$$

Additionally, $F(z) = 0$ for $z \le 0$ and $F(z) = 1$ for $z \ge 1$. Prove that the return distribution for the MDP in Example 2.11 is the Cantor distribution. △

Exercise 2.8. The *n-step Bellman equation* relates the value function at a state x with the discounted sum of $n \in \mathbb{N}^+$ future rewards and the value of the nth successor state. Prove that, for each $n \in \mathbb{N}^+$,

$$V^\pi(x) = \mathbb{E}_\pi\Big[\sum_{t=0}^{n-1} \gamma^t R_t + \gamma^n V^\pi(X_n) \mid X_0 = x\Big].$$ △

Exercise 2.9. In Section 2.5, we argued that the number of partial trajectories of length T can grow exponentially with T. Give examples of Markov decision processes with $N_{\mathcal{X}}$ states and $N_{\mathcal{A}}$ actions where the number of possible length T realizations $(x_t, a_t, r_t)_{t=0}^{T}$ of the random trajectory is

(i) Bounded by a constant.
(ii) Linear in T. △

Exercise 2.10. In Section 2.7, we argued that the equation

$$\tilde{G}^\pi(x) = R + \gamma \tilde{G}^\pi(X'), \quad X = x$$

leads to the wrong probability distribution over returns. Are there scenarios in which this is a sensible model? △

Exercise 2.11. The purpose of this exercise is to familiarize yourself with the transformations of the pushforward operation applied to the bootstrap function b. Let $\nu \in \mathscr{P}(\mathbb{R})$ be a probability distribution and let Z be a random variable with distribution ν. Let $r = 1/2$, $\gamma = 1/3$, and let R be a Bernoulli($1/4$) random variable independent of Z. For each of the following probability distributions:

(i) $\nu = \delta_1$;
(ii) $\nu = 1/2\delta_{-1} + 1/2\delta_1$;
(iii) $\nu = \mathcal{N}(2, 1)$,

express the probability distributions produced by the following operations:

(i) $(\mathrm{b}_{r,1})_\# \nu = \mathcal{D}(r + Z)$;
(ii) $(\mathrm{b}_{0,\gamma})_\# \nu = \mathcal{D}(\gamma Z)$;
(iii) $(\mathrm{b}_{r,\gamma})_\# \nu = \mathcal{D}(r + \gamma Z)$; and
(iv) $\mathbb{E}[(\mathrm{b}_{R,\gamma})_\# \nu] = \mathcal{D}(R + \gamma Z)$. △

Exercise 2.12. Repeat the preceding exercise, now with $R \sim \mathcal{N}(0, 1)$. Conclude on the distribution of the random return

$$G = \sum_{t=0}^{\infty} \gamma^t R_t ,$$

where $R_t \sim \mathcal{N}(1, 1)$ for all $t \geq 0$. △

Exercise 2.13. Let $\eta \in \mathscr{P}(\mathbb{R})^{\mathcal{X}}$ be a return-distribution function, and let X' be a $\mathcal{X}$-valued random variable. Show that for any $r, \gamma \in \mathbb{R}$, the pushforward operation combined with the bootstrap function forms an *affine map*, in the sense that

$$(\mathrm{b}_{r,\gamma})_\# \, \mathbb{E}\left[\eta(X')\right] = \mathbb{E}\left[(\mathrm{b}_{r,\gamma})_\# \eta(X')\right] .$$

△

Exercise 2.14. Consider the standard discounted return $\sum_{t=0}^{\infty} \gamma^t R_t$. Recall from Section 2.9 the random-horizon return, $\sum_{t=0}^{T} R_t$, where T has a geometric probability distribution with parameter γ:

$$\mathbb{P}_\pi(T = k) = \gamma^k (1 - \gamma) .$$

Show that

$$\mathbb{E}_\pi\left[\sum_{t=0}^{\infty} \gamma^t R_t\right] = \mathbb{E}_\pi\left[\sum_{t=0}^{T} R_t\right] . \qquad \triangle$$

Exercise 2.15. Consider again the random-horizon return described in Section 2.9.

(i) Show that if the termination time T has a geometric distribution with parameter γ, then T has the memoryless property: that is, $\mathbb{P}_\pi(T \geq k + l \,|\, T \geq k) = \mathbb{P}_\pi(T \geq l)$, for all $k, l \in \mathbb{N}$.

(ii) Hence, show that the return-variable function corresponding to this alternative definition of return satisfies the Bellman equation

$$G^\pi(x) \overset{\mathcal{D}}{=} R + \mathbb{1}_{\{T>0\}} G^\pi(X'), \quad X = x .$$

for each $x \in \mathcal{X}$.

Consider now a further alternative notion of return, given by considering a sequence of identically and independent distributed (i.i.d.) nonnegative random variables $(I_t)_{t \geq 0}$, independent from all other random variables in the MDP. Suppose $\mathbb{E}_\pi[I_t] = \gamma$. Define the return to be

$$\sum_{t=0}^{\infty} \Big(\prod_{s=0}^{t-1} I_s \Big) R_t .$$

(iii) Show that both the standard definition of the return and the random-horizon return can be viewed as special cases of this more general notion of return by particular choices of the distribution of the $(I_t)_{t \geq 0}$ variables.

(iv) Verify that

$$\mathbb{E}_\pi\Big[\sum_{t=0}^{\infty} \Big(\prod_{s=0}^{t-1} I_s \Big) R_t\Big] = \mathbb{E}_\pi\Big[\sum_{t=0}^{\infty} \gamma^t R_t\Big].$$

(v) Show that the random-variable Bellman equation associated with this notion of return is given by

$$G^\pi(x) = R + I_0 G^\pi(X') , \quad X = x$$

for each $x \in \mathcal{X}$. $\triangle$

Exercise 2.16. For a state $x \in \mathcal{X}$, write

$$T_x = \min\{t \geq 1 : X_t = x\}$$

with $T_x = \infty$ if x is never reached from time $t = 1$ onward. When $X_0 = x$, this is the random time of first return to x. Prove that the following alternative random-variable Bellman equation holds:

$$G^\pi(x) \overset{\mathcal{D}}{=} \sum_{t=0}^{T-1} \gamma^t R_t + \gamma^T G^\pi(x), \quad X_0 = x\,. \qquad \triangle$$

Exercise 2.17. The aim of the following two exercises is to explore under what conditions the random return $G = \sum_{t=0}^{\infty} \gamma^t R_t$ exists. As in Remark 2.2, denote the truncated return by

$$G_{0:T} = \sum_{t=0}^{T} \gamma^t R_t.$$

Show that if all reward distributions are supported on the interval $[R_{\text{MIN}}, R_{\text{MAX}}]$, then for any values of the random variables $(R_t)_{t\geq 0}$, the sequence $(G_{0:T})_{T\geq 0}$ converges, and hence G is well defined. △

Exercise 2.18 (*). The random-variable Bellman equation is effectively a system of *recursive distributional equations* (Aldous and Bandyopadhyay 2005). This example asks you to characterize the normal distribution as the solution to another such equation. Let Z be a normally distributed random variable with mean zero and unit variance (that is, $Z \sim \mathcal{N}(0, 1)$). Recall that the probability density p_Z and characteristic function χ_Z of Z are

$$p_Z(z) = \frac{1}{\sqrt{2\pi}} \exp\Big(-\frac{z^2}{2}\Big), \qquad \chi_Z(s) = \exp\Big(-\frac{s^2}{2}\Big).$$

Suppose that Z' is an independent copy of Z. Show that

$$Z \overset{\mathcal{D}}{=} 2^{-1/2} Z + 2^{-1/2} Z'$$

by considering

(i) the probability density of each side of the equation above;
(ii) the characteristic function of each side of the equation above. △

Exercise 2.19 (*). Under the overarching assumption of this chapter and the remainder of the book that there are finitely many states and all reward distributions have finite first moment, we can show that the limit

$$\lim_{T\to\infty} \sum_{t=0}^{T} \gamma^t R_t \tag{2.22}$$

exists with probability 1. First, show that the sequence of truncated returns $(G_{0:T})_{T\geq 0}$ forms a semi-martingale with respect to the filtration $(\mathcal{F}_t)_{t\geq 0}$, defined by

$$\mathcal{F}_t = \sigma(X_{0:t}, A_{0:t}, R_{0:t-1})\,;$$

the relevant decomposition is

$$G_{0:T} = G_{0:T-1} + \underbrace{\gamma^t \mathbb{E}_\pi[R_T \mid X_T, A_T]}_{\text{predictable increment}} + \underbrace{\gamma^t (R_T - \mathbb{E}_\pi[R_T \mid X_T, A_T])}_{\text{martingale noise}} .$$

Writing

$$C_T = \sum_{t=0}^{T} \gamma^t \mathbb{E}_\pi[R_t \mid X_t, A_t]$$

for the sum of predictable increments, show that this sequence converges for all realizations of the random trajectory. Next, show that the sequence $(\bar{C}_T)_{T\geq 0}$ defined by $\bar{C}_T = G_{0:T} - C_T$ is a martingale, and by using the assumption of a finite state space $\mathcal{X}$, show that it satisfies

$$\sup_{T\geq 0} \mathbb{E}[|\bar{C}_T|] < \infty .$$

Hence, use the martingale convergence theorem to argue that $(\bar{C}_T)_{T\geq 0}$ converges with probability 1 (w.p. 1). From there, deduce that the limit of Equation 2.22 exists w.p. 1. △

Exercise 2.20 (*). The second Borel–Cantelli lemma (Billingsley 2012, Section 4) states that if $(E_t)_{t\geq 0}$ is a sequence of independent events and $\sum_{t\geq 0} \mathbb{P}(E_t) = \infty$, then

$$\mathbb{P}(\text{infinitely many } E_t \text{ occur}) = 1 .$$

Let $\varepsilon > 0$. Exhibit a Markov decision process and policy for which the events $\{|\gamma^t R_t| > \varepsilon\}$ are independent, and satisfy

$$\mathbb{P}_\pi(|\gamma^t R_t| > \varepsilon \text{ for infinitely many } t) = 1 .$$

Deduce that

$$\mathbb{P}_\pi\Big(\sum_{t=0}^{T} \gamma^t R_t \text{ converges as } T \to \infty\Big) = 0$$

in this case. *Hint.* Construct a sufficiently heavy-tailed reward distribution from a suitable cumulative distribution function. △

3 Learning the Return Distribution

Reinforcement learning provides a computational model of what and how an animal, or more generally an agent, learns to *predict* on the basis of received experience. Often, the prediction of interest is about the expected return under the initial state distribution or, counterfactually, from a query state $x \in \mathcal{X}$. In the latter case, making good predictions is equivalent to having a good estimate of the value function V^π. From these predictions, reinforcement learning also seeks to explain how the agent might best *control* its environment: for example, to maximize its expected return.

By *learning from experience*, we generally mean from data rather than from the Markov decision process description of the environment (its transition kernel, reward distribution, and so on). In many settings, these data are taken from sample interactions with the environment. In their simplest forms, these interactions are realizations of the random trajectory $(X_t, A_t, R_t)_{t \geq 0}$. The record of a particular game of chess and the fluctuations of an investment account throughout the month are two examples of sample trajectories. Learning from experience is a powerful paradigm, because it frees us from needing a complete description of the environment, an often impractical if not infeasible requirement. It also enables incremental algorithms whose run time does not depend on the size of the environment and that can easily be implemented and parallelized.

The aim of this chapter is to introduce a concrete algorithm for learning return distributions from experience, called *categorical temporal-difference learning*. In doing so, we will provide an overview of the design choices that must be made when creating distributional reinforcement learning algorithms. In classical reinforcement learning, there are well-established (and in some cases, definitive) methods for learning to predict the expected return; in the distributional setting, choosing the right algorithm requires balancing multiple, sometimes conflicting considerations. In great part, this is because of the unique and exciting challenges that arise when one wishes to estimate sometimes intricate probability distributions, rather than scalar expectations.

3.1 The Monte Carlo Method

Birds such as the pileated woodpecker follow a feeding routine that regularly takes them back to the same foraging grounds. The success of this routine can be measured in terms of the total amount of food obtained during a fixed period of time, say a single day. As part of a field study, it may be desirable to predict the success of a particular bird's routine on the basis of a limited set of observations: for example, to assess its survival chances at the beginning of winter based on feeding observations from the summer months. In reinforcement learning terms, we view this as the problem of learning to predict the expected return (total food per day) of a given policy π (the feeding routine). Here, variations in weather, human activity, and other foraging animals are but a few of the factors that affect the amount of food obtained on any particular day.

In our example, the problem of learning to predict is abstractly a problem of statistical estimation. To this end, let us model the woodpecker's feeding routine as a Markov decision process.[17] We associate each day with a sample trajectory or *episode*, corresponding to measurements made at regular intervals about the bird's location x, behavior a, and per-period food intake r. Suppose that we have observed a set of K sample trajectories,

$$\left\{(x_{k,t}, a_{k,t}, r_{k,t})_{t=0}^{T_k-1}\right\}_{k=1}^{K}, \tag{3.1}$$

where we use k to index the trajectory and t to index time, and where T_k denotes the number of measurements taken each day. In this example, it is most sensible to assume a fixed number of measurements $T_k = T$, but in the general setting, T_k may be random and possibly dependent on the trajectory, often corresponding to the time when a terminal state is first reached. For now, let us also assume that there is a unique starting state x_0, such that $x_{k,0} = x_0$ for all k. We are interested in the problem of estimating the expected return

$$\mathbb{E}_\pi\Big[\sum_{t=0}^{T-1} \gamma^t R_t\Big] = V^\pi(x_0),$$

corresponding to the expected per-day food intake of our bird.[18]

Monte Carlo methods estimate the expected return by averaging the outcomes of observed trajectories. Let us denote by g_k the *sample return* for the kth

17. Although this may be a reasonable approximation of reality, it is useful to remember that concepts such as the Markov property are in this case *modeling assumptions*, rather than actual facts.

18. In this particular example, a discount factor of $\gamma = 1$ is reasonable given that T is fixed and we should have no particular preference for mornings over evenings.

trajectory:

$$g_k = \sum_{t=0}^{T-1} \gamma^t r_{k,t} \,. \tag{3.2}$$

The sample-mean Monte Carlo estimate is the average of these K sample returns:

$$\hat{V}^\pi(x_0) = \frac{1}{K} \sum_{k=1}^{K} g_k \,. \tag{3.3}$$

This a sensible procedure that also has the benefit of being simple to implement and broadly applicable. In our example above, the Monte Carlo method corresponds to estimating the expected per-day food intake by simply averaging the total intake measured over different days.

Example 3.1. *Monte Carlo tree search* is a family of approaches that have proven effective for planning in stochastic environments and have been used to design state-of-the-art computer programs that play the game of Go. Monte Carlo tree search combines a search procedure with so-called Monte Carlo rollouts (simulations). In a typical implementation, a computer Go algorithm will maintain a partial search tree whose nodes are Go positions, rooted in the current board configuration. Nodes at the fringe of this tree are evaluated by performing a large number of rollouts of a fixed rollout policy (often uniformly random) from the nodes' position to the game's end.

By defining the return of one rollout to be 1 if the game is won and 0 if the game is lost, the expected undiscounted return from a leaf node corresponds to the probability of winning the game (under the rollout policy) from that position. By estimating the expected return from complete rollouts, the Monte Carlo method avoids the challenges associated with devising a heuristic to determine the value of a given position. △

We can use the Monte Carlo method to estimate the value function V^π rather than only the expected return from the initial state. Suppose now that the sample trajectories (Equation 3.1) have different starting states (that is, $x_{k,0}$ varies across trajectories). The Monte Carlo method constructs the value function estimate

$$\hat{V}^\pi(x) = \frac{1}{N(x)} \sum_{k=1}^{K} \mathbb{1}_{\{x_{k,0}=x\}} g_k \,,$$

where

$$N(x) = \sum_{k=1}^{K} \mathbb{1}_{\{x_{k,0}=x\}}$$

is the number of trajectories whose starting state is x. For simplicity of exposition, we assume here that $N(x) > 0$ for all $x \in \mathcal{X}$.

Learning the value function is useful because it provides finer-grained detail about the environment and chosen policy compared to only learning about the expected return from the initial state. Often, estimating the full value function can be done at little additional cost, because sample trajectories contain information about the expected return from multiple states. For example, given a collection of K sample trajectories drawn independently from the generative equations, the *first-visit Monte Carlo estimate* is

$$\hat{V}^{\pi}_{\text{FV}}(x) = \frac{1}{N_{\text{FV}}(x)} \sum_{k=1}^{K} \sum_{t=0}^{T_k-1} \underbrace{\mathbb{1}\{x_{k,t} = x, x_{k,0}, \dots, x_{k,t-1} \neq x\}}_{c_{k,t}} \Big(\sum_{s=t}^{T_k-1} \gamma^{s-t} r_{k,s} \Big)$$

$$N_{\text{FV}}(x) = \sum_{k=1}^{K} \sum_{t=0}^{T_k-1} c_{k,t} .$$

The first-visit Monte Carlo estimate treats each time step as the beginning of a new trajectory; this is justified by the Markov property and time-homogeneity of the random trajectory $(X_t, A_t, R_t)_{t\geq 0}$ (Section 2.6). The restriction to the first occurrence of x in a particular trajectory guarantees that $\hat{V}^{\pi}_{\text{FV}}$ behaves like the sample-mean estimate.

3.2 Incremental Learning

Both in practice and in theory, it is useful to consider a learning model under which sample trajectories are processed sequentially, rather than all at once. Algorithms that operate in this fashion are called *incremental algorithms*, as they maintain a running value function estimate $V \in \mathbb{R}^{\mathcal{X}}$ that they improve with each sample.[19] Under this model, we now consider an infinite sequence of sample trajectories

$$\Big((x_{k,t}, a_{k,t}, r_{k,t})_{t=0}^{T_k-1} \Big)_{k\geq 0} , \tag{3.4}$$

presented one at a time to the learning algorithm. In addition, we consider the more general setting in which the initial states $(x_{k,0})_{k\geq 0}$ may be different; we call these states the *source states*, as with the sample transition model (Section 2.6). As in the previous section, a minimum requirement for learning V^{π} is that every state $x \in \mathcal{X}$ should be the source state of some trajectories.

The incremental Monte Carlo algorithm begins by initializing its value function estimate:

$$V(x) \leftarrow V_0(x), \quad \text{for all } x \in \mathcal{X} .$$

19. Incremental methods are also sometimes called *stochastic* or *sample based*. We prefer the term "incremental" as these methods can be applied even in the absence of randomness, and because other estimation methods – including the sample-mean method of the preceding section – are also based on samples.

Algorithm 3.1: Online incremental first-visit Monte Carlo

```
Algorithm parameters:  step size α ∈ (0, 1],
                       policy π : X → P(A),
                       initial estimate V_0 ∈ R^X

V(x) ← V_0(x) for all x ∈ X
Loop for each episode:
  | Observe initial state x_0
  | T ← 0
  | Loop for t = 0, 1, ...
  |   | Draw a_t from π( · | x_t)
  |   | Take action a_t, observe r_t, x_{t+1}
  |   | T ← T + 1
  | until x_{t+1} is terminal
  | g ← 0
  | for t = T − 1, ..., 0 do
  |   | g ← r_t + γg
  |   | if x_t is not in {x_0, ..., x_{t−1}} then
  |   |   | V(x_t) ← (1 − α)V(x_t) + αg
  | end for
end
```

Given the kth trajectory, the algorithm computes the sample return g_k (Equation 3.2), called the *Monte Carlo target* in this context. It then adjusts its value function estimate for the initial state $x_{k,0}$ toward this target, according to the *update rule*

$$V(x_{k,0}) \leftarrow (1 - \alpha_k)V(x_{k,0}) + \alpha_k g_k \,.$$

The parameter $\alpha_k \in [0, 1)$ is a time-varying step size that controls the impact of a single sample trajectory on the value estimate V. Because the incremental Monte Carlo update rule only depends on the starting state and the sample return, we can more generally consider learning V^π on the basis of the sequence of state-return pairs

$$(x_k, g_k)_{k \geq 0} \,. \tag{3.5}$$

Under this simplified model, the sample return g_k is assumed drawn from the return distribution $\eta^\pi(x_k)$ (rather than constructed from the sample trajectory; of course, these two descriptions are equivalent). The initial state $x_{k,0}$ is substituted for x_k to yield

$$V(x_k) \leftarrow (1 - \alpha_k)V(x_k) + \alpha_k g_k \,. \tag{3.6}$$

By choosing a step size that is inversely proportional to the number of times $N_k(x_k)$ that the value of state x_k has been previously encountered, we recover the sample-mean estimate of the previous section (see Exercise 3.1). In practice, it is also common to use a constant step size α to avoid the need to track N_k. We will study the effect of step sizes in greater detail in Chapter 6.

As given, the Monte Carlo update rule is agnostic to the mechanism by which the sample trajectories are generated or when learning occurs in relation to the agent observing these trajectories. A frequent and important setting in reinforcement learning is when each trajectory is consumed by the learning algorithm immediately after being experienced, which we call the *online* setting.[20] Complementing the abstract presentation given in this section, Algorithm 3.1 gives pseudo-code for an incremental implementation of the first-visit Monte Carlo algorithm in the online, episodic[21] setting.

In Equation 3.6, the notation $A \leftarrow B$ indicates that the value B (which may be a scalar, a vector, or a distribution) should be stored in the variable A. In the case of Equation 3.6, V is a collection of scalars – one per state – and the update rule describes the process of modifying the value of one of these variables "in place." This provides a succinct way of highlighting the incremental nature of the process. On the other hand, it is often useful to consider the value of the variable after a given number of iterations. For $k > 0$, we express this with the notation

$$
\begin{aligned}
V_{k+1}(x_k) &= (1 - \alpha_k)V_k(x_k) + \alpha_k g_k \\
V_{k+1}(x) &= V_k(x) \text{ for } x \neq x_k\,,
\end{aligned}
\tag{3.7}
$$

where the second line reflects the fact that only the variable associated to state x_k is modified at time k.

3.3 Temporal-Difference Learning

Incremental algorithms are useful because they do not need to maintain the entire set of sample trajectories in memory. In addition, they are often simpler to implement and enable distributed and approximate computation. *Temporal-difference learning* (TD learning, or simply TD) is particularly well suited to the incremental setting, because it can learn from sample transitions, rather than entire trajectories. It does so by leveraging the relationship between the value

20. The online setting is sometimes defined in terms of individual transitions, which are to be consumed immediately after being experienced. Our use of the term is broader and related to the *streaming* setting considered in other fields of research (see, e.g., Cormode and Muthukrishnan 2005).

21. A environment is said to be episodic when all trajectories eventually reach a terminal state.

Algorithm 3.2: Online temporal-difference learning

Algorithm parameters: step size $\alpha \in (0, 1]$,
policy $\pi : \mathcal{X} \to \mathscr{P}(\mathcal{A})$
initial estimate $V_0 \in \mathbb{R}^{\mathcal{X}}$

$V(x) \leftarrow V_0(x)$ for all $x \in \mathcal{X}$
Loop for each episode:
  Observe initial state x_0
  Loop for $t = 0, 1, \ldots$
    Draw a_t from $\pi(\cdot \mid x_t)$
    Take action a_t, observe r_t, x_{t+1}
    if x_{t+1} is terminal **then**
      $V(x_t) \leftarrow (1 - \alpha)V(x_t) + \alpha r_t$
    else
      $V(x_t) \leftarrow (1 - \alpha)V(x_t) + \alpha(r_t + \gamma V(x_{t+1}))$
    end if
  until x_{t+1} is terminal
end

function at successive states – effectively, it takes advantage of the Bellman equation.

To begin, let us abstract away the sequential nature of the trajectory and consider the sample transition model (X, A, R, X') defined by Equation 2.12. Here, the distribution ξ of the source state X may correspond, for example, to the relative frequency at which states are visited over the random trajectory. We consider a sequence of sample transitions drawn independently according to this model, denoted

$$(x_k, a_k, r_k, x'_k)_{k \geq 0} \,. \tag{3.8}$$

As with the incremental Monte Carlo algorithm, the update rule of temporal-difference learning is parameterized by a time-varying step size α_k. For the kth sample transition, this update rule is given by

$$V(x_k) \leftarrow (1 - \alpha_k)V(x_k) + \alpha_k(r_k + \gamma V(x'_k)) \,. \tag{3.9}$$

Algorithm 3.2 instantiates this update rule in the online, episodic setting. We call the term $r_k + \gamma V(x'_k)$ in Equation 3.9 the *temporal-difference target*. If we again write the Bellman equation

$$V^\pi(x) = \mathbb{E}_\pi \left[R + \gamma V^\pi(X') \mid X = x \right],$$

we see that the temporal-difference target can be thought of as a realization of the random variable whose expectation forms the right-hand side of the Bellman equation, with the exception that the TD target uses the estimated value function V in place of the true value function V^π. This highlights the fact that temporal-difference learning adjusts its value function estimate to be more like the right-hand side of the Bellman equation; we will study this relationship more formally in Chapter 6.

By rearranging terms, we can also express the temporal-difference update rule in terms of the *temporal-difference error* $r_k + \gamma V(x'_k) - V(x_k)$:

$$V(x_k) \leftarrow V(x_k) + \alpha_k(r_k + \gamma V(x'_k) - V(x_k));$$

this form highlights the direction of change of the value function estimate (positive if the target is greater than our estimate, negative if it is not) and is needed to express certain reinforcement learning algorithms, as we will see in Chapter 9.

The incremental Monte Carlo algorithm updates its value function estimate toward a fixed (but noisy) target. By contrast, Equation 3.9 describes a recursive process, without such a fixed target. Temporal-difference learning instead depends on the value function at the next state $V(x'_k)$ being approximately correct. As such, it is said to *bootstrap* from its own value function estimate. Because of this recursive dependency, the dynamics of temporal-difference learning are usually different from those of Monte Carlo methods, and are more challenging to analyze (Chapter 6).

On the other hand, temporal-difference learning offers some important advantages over Monte Carlo methods. One is the way in which value estimates are naturally shared between states, so that once a value has been estimated accurately at one state, this can often be used to improve the value estimates at other states. In many situations, the estimates produced by temporal-difference learning are consequently more accurate than their Monte Carlo counterparts. A full treatment of the statistical properties of temporal-difference learning is beyond the scope of this book, but we provide references on the topic in the bibliographical remarks.

3.4 From Values to Probabilities

In distributional reinforcement learning, we are interested in understanding the random return as it arises from interactions with the environment. In the context of this chapter, we are specifically interested in how we can learn the return-distribution function η^π.

As a light introduction, consider the scenario in which rewards are binary ($R_t \in \{0, 1\}$) and where we are interested in learning the distribution of the undiscounted finite-horizon return function

$$G^\pi(x) = \sum_{t=0}^{H-1} R_t, \quad X_0 = x. \tag{3.10}$$

Here, $H \in \mathbb{N}^+$ denotes the *horizon* of the learner – how far it predicts into the future. By enumeration, we see that $G^\pi(x)$ takes on one of $H+1$ possible values, integers ranging from 0 to H. These form the support of the probability distribution $\eta^\pi(x)$. To learn η^π, we will construct an incremental algorithm that maintains a return-distribution function estimate η, the distributional analogue of V from the previous sections. This estimate assigns a probability $p_i(x)$ to each possible return $i \in \{0, \dots, H\}$:

$$\eta(x) = \sum_{i=0}^{H} p_i(x)\delta_i, \tag{3.11}$$

where $p_i(x) \geq 0$ and $\sum_{i=0}^{H} p_i(x) = 1$. We call Equation 3.11 a *categorical representation*, since each possible return can now be thought of as one of $H+1$ categories. Under this perspective, we can think of the problem of learning the return distribution for a given state x as a classification problem – assigning probabilities to each of the possible categories. We may then view the problem of learning the return function η^π as a collection of classification problems (one per state).

As before, let us consider a sequence of state-return pairs $(x_k, g_k)_{k\geq 0}$, where each g_k is drawn from the distribution $\eta^\pi(x_k)$. As in Section 3.2, the sample return g_k provides a target for an update rule except that now we want to adjust the *probability* of observing g_k rather than an estimate of the expected return. For a step size $\alpha_k \in (0, 1]$, the *categorical update rule* is

$$\begin{aligned} p_{g_k}(x_k) &\leftarrow (1-\alpha_k)p_{g_k}(x_k) + \alpha_k \\ p_i(x_k) &\leftarrow (1-\alpha_k)p_i(x_k), \quad i \neq g_k. \end{aligned} \tag{3.12}$$

The adjustment of the probabilities for returns other than g_k ensures that the return-distribution estimate at x_k continues to sum to 1 after the update. Expressed as an operation over probability distributions, this update rule corresponds to changing $\eta(x_k)$ to be a *mixture* between itself and a distribution that puts all of its mass on g_k:

$$\eta(x_k) \leftarrow (1-\alpha_k)\eta(x_k) + \alpha_k\delta_{g_k}. \tag{3.13}$$

We call this the *undiscounted finite-horizon categorical Monte Carlo algorithm*. It is instantiated in the online, episodic setting in Algorithm 3.3. Similar to the other incremental algorithms presented thus far, it is possible to demonstrate that under the right conditions, this algorithm learns a good approximation to the distribution of the binary-reward, undiscounted, finite-horizon return. In the next section, we will see how this idea can be carried over to the more general setting.

Algorithm 3.3: Undiscounted finite-horizon categorical Monte Carlo

Algorithm parameters: step size $\alpha \in (0, 1]$,
horizon H,
policy $\pi : \mathcal{X} \to \mathscr{P}(\mathcal{A})$,
initial probabilities $((p_i^0(x))_{i=0}^H : x \in \mathcal{X})$

$p_i(x) \leftarrow p_i^0(x)$ for all $0 \le i \le H$, $x \in \mathcal{X}$
Loop for each episode:
  Observe initial state x_0
  $T \leftarrow 0$
  Loop for $t = 0, 1, \ldots$
    Draw a_t from $\pi(\cdot \mid x_t)$
    Take action a_t, observe r_t, x_{t+1}
    $T \leftarrow T + 1$
  until x_{t+1} is terminal
  $g \leftarrow 0$
  for $t = T - 1, \ldots, 0$ **do**
    $g \leftarrow g + r_t$
    if $t < T - H$ **then**
      $g \leftarrow g - r_{t+H}$
    if x_t is not in $\{x_0, \ldots, x_{t-1}\}$ **then**
      $p_g(x_t) \leftarrow (1 - \alpha)p_g(x_t) + \alpha$
      $p_i(x_t) \leftarrow (1 - \alpha)p_i(x_t)$, for $i \neq g$
  end for
end

3.5 The Projection Step

The finite, undiscounted algorithm of the previous section is a sensible approach when the random return takes only a few distinct values. In the undiscounted setting, we already saw that the number of possible returns is $N_G = H + 1$ when there are two possible rewards. However, this small number is the exception, rather than the rule. If there are $N_\mathcal{R}$ possible rewards, then N_G can be as large as

$\binom{N_{\mathcal{R}}+H-1}{H}$, already a potentially quite large number for $N_{\mathcal{R}} > 2$ (see Exercise 3.5). Worse, when a discount factor γ is introduced, the number of possible returns depends exponentially on H. To see why, recall the single-state, single-action Markov decision process of Example 2.10 with reward distribution

$$\mathbb{P}_\pi(R_t = 0) = \mathbb{P}_\pi(R_t = 1) = 1/2\,.$$

With a discount factor $\gamma = 1/2$, we argued that the set of possible returns for this MDP corresponds to the binary expansion of all numbers in the $[0, 2]$ interval, from which we concluded that the random return is uniformly distributed on that interval. By the same argument, we can show that the H-horizon return

$$G = \sum_{t=0}^{H-1} \gamma^t R_t$$

has support on all numbers in the $[0, 2]$ interval that are described by H binary digits; there are 2^H such numbers. Of course, if we take H to be infinite, there may be uncountably many possible returns.

To deal with the issue of a large (or even infinite) set of possible returns, we insert a *projection step* prior to the mixture update in Equation 3.13.[22] The purpose of the projection step is to transform an arbitrary target return g into a modified target taking one of m values, for m reasonably small. From a classification perspective, we can think of the projection step as assigning a label (from a small set of categories) to each possible return.

We will consider return distributions that assign probability mass to $m \geq 2$ evenly spaced values or *locations*. In increasing order, we denote these locations by $\theta_1 \leq \theta_2 \leq \cdots \leq \theta_m$. We write ς_m for the gap between consecutive locations, so that for $i = 1, \ldots, m-1$, we have

$$\varsigma_m = \theta_{i+1} - \theta_i\,.$$

A common design takes $\theta_1 = V_{\text{MIN}}$ and $\theta_m = V_{\text{MAX}}$, in which case

$$\varsigma_m = \frac{V_{\text{MAX}} - V_{\text{MIN}}}{m-1}\,.$$

However, other choices are possible and sometimes desirable. Note that the undiscounted algorithm of the previous section corresponds to $m = H + 1$ and $\theta_i = i - 1$.

22. When there are relatively few sample trajectories, a sensible alternative is to construct a *nonparametric* estimate of the return distribution that simply puts equal probability mass on all observed returns. The return distributions described in Section 1.2 and Example 2.9, in particular, were estimated in this manner. See Remark 3.1 for further details.

We express the corresponding return distribution as a weighted sum of Dirac deltas at these locations:

$$\eta(x) = \sum_{i=1}^{m} p_i(x)\delta_{\theta_i} .$$

For mathematical convenience, let us write $\theta_0 = -\infty$ and $\theta_{m+1} = \infty$. Consider a sample return g, and denote by $i^*(g)$ the index of the largest element of the support (extended with $-\infty$) that is no greater than g:

$$i^* = \arg\max_{i \in \{0,\ldots,m\}} \{\theta_i : \theta_i \leq g\} .$$

For this sample return, we write

$$\Pi_-(g) = \theta_{i^*} \qquad \Pi_+(g) = \theta_{i^*+1}$$

to denote the corresponding element of the support and its immediate successor; when $\theta_1, \ldots, \theta_m$ are consecutive integers and $g \in [\theta_1, \theta_m]$, these are the floor and ceiling of g, respectively.

The projection step begins by computing

$$\zeta(g) = \frac{g - \Pi_-(g)}{\Pi_+(g) - \Pi_-(g)} ,$$

with the convention that $\zeta(g) = 1$ if $\Pi_-(g) = -\infty$ and $\zeta(g) = 0$ if $\Pi_+(g) = \infty$. The $\zeta(g)$ term corresponds to the distance of g to the two closest elements of the support, scaled to lie in the interval $[0, 1]$ – effectively, the fraction of the distance between $\Pi_-(g)$ and $\Pi_+(g)$ at which g lies. The *stochastic projection* of g is

$$\Pi_\pm(g) = \begin{cases} \Pi_-(g) & \text{with probability } 1 - \zeta(g) \\ \Pi_+(g) & \text{with probability } \zeta(g). \end{cases} \tag{3.14}$$

We use this projection to construct the update rule

$$\eta(x) \leftarrow (1 - \alpha)\eta(x) + \alpha\delta_{\Pi_\pm g} ,$$

where as before, x is a source state and g a sample return.[23] Similar to Equation 3.12, this update rule is implemented by adjusting the probabilities according to

$$p_{i^\pm}(x) \leftarrow (1 - \alpha)p_{i^\pm}(x) + \alpha$$

$$p_i(x) \leftarrow (1 - \alpha)p_i(x) , \quad i \neq i^\pm .$$

where $i^\pm$ is the index of location $\Pi_\pm g$.

23. From here onward, we omit the iteration index k from the notation as it is not needed in the definition of the update rule. The algorithm proper should still be understood as applying this update rule to a sequence of state-return pairs $(x_k, g_k)_{k\geq 0}$, possibly with a time-varying step size α_k.

We can improve on the stochastic projection by constructing a modified target that contains information about both $\Pi_-(g)$ and $\Pi_+(g)$. In classification terms, this corresponds to using soft labels: the target is a probability distribution over labels, rather than a single label. This *deterministic projection* of g results in the update rule

$$\eta(x) \leftarrow (1-\alpha)\eta(x) + \alpha\big[(1-\zeta(g))\delta_{\Pi_-(g)} + \zeta(g)\delta_{\Pi_+(g)}\big]. \tag{3.15}$$

We denote the deterministic projection by Π_c. Statistically speaking, the deterministic projection produces return-distribution estimates that are on average the same as those produced by the stochastic projection but are comparatively more concentrated around their mean. Going forward, we will see that it is conceptually simpler to apply this projection to probability distributions, rather than to sample returns. Rather than $\Pi_c(g)$, we therefore write

$$\Pi_c\delta_g = (1-\zeta(g))\delta_{\Pi_-(g)} + \zeta(g)\delta_{\Pi_+(g)}.$$

We call this method the *categorical Monte Carlo algorithm*. This algorithm can be used to learn to predict infinite-horizon, discounted returns and is applicable even when there are a large number of possible rewards.

Example 3.2. *Montezuma's Revenge* is a 1984 platform game designed by then-sixteen-year-old Robert Jaeger. As part of the Arcade Learning Environment (Bellemare et al. 2013a), the Atari 2600 version of the game poses a challenging reinforcement learning problem due to the rare occurrence of positive rewards.

The very first task in *Montezuma's Revenge* consists in collecting a key, an act that rewards the agent with 100 points. Let us consider the integer support $\theta_i = i-1$, for $i = 1, 2, \ldots, 101$. For a discount factor $\gamma = 0.99$, the discounted H-horizon return from the initial game state is

$$\begin{aligned} G &= \sum_{t=0}^{H-1} \gamma^t R_t \\ &= \mathbb{1}_{\{\tau < H\}} 0.99^\tau \times 100, \end{aligned}$$

where τ denotes the time at which the key is obtained. For a fixed $\tau \in \mathbb{R}$, write

$$g = 0.99^\tau \times 100 \qquad \zeta = g - \lfloor g \rfloor.$$

For $\tau < H$, the deterministic projection of the return 0.99^τ puts probability mass $1-\zeta$ on $\lfloor g \rfloor$ and ζ on $\lceil g \rceil$. For example, if $\tau = 60$, then $100 \times 0.99^\tau \approx 54.72$ and the deterministic projection is

$$\Pi_c\delta_g = 0.28 \times \delta_{54} + 0.72 \times \delta_{55}. \qquad \triangle$$

By introducing a projection step, we typically lose some information about the target return. This is by necessity: we are asking the learning algorithm to

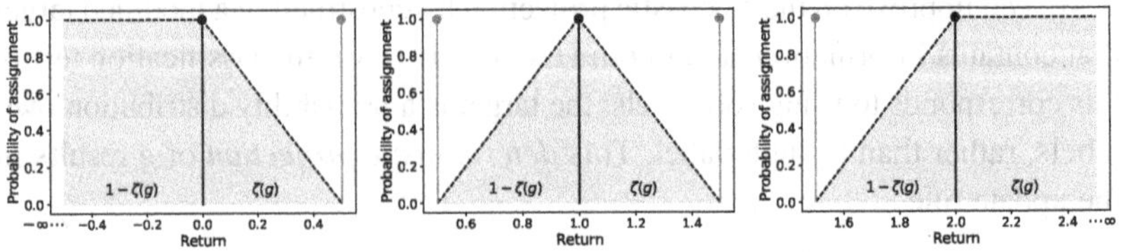

Figure 3.1
Illustration of the projection for the five-location grid used in Example 3.3. The middle panel depicts the proportion of probability mass assigned to the location $\theta_3 = 1$ in terms of the sample return g. The left and right panels illustrate this probability assignment at the boundary locations $\theta_1 = 0$ and $\theta_m = 2$.

approximate complex distributions using a small, finite set of possible returns. Under the right conditions, Equation 3.15 gives rise to a convergent algorithm. The point of convergence of this algorithm is the return function $\hat{\eta}^\pi(x)$ for which, for all $x \in \mathcal{X}$, we have

$$\hat{\eta}^\pi(x) = \mathbb{E}[\Pi_c \delta_{G^\pi(x)}] = \mathbb{E}\left[(1 - \zeta(G^\pi(x)))\delta_{\Pi_-(G^\pi(x))} + \zeta(G^\pi(x))\delta_{\Pi_+(G^\pi(x))}\right]. \quad (3.16)$$

In fact, we may write

$$\mathbb{E}[\Pi_c \delta_{G^\pi(x)}] = \Pi_c \eta^\pi(x),$$

where $\Pi_c \eta^\pi(x)$ denotes distribution supported on $\{\theta_1, \ldots, \theta_m\}$ produced by projecting each possible outcome under the distribution $\eta^\pi(x)$; we will discuss this definition in further detail in Chapter 5.

Example 3.3. Recall the single-state Markov decision process of Example 2.10, whose return is uniformly distributed on the interval $[0, 2]$. Suppose that we take our support to be the uniform grid $\{0, 0.5, 1, 1.5, 2\}$. Let us write $\hat{p}_0, \hat{p}_{0.5}, \ldots, \hat{p}_2$ for the probabilities assigned to these locations by the projected distribution $\hat{\eta}^\pi(x) = \Pi_c \eta^\pi(x)$, where x is the unique state of the MDP. The probability density of the return on the interval $[0, 2]$ is 0.5. We thus have

$$\begin{aligned}
\hat{p}_0 &\overset{(a)}{=} 0.5 \int_{g \in [0,2]} [\mathbb{1}_{\{\Pi_-(g) = 0\}}(1 - \zeta(g)) + \mathbb{1}_{\{\Pi_+(g) = 0\}}\zeta(g)] dg \\
&= 0.5 \int_{g \in [0,0.5]} (1 - \zeta(g)) dg \\
&= 0.5 \int_{g \in [0,0.5]} (1 - 2g) dg \\
&= 0.125.
\end{aligned}$$

In (a), we reexpressed Equation 3.16 in terms of the probability assigned to $\hat{p}_0$. A similar computation shows that $\hat{p}_{0.5} = \hat{p}_1 = \hat{p}_{1.5} = 0.25$, while $\hat{p}_2 = 0.125$. Figure 3.1 illustrates the process of assigning probability mass to different locations. Intuitively, the solution makes sense: there is less probability mass near the boundaries of the interval $[0, 2]$. △

3.6 Categorical Temporal-Difference Learning

With the use of a projection step, the categorical Monte Carlo method allows us to approximate the return-distribution function of a given policy from sample trajectories and using a fixed amount of memory. Like the Monte Carlo method for value functions, however, it ignores the relationship between successive states in the trajectory. To leverage this relationship, we turn to *categorical temporal-difference learning* (CTD).

Consider now a sample transition (x, a, r, x'). Like the categorical Monte Carlo algorithm, CTD maintains a return function estimate $\eta(x)$ supported on the evenly spaced locations $\{\theta_1, \ldots, \theta_m\}$. Like temporal-difference learning, it learns by bootstrapping from its current return function estimate. In this case, however, the update target is a probability distribution supported on $\{\theta_1, \ldots, \theta_m\}$. The algorithm first constructs an intermediate target by scaling the return distribution $\eta(x')$ at the next state by the discount factor γ, then shifting it by the sample reward r. That is, if we write

$$\eta(x') = \sum_{i=1}^{m} p_i(x')\delta_{\theta_i} ,$$

then the intermediate target is

$$\tilde{\eta}(x) = \sum_{i=1}^{m} p_i(x')\delta_{r+\gamma\theta_i} ,$$

which can also be expressed in terms of a pushforward distribution:

$$\tilde{\eta}(x) = (\mathrm{b}_{r,\gamma})_{\#}\eta(x') . \tag{3.17}$$

Observe that the shifting and scaling operations are applied to each particle in isolation. After shifting and scaling, however, these particles in general no longer lie in the support of the original distribution. This motivates the use of the projection step described in the previous section. Let us denote the intermediate particles by

$$\tilde{\theta}_i = \mathrm{b}_{r,\gamma}(\theta_i) = r + \gamma\theta_i .$$

The CTD target is formed by individually projecting each of these particles back onto the support and combining their probabilities. That is,

$$\Pi_c(\mathrm{b}_{r,\gamma})_\# \eta(x') = \sum_{j=1}^m p_j(x')\Pi_c \delta_{r+\gamma\theta_j} .$$

More explicitly, this is

$$\begin{aligned}\Pi_c(\mathrm{b}_{r,\gamma})_\# \eta(x') &= \sum_{j=1}^m p_j(x')[(1-\zeta(\tilde\theta_j))\delta_{\Pi_-(\tilde\theta_j)} + \zeta(\tilde\theta_j)\delta_{\Pi_+(\tilde\theta_j)}] \\ &= \sum_{i=1}^m \delta_{\theta_i}\Big(\sum_{j=1}^m p_j(x')\zeta_{i,j}(r)\Big), \end{aligned} \tag{3.18}$$

where

$$\zeta_{i,j}(r) = \big((1-\zeta(\tilde\theta_j))\mathbb{1}_{\{\Pi_-(\tilde\theta_j)=\theta_i\}} + \zeta(\tilde\theta_j)\mathbb{1}_{\{\Pi_+(\tilde\theta_j)=\theta_i\}}\big).$$

In Equation 3.18, the second line highlights that the CTD target is a probability distribution supported on $\{\theta_1, \dots, \theta_m\}$. The probabilities assigned to specific locations are given by the inner sum; as shown here, this assignment is obtained by weighting the next-state probabilities $p_j(x')$ by the coefficients $\zeta_{i,j}(r)$.

The CTD update adjusts the return function estimate $\eta(x)$ toward this target:

$$\eta(x) \leftarrow (1-\alpha)\eta(x) + \alpha(\Pi_c(\mathrm{b}_{r,\gamma})_\# \eta(x')) . \tag{3.19}$$

Expressed in terms of the probabilities of the distributions $\eta(x)$ and $\eta(x')$, this is (for $i = 1, \dots, m$)

$$p_i(x) \leftarrow (1-\alpha)p_i(x) + \alpha \sum_{j=1}^m \zeta_{i,j}(r)p_j(x') . \tag{3.20}$$

With this form, we see that the CTD update rule adjusts each probability $p_i(x)$ of the return distribution at state x toward a mixture of the probabilities of the return distribution at the next state (see Algorithm 3.4).

The definition of the intermediate target in pushforward terms (Equation 3.17) illustrates that categorical temporal-difference learning relates to the distributional Bellman equation

$$\eta^\pi(x) = \mathbb{E}_\pi[(\mathrm{b}_{R,\gamma})_\# \eta^\pi(X') \mid X = x] ,$$

analogous to the relationship between TD learning and the classical Bellman equation. We will continue to explore this relationship in later chapters. However, the two algorithms usually learn different solutions, due to the introduction of approximation error from the bootstrapping process.

Example 3.4. We can study the behavior of the categorical temporal-difference learning algorithm by visualizing how its predictions vary over the course

Algorithm 3.4: Online categorical temporal-difference learning

Algorithm parameters: step size $\alpha \in (0, 1]$,
policy $\pi : \mathcal{X} \to \mathscr{P}(\mathcal{A})$,
evenly spaced locations $\theta_1, \ldots, \theta_m \in \mathbb{R}$,
initial probabilities $((p_i^0(x))_{i=0}^H : x \in \mathcal{X})$

$(p_i(x))_{i=0}^H \leftarrow (p_i^0(x))_{i=0}^H$ for all $x \in \mathcal{X}$
Loop for each episode:
  Observe initial state x_0
  Loop for $t = 0, 1, \ldots$
    Draw a_t from $\pi(\cdot \mid x_t)$
    Take action a_t, observe r_t, x_{t+1}
    $\hat{p}_i \leftarrow 0$ for $i = 1, \ldots, m$
    for $j = 1, \ldots, m$ **do**
      if x_{t+1} is terminal **then**
        $g \leftarrow r_t$
      else
        $g \leftarrow r_t + \gamma\theta_j$
      if $g \leq \theta_1$ **then**
        $\hat{p}_1 \leftarrow \hat{p}_1 + p_j(x_{t+1})$
      else if $g \geq \theta_m$ **then**
        $\hat{p}_m \leftarrow \hat{p}_m + p_j(x_{t+1})$
      else
        $i^* \leftarrow$ largest i s.t. $\theta_i \leq g$
        $\zeta \leftarrow \frac{g - \theta_{i^*}}{\theta_{i^*+1} - \theta_{i^*}}$
        $\hat{p}_{i^*} \leftarrow \hat{p}_{i^*} + (1 - \zeta)p_j(x_{t+1})$
        $\hat{p}_{i^*+1} \leftarrow \hat{p}_{i^*+1} + \zeta p_j(x_{t+1})$
    end for
    for $i = 1, \ldots, m$ **do**
      $p_i(x_t) \leftarrow (1 - \alpha)p_i(x_t) + \alpha\hat{p}_i$
    end for
  until x_{t+1} is terminal
end

of learning. We apply CTD to approximate the return function of the safe policy in the Cliffs domain (Example 2.9). Learning takes place online (as per Algorithm 3.4), using a constant step size of $\alpha = 0.05$, and return distributions are approximated with $m = 31$ locations evenly spaced between -1 and 1.

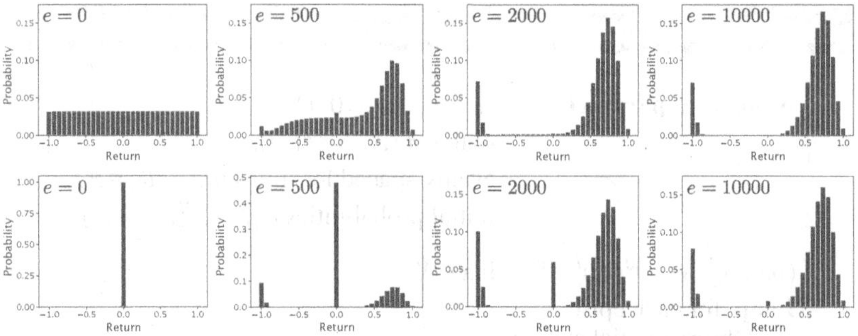

Figure 3.2
The return distribution at the initial state in the Cliffs domain, as predicted by categorical temporal-difference learning over the course of learning. **Top panels.** The predictions after $e \in \{0, 1000, 2000, 10000\}$ episodes ($\alpha = 0.05$; see Algorithm 3.4). Here, the return distributions are initialized by assigning equal probability to all locations. **Bottom panels.** Corresponding predictions when the return distributions initially put all probability mass on the zero return.

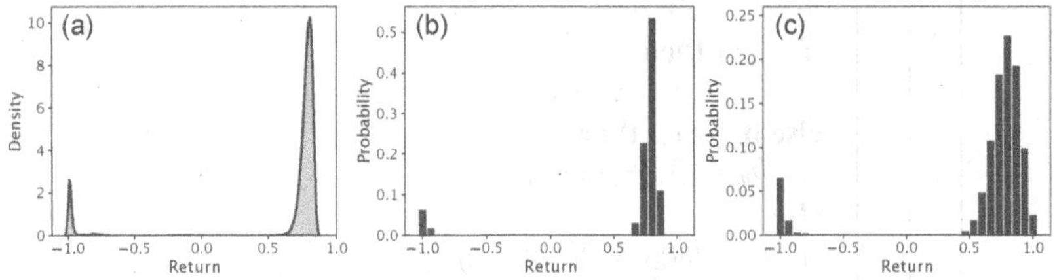

Figure 3.3
(a) The return distribution at the initial state in the Cliffs domain, visualized using kernel density estimation (see Figure 2.2). **(b)** The return-distribution estimate learned by the categorical Monte Carlo algorithm with $m = 31$. **(c)** The estimate learned by the categorical temporal-difference learning algorithm.

Figure 3.2 illustrates that the initial return function plays a substantial role in the speed at which CTD learns a good approximation. Informally, this occurs because the uniform distribution is closer to the final approximation than the distribution that puts all of its probability mass on zero (what "close" means in this context will be the topic of Chapter 4). In addition, we see that categorical temporal-difference learning learns a different approximation to the true return function, compared to the categorical Monte Carlo algorithm (Figure 3.3). This is due to a phenomenon we call *diffusion*, which arises from the combination of the bootstrapping step and projection; we will study diffusion in Chapter 5. △

3.7 Learning to Control

A large part of this book considers the problem of learning to predict the distribution of an agent's returns. In Chapter 7, we will discuss how one might instead learn to maximize or *control* these returns and the role that distributional reinforcement learning plays in this endeavor. By learning to control, we classically mean obtaining (from experience) a policy π^* that maximizes the expected return:

$$\mathbb{E}_{\pi^*}\Big[\sum_{t=0}^{\infty}\gamma^t R_t\Big] \geq \mathbb{E}_{\pi}\Big[\sum_{t=0}^{\infty}\gamma^t R_t\Big], \text{ for all } \pi .$$

Such a policy is called an *optimal policy*. From Section 2.5, recall that the state-action value function Q^π is given by

$$Q^\pi(x,a) = \mathbb{E}_\pi\Big[\sum_{t=0}^{\infty}\gamma^t R_t \,\Big|\, X_0 = x, A_0 = a\Big].$$

Any optimal policy π^* has the property that its state-action value function also satisfies the *Bellman optimality equation*:

$$Q^{\pi^*}(x,a) = \mathbb{E}_\pi\big[R + \gamma \max_{a' \in \mathcal{A}} Q^{\pi^*}(X', a') \,|\, X = x, A = a\big].$$

Similar in spirit to temporal-difference learning, *Q-learning* is an incremental algorithm that finds an optimal policy. Q-learning maintains a state-action value function estimate, Q, which it updates according to

$$Q(x,a) \leftarrow (1-\alpha)Q(x,a) + \alpha\big(r + \gamma \max_{a' \in \mathcal{A}} Q(x', a')\big). \tag{3.21}$$

The use of the maximum in the Q-learning update rule results in different behavior than TD learning, as the selected action depends on the current value estimate. We can think of this difference as constructing one target for each action a' and updating the value estimate toward the largest of such targets. It can be shown that under the right conditions, Q-learning converges to the *optimal state-action value function* Q^*, corresponding to the expected return under any optimal policy. We extract an optimal policy from Q^* by acting *greedily* with respect to Q^*: that is, choosing a policy π^* that selects maximally valued actions according to Q^*.

The simplest way to extend Q-learning to the distributional setting is to express the maximal action in Equation 3.21 as a greedy policy. Denote by η a return-function estimate over state-action pairs, such that $\eta(x,a)$ is the return distribution associated with the state-action pair $(x,a) \in \mathcal{X} \times \mathcal{A}$. Define the greedy action

$$\mathrm{a}_\eta(x) = \arg\max_{a \in \mathcal{A}} \mathop{\mathbb{E}}_{Z \sim \eta(x,a)}[Z],$$

breaking ties arbitrarily. The *categorical Q-learning* update rule is

$$\eta(x, a) \leftarrow (1 - \alpha)\eta(x, a) + \alpha\Big(\Pi_c(\mathrm{b}_{r,\gamma})_\# \eta(x', \mathrm{a}_\eta(x'))\Big).$$

It can be shown that, under the same conditions as Q-learning, the mean of the return-function estimates also converges to Q^*. The behavior of the distributions themselves, however, may be surprisingly complicated. We can also put the learned distributions to good use and make decisions on the basis of their full characterization, rather than from their mean alone. This forms the topic of risk-sensitive reinforcement learning. We return to both of these points in Chapter 7.

3.8 Further Considerations

Categorical temporal-difference learning learns to predict return distributions from sample experience. As we will see in subsequent chapters, the choices that we made in designing CTD are not unique, and the algorithm is best thought of as a jumping-off point into a broad space of methods. For example, an important question in distributional reinforcement learning asks how we should represent probability distributions, given a finite memory budget. One issue with the categorical representation is that it relies on a fixed grid of locations to cover the range $[\theta_1, \theta_m]$, which lacks flexibility and is in many situations inefficient. We will take a closer look at this issue in Chapter 5. In many practical situations, we also need to deal with a few additional considerations, including the use of function approximation to deal with very large state spaces (Chapters 9 and 10).

3.9 Technical Remarks

Remark 3.1 (Nonparametric distributional Monte Carlo algorithm). In Section 3.4, we saw that the (unprojected) finite-horizon categorical Monte Carlo algorithm can in theory learn finite-horizon return-distribution functions when there are only a small number of possible returns. It is possible to extend these ideas to obtain a straightforward, general-purpose algorithm that can be sometimes be used to learn an accurate approximation to the return distribution.

Like the sample-mean Monte Carlo method, the *nonparametric distributional Monte Carlo algorithm* takes as input K finite-length trajectories with a common source state x_0. After computing the sample returns $(g_k)_{k=1}^K$ from these trajectories, it constructs the estimate

$$\hat{\eta}^\pi(x_0) = \frac{1}{K}\sum_{k=1}^K \delta_{g_k} \tag{3.22}$$

of the return distribution $\eta^\pi(x_0)$. Here, *nonparametric* refers to the fact that the approximating distribution in Equation 3.22 is not described by a finite collection of parameters; in fact, the memory required to represent this object may grow linearly with K. Although this is not an issue when K is relatively small, this can be undesirable when working with large amounts of data and moreover precludes the use of function approximation (see Chapters 9 and 10).

However, unlike the categorical Monte Carlo and temporal-difference learning algorithms presented in this chapter, the accuracy of this estimate is only limited by the number of trajectories K; we describe various ways to quantify this accuracy in Remark 4.3. As such, it provides a useful baseline for measuring the quality of other distributional algorithms. In particular, we used this algorithm to generate the ground-truth return-distribution estimates in Example 2.9 and in Figure 3.3. △

3.10 Bibliographical Remarks

The development of a distributional algorithm in this chapter follows our own development of the distributional perspective, beginning with our work on using compression algorithms in reinforcement learning (Veness et al. 2015).

3.1. The first-visit Monte Carlo estimate is studied by Singh and Sutton (1996), where it is used to characterize the properties of replacing eligibility traces (see also Sutton and Barto 2018). Statistical properties of model-based estimates (which solve for the Markov decision process's parameters as an intermediate step) are analyzed by Mannor et al. (2007). Grünewälder and Obermayer (2011) argue that model-based methods must incur statistical bias, an argument that also extends to temporal-difference algorithms. Their work also introduces a refined sample-mean Monte Carlo method that yields a minimum-variance unbiased estimator (MVUE) of the value function. See Browne et al. (2012) for a survey of Monte Carlo tree search methods and Liu (2001), Robert and Casella (2004), and Owen (2013) for further background on Monte Carlo methods more generally.

3.2. Incremental algorithms are a staple of reinforcement learning and have roots in stochastic approximation (Robbins and Monro 1951; Widrow and Hoff 1960; Kushner and Yin 2003) and psychology (Rescorla and Wagner 1972). In the control setting, these are also called *optimistic policy iteration* methods and exhibit fairly complex behavior (Sutton 1999; Tsitsiklis 2002).

3.3. Temporal-difference learning was introduced by Sutton (1984, 1988). The sample transition model presented here differs from the standard algorithmic presentation but allows us to separate concerns of behavior (data collection) from learning. A similar model is used in Bertsekas and Tsitsiklis (1996) to prove convergence of a broad class of TD methods and by Azar et al. (2012) to provide sample efficiency bounds for model-based control.

3.4. What is effectively the undiscounted finite-horizon categorical Monte Carlo algorithm was proposed by Veness et al. (2015). There, the authors demonstrate that by means of Bayes's rule, one can learn the return distribution by first learning the joint distribution over returns and states (in our notation, $\mathbb{P}_\pi(X, G)$) by means of a compression algorithm and subsequently using Bayes's rule to extract $\mathbb{P}_\pi(G \mid X)$. The method proved surprisingly effective at learning to play Atari 2600 games. Toussaint and Storkey (2006) consider the problem of control as probabilistic inference, where the reward and trajectory length are viewed as random variables to be optimized over, again obviating the need to deal with a potentially large support for the return.
3.5–3.6. Categorical temporal-difference learning for both prediction and control was introduced by Bellemare et al. (2017a), in part to address the shortcomings of the undiscounted algorithm. Its original form contains both the projection step and the categorical representation as given here. The mixture update that we study in this chapter is due to Rowland et al. (2018).
3.7. The Q-learning algorithm is due to Watkins (1989); see also Watkins and Dayan (1992). The explicit construction of a greedy policy is commonly found in more complex reinforcement learning algorithms, including modified policy iteration (Puterman and Shin 1978), λ-policy iteration (Bertsekas and Ioffe 1996), and nonstationary policy iteration (Scherrer and Lesner 2012; Scherrer 2014).
3.9. The algorithm introduced in Remark 3.1 is essentially an application of the standard Monte Carlo method to the return distribution and is a special case of the framework set out by Chandak et al. (2021), who also analyze the statistical properties of the approach.

3.11 Exercises

Exercise 3.1. Suppose that we begin with the initial value function estimate $V(x) = 0$ for all $x \in \mathcal{X}$.

(i) Consider first the setting in which we are given sample returns for a single state x. Show that in this case, the incremental Monte Carlo algorithm (Equation 3.6), instantiated with $\alpha_k = \frac{1}{k+1}$, is equivalent to computing the sample-mean Monte Carlo estimate for x. That is, after processing the sample returns $g_1, \ldots, g_K$, we have

$$V_K(x) = \frac{1}{K} \sum_{i=1}^{K} g_i.$$

(ii) Now consider the case where source states are drawn from a distribution ξ, with $\xi(x) > 0$ for all x, and let $N_k(x_k)$ be the number of times x_k has been updated up to but excluding time k. Show that the appropriate step size to match the sample-mean Monte Carlo estimate is $\alpha_k = \frac{1}{N_k(x_k)+1}$. △

Exercise 3.2. Recall from Exercise 2.8 the n-step Bellman equation:

$$V^\pi(x) = \mathbb{E}_\pi\Big[\sum_{t=0}^{n-1} \gamma^t R_t + \gamma^n V^\pi(X_n) \mid X_0 = x\Big].$$

Explain what a sensible n-step temporal-difference learning update rule might look like. △

Exercise 3.3. The *Cart–Pole* domain is a small, two-dimensional reinforcement learning problem (Barto et al. 1983). In this problem, the learning agent must balance a swinging pole that is a attached to a moving cart. Using an open-source implementation of Cart–Pole, implement the undiscounted finite-horizon categorical Monte Carlo algorithm and evaluate its behavior. Construct a finite state space by discretising the four-dimensional state space using a uniform grid of size $10 \times 10 \times 10 \times 10$. Plot the learned return distribution for a fixed initial state and other states of your choice when

(i) the policy chooses actions uniformly at random;
(ii) the policy moves the cart in the direction that the pole is leaning toward. △

You may want to pick the horizon H to be the maximum length of an episode.

Exercise 3.4. Implement the categorical Monte Carlo (CMC), nonparametric categorical Monte Carlo (Remark 3.1), and categorical temporal-difference learning (CTD) algorithms. For a discount factor $\gamma = 0.99$, compare the return distributions learned by these three algorithms on the Cart–Pole domain of the previous exercise. For CMC and CTD, vary the number of particles m and the range of the support $[\theta_1, \theta_m]$. How do the approximations vary as a function of these parameters? △

Exercise 3.5. Suppose that the rewards R_t take on one of $N_{\mathcal{R}}$ values. Consider the undiscounted finite-horizon return

$$G = \sum_{t=0}^{H-1} R_t .$$

Denote by N_G the number of possible realizations of G.

(i) Show that N_G can be as large as $\binom{N_{\mathcal{R}}+H-1}{H}$.
(ii) Derive a better bound when $R_t \in \{0, 1, \ldots, N_{\mathcal{R}} - 1\}$.
(iii) Explain, in words, why the bound is better in this case. Are there other sets of rewards for which N_G is smaller than the worst-case from (i)? △

Exercise 3.6. Recall from Section 3.5 that the categorical Monte Carlo algorithm aims to find the approximation

$$\hat{\eta}^\pi_m(x) = \Pi_c \eta^\pi(x), \quad x \in \mathcal{X},$$

where we use the subscript m to more explicitly indicate that the quality of this approximation depends on the number of locations.

Consider again the problem setting of Example 3.3. For $m \geq 2$, suppose that we take $\theta_1 = 0$ and $\theta_m = 2$, such that

$$\theta_i = 2\frac{i-1}{m-1} \quad i = 1, \ldots m.$$

Show that in this case, $\hat{\eta}^\pi_m(x)$ converges to the uniform distribution on the interval $[0, 2]$, in the sense that

$$\lim_{m \to \infty} \hat{\eta}^\pi_m(x)([a, b]) \to \frac{b-a}{2}$$

for all $a < b$; recall that $\nu([a, b])$ denotes the probability assigned to the interval $[a, b]$ by a probability distribution ν. △

Exercise 3.7. The purpose of this exercise is to demonstrate that the categorical Monte Carlo algorithm is what we call *mean-preserving*. Consider a sequence of state-return pairs $(x_k, g_k)_{k=1}^K$ and an evenly spaced grid $\{\theta_1, \ldots, \theta_m\}$, $m \geq 2$. Suppose that rewards are bounded in $[R_{\text{MIN}}, R_{\text{MAX}}]$.

(i) Suppose that $V_{\text{MIN}} \geq \theta_1$ and $V_{\text{MAX}} \leq \theta_m$. For a given $g \in [V_{\text{MIN}}, V_{\text{MAX}}]$, show that the distribution $\nu = \Pi_c \delta_g$ satisfies

$$\mathop{\mathbb{E}}_{Z \sim \nu}[Z] = g.$$

(ii) Based on this, argue that if $\eta(x)$ is a distribution with mean V, then after applying the update

$$\eta(x) \leftarrow (1 - \alpha)\eta(x) + \alpha \Pi_c \delta_g,$$

the mean of $\eta(x)$ is $(1 - \alpha)V + \alpha g$.

(iii) By comparing with the incremental Monte Carlo update rule, explain why categorical Monte Carlo can be said to be mean-preserving.

(iv) Now suppose that $[V_{\text{MIN}}, V_{\text{MAX}}] \not\subseteq [\theta_1, \theta_m]$. How are the preceding results affected? △

Exercise 3.8. Following the notation of Section 3.5, consider the *nearest neighbor* projection method

$$\Pi_{\text{NN}} g = \begin{cases} \Pi_-(g) & \text{if } \zeta(g) \leq 0.5 \\ \Pi_+(g) & \text{otherwise.} \end{cases}$$

Show that this projection is not mean-preserving in the sense of the preceding exercise. Implement and evaluate on the Cart–Pole domain. What do you observe? △

4 Operators and Metrics

Anyone who has learned to play a musical instrument knows that practice makes perfect. Along the way, however, one's ability at playing a difficult passage usually varies according to a number of factors. On occasion, something that could be played easily the day before now seems insurmountable. The adage expresses an abstract notion – that practice improves performance, on average or over a long period of time – rather than a concrete statement about instantaneous ability.

In the same way, reinforcement learning algorithms deployed in real situations behave differently from moment to moment. Variations arise due to different initial conditions, specific choices of parameters, hardware nondeterminism, or simply because of randomness in the agent's interactions with its environment. These factors make it hard to make precise predictions, for example, about the magnitude of the value function estimate learned by TD learning at a particular state x and step k, other than by extensive simulations. Nevertheless, the large-scale behavior of TD learning is relatively predictable, sufficiently so that convergence can be established under certain conditions, and convergence rates can be derived.

This chapter introduces the language of operators as an effective abstraction with which to study such long-term behavior, characterize the asymptotic properties of reinforcement learning algorithms, and eventually explain what makes an effective algorithm. In addition to being useful in the study of existing algorithms, operators also serve as a kind of blueprint when designing new algorithms, from which incremental methods such as categorical temporal-difference learning can then be derived. In parallel, we will also explore *probability metrics* – essentially distance functions between probability distributions. These metrics play an immediate role in our analysis of the distributional Bellman operator, and will recur in later chapters as we design algorithms for approximating return distributions.

4.1 The Bellman Operator

The value function V^π characterizes the expected return obtained by following a policy π, beginning in a given state x:

$$V^\pi(x) = \mathbb{E}_\pi\Big[\sum_{t=0}^{\infty} \gamma^t R_t \mid X_0 = x\Big].$$

The Bellman equation establishes a relationship between the expected return from one state and from its successors:

$$V^\pi(x) = \mathbb{E}_\pi\left[R + \gamma V^\pi(X') \mid X = x\right].$$

Let us now consider a state-indexed collection of real variables, written $V \in \mathbb{R}^{\mathcal{X}}$, which we call a value function estimate. By substituting V^π for V in the original Bellman equation, we obtain the system of equations

$$V(x) = \mathbb{E}_\pi\left[R + \gamma V(X') \mid X = x\right], \quad \text{for all } x \in \mathcal{X}. \tag{4.1}$$

From Chapter 2, we know that V^π is one solution to the above.

Are there other solutions to Equation 4.1? In this chapter, we answer this question (negatively) by interpreting the right-hand side of the equation as applying a transformation on the estimate V. For a given realization (x, a, r, x') of the random transition, this transformation indexes V by x', multiplies it by the discount factor, and adds it to the immediate reward (this yields $r + \gamma V(x')$). The actual transformation returns the value that is obtained by following these steps, in expectation. Functions that map elements of a space onto itself, such as this one (from estimates to estimates), are called *operators*.

Definition 4.1. The *Bellman operator* is the mapping $T^\pi : \mathbb{R}^{\mathcal{X}} \to \mathbb{R}^{\mathcal{X}}$ defined by

$$(T^\pi V)(x) = \mathbb{E}_\pi[R + \gamma V(X') \mid X = x]. \tag{4.2}$$

Here, the notation $T^\pi V$ should be understood as "T^π, applied to V." △

The Bellman operator gives a particularly concise way of expressing the transformations implied in Equation 4.1:

$$V = T^\pi V.$$

As we will see in later chapters, it also serves as the springboard for the design and analysis of algorithms for learning V^π.

When working with the Bellman operator, it is often useful to treat V as a finite-dimensional vector in $\mathbb{R}^{\mathcal{X}}$ and to express the Bellman operator in terms of vector operations. That is, we write

$$T^\pi V = r^\pi + \gamma P^\pi V, \tag{4.3}$$

where $r^\pi(x) = \mathbb{E}_\pi[R \mid X = x]$ and P^π is the transition operator[24] defined as

$$(P^\pi V)(x) = \sum_{a \in \mathcal{A}} \pi(a \mid x) \sum_{x' \in \mathcal{X}} P_{\mathcal{X}}(x' \mid x, a) V(x') .$$

Equation 4.3 follows from these definitions and the linearity of expectations.

A vector $\tilde{V} \in \mathbb{R}^{\mathcal{X}}$ is a solution to Equation 4.1 if it remains unchanged by the transformation corresponding to the Bellman operator T^π; that is, if it is a *fixed point* of T^π. This means that the value function V^π is a fixed point of T^π:

$$V^\pi = T^\pi V^\pi .$$

To demonstrate that V^π is the only fixed point of the Bellman operator, we will appeal to the notion of *contraction mappings*.

4.2 Contraction Mappings

When we apply the operator T^π to a value function estimate $V \in \mathbb{R}^{\mathcal{X}}$, we obtain a new estimate $T^\pi V \in \mathbb{R}^{\mathcal{X}}$. A characteristic property of the Bellman operator is that this new estimate is guaranteed to be closer to V^π than V (unless $V = V^\pi$, of course). In fact, as we will see in this section, applying the operator to any two estimates must bring them closer together.

To formalize what we mean by "closer," we need a way of measuring distances between value function estimates. Because these estimates can be viewed as finite-dimensional vectors, there are many well-established ways of doing so: the reader may have come across the Euclidean (L^2) distance, the Manhattan (L^1) distance, and curios such as the British Rail distance. We use the term *metric* to describe distances that satisfy the following standard definition.

Definition 4.2. Given a set M, a metric $d : M \times M \to \mathbb{R}$ is a function that satisfies, for all $U, V, W \in M$,

(a) $d(U, V) \geq 0$,
(b) $d(U, V) = 0$ if and only if $U = V$,
(c) $d(U, V) \leq d(U, W) + d(W, V)$,
(d) $d(U, V) = d(V, U)$.

We call the pair (M, d) a *metric space*. △

In our setting, M is the space of value function estimates, $\mathbb{R}^{\mathcal{X}}$. Because we assume that there are finitely many states, this space can be equivalently thought of as the space of real-valued vectors with $N_{\mathcal{X}}$ entries, where $N_{\mathcal{X}}$ is the number

24. It is also possible to express P^π as a stochastic matrix, in which case $P^\pi V$ describes a matrix–vector multiplication. We will return to this point in Chapter 5.

of states. On this space, we measure distances in terms of the L^∞ metric, defined by

$$\|V - V'\|_\infty = \max_{x \in \mathcal{X}} |V(x) - V'(x)|, \quad V, V' \in \mathbb{R}^{\mathcal{X}}. \tag{4.4}$$

A key result is that the Bellman operator T^π is a *contraction mapping* with respect to this metric. Informally, this means that its application to different value function estimates brings them closer by at least a constant multiplicative factor, called its *contraction modulus*.

Definition 4.3. Let (M, d) be a metric space. A function $\mathcal{O}: M \to M$ is a contraction mapping with respect to d and with contraction modulus $\beta \in [0, 1)$, if for all $U, U' \in M$,

$$d(\mathcal{O}U, \mathcal{O}U') \leq \beta d(U, U'). \qquad \triangle$$

Proposition 4.4. The operator $T^\pi : \mathbb{R}^{\mathcal{X}} \to \mathbb{R}^{\mathcal{X}}$ is a contraction mapping with respect to the L^∞ metric on $\mathbb{R}^{\mathcal{X}}$ with contraction modulus given by the discount factor γ. That is, for any two value functions $V, V' \in \mathbb{R}^{\mathcal{X}}$,

$$\|T^\pi V - T^\pi V'\|_\infty \leq \gamma \|V - V'\|_\infty. \qquad \triangle$$

Proof. The proof is most easily stated in vector notation. Here, we make use of two properties of the operator P^π. First, P^π is *linear*, in the sense that for any V, V',

$$P^\pi V + P^\pi V' = P^\pi (V + V').$$

Second, because $(P^\pi V)(x)$ is a convex combination of elements from V, it must be that

$$\|P^\pi V\|_\infty \leq \|V\|_\infty.$$

From here, we have

$$\begin{aligned}
\|T^\pi V - T^\pi V'\|_\infty &= \|(r^\pi + \gamma P^\pi V) - (r^\pi + \gamma P^\pi V')\|_\infty \\
&= \|\gamma P^\pi V - \gamma P^\pi V'\|_\infty \\
&= \gamma \|P^\pi (V - V')\|_\infty \\
&\leq \gamma \|V - V'\|_\infty,
\end{aligned}$$

as desired. □

The fact that T^π is a contraction mapping guarantees the uniqueness of V^π as a solution to the equation $V = T^\pi V$. As made formal by the following proposition, because the operator T^π brings any two value functions closer together, it cannot keep more than one value function fixed.

Proposition 4.5. Let (M, d) be a metric space and $O: M \to M$ be a contraction mapping. Then O has at most one fixed point in M. △

Proof. Let $\beta \in [0, 1)$ be the contraction modulus of O, and suppose $U, U' \in M$ are distinct fixed points of O, so that $d(U, U') > 0$ (following Definition 4.2). Then we have

$$d(U, U') = d(OU, OU') \leq \beta d(U, U'),$$

which is a contradiction. □

Because we know that V^π is a fixed point of the Bellman operator T^π, following Proposition 4.5, we deduce that there are no other such fixed points – and hence no other solutions to the Bellman equation. As the phrasing of Proposition 4.5 suggests, in some metric spaces, it is possible for O to be a contraction mapping yet to not possess a fixed point. This can matter when dealing with return functions, as we will see in the second half of this chapter and in Chapter 5.

Example 4.6. Consider the *no-loop operator*

$$(T^\pi_{\text{NL}} V)(x) = \mathbb{E}_\pi \left[R + \gamma V(X')\mathbb{1}_{\{X' \neq x\}} \mid X = x\right],$$

where the name denotes the fact that we omit the next-state value whenever a transition from x to itself occurs. By inspection, we can determine that the fixed point of this operator is

$$V^\pi_{\text{NL}}(x) = \mathbb{E}_\pi \Big[\sum_{t=0}^{T-1} \gamma^t R_t \mid X_0 = x\Big],$$

where T denotes the (random) first time at which $X_T = X_{T-1}$. In words, this fixed point describes the discounted sum of rewards obtained until the first time that an action leaves the state unchanged.[25]

Exercise 4.1 asks you to show that T^π_{NL} is a contraction mapping with modulus

$$\beta = \gamma \max_{x \in \mathcal{X}} \mathbb{P}_\pi(X' \neq x \mid X = x).$$

Following Proposition 4.5, we deduce that this is the unique fixed point to T^π_{NL}. △

25. The reader is invited to consider the kind of environments in which policies that maximize the no-loop return are substantially different from those that maximize the usual expected return.

When an operator O is contractive, we can also straightforwardly construct a mathematical approximation to its fixed point.[26] This approximation is given by the sequence $(U_k)_{k\geq 0}$, defined by an initial value $U_0 \in M$, and the recursive relationship

$$U_{k+1} = OU_k .$$

By contractivity, successive iterates of this sequence must come progressively closer to the operator's fixed point. This is formalized by the following.

Proposition 4.7. Let (M, d) be a metric space, and let $O : M \to M$ be a contraction mapping with contraction modulus $\beta \in [0, 1)$ and fixed point $U^* \in M$. Then for any initial point $U_0 \in M$, the sequence $(U_k)_{k\geq 0}$ defined by $U_{k+1} = OU_k$ is such that

$$d(U_k, U^*) \leq \beta^k d(U_0, U^*). \tag{4.5}$$

and in particular $d(U_k, U^*) \to 0$ as $k \to \infty$. △

Proof. We will prove Equation 4.5 by induction, from which we obtain convergence of $(U_k)_{k\geq 0}$ in d. For $k = 0$, Equation 4.5 trivially holds. Now suppose for some $k \geq 0$, we have

$$d(U_k, U^*) \leq \beta^k d(U_0, U^*) .$$

Then note that

$$d(U_{k+1}, U^*) \overset{(a)}{=} d(OU_k, OU^*) \overset{(b)}{\leq} \beta d(U_k, U^*) \overset{(c)}{\leq} \beta^{k+1} d(U_0, U^*) ,$$

where (a) follows from the definition of the sequence $(U_k)_{k\geq 0}$ and the fact that U^* is fixed by O, (b) follows from the contractivity of O, and (c) follows from the inductive hypothesis. By induction, we conclude that Equation 4.5 holds for all $k \in \mathbb{N}$. □

In the case of the Bellman operator T^π, Proposition 4.7 means that repeated application of T^π to any initial value function estimate $V_0 \in \mathbb{R}^{\mathcal{X}}$ produces a sequence of estimates $(V_k)_{k\geq 0}$ that are progressively closer to V^π. This observation serves as the starting point for a number of computational approaches that approximate V^π, including dynamic programming (Chapter 5) and temporal-difference learning (Chapter 6).

26. We use the term *mathematical approximation* to distinguish it from an approximation that can be computed. That is, there may or may not exist an algorithm that can determine the elements of the sequence $(U_k)_{k\geq 0}$ given the initial estimate U_0.

4.3 The Distributional Bellman Operator

Designing distributional reinforcement learning algorithms such as categorical temporal-difference learning involves a few choices – such as how to represent probability distributions in a computer's memory – that do not have an equivalent in classical reinforcement learning. Throughout this book, we will make use of the *distributional Bellman operator* to understand and characterize many of these choices. To begin, recall the random-variable Bellman equation:

$$G^\pi(x) \overset{\mathcal{D}}{=} R + \gamma G^\pi(X'), \quad X = x. \tag{4.6}$$

As in the expected-value setting, we construct a random-variable operator by viewing the right-hand side of Equation 4.6 as a transformation of G^π. In this case, we break down the transformation of G^π into three operations, each of which produces a new random variable (Figure 4.1):

(a) $G^\pi(X')$: the indexing of the collection of random variables G^π by X';
(b) $\gamma G^\pi(X')$: the multiplication of the random variable $G^\pi(X')$ with the scalar γ;
(c) $R + \gamma G^\pi(X')$: the addition of two random variables (R and $\gamma G^\pi(X')$).

More generally, we may apply these operations to any state-indexed collection of random variables $G = (G(x) : x \in \mathcal{X})$, taken to be independent of the random transitions used to define the transformation. With some mathematical caveats discussed below, let us introduce the *random-variable Bellman operator*

$$(\mathcal{T}^\pi G)(x) \overset{\mathcal{D}}{=} R + \gamma G(X'), \quad X = x. \tag{4.7}$$

Equation 4.7 states that the application of the Bellman operator to G (evaluated at x; the left-hand side) produces a random variable that is equal in distribution to the random variable constructed on the right-hand side. Because this holds for all x, we think of $\mathcal{T}^\pi$ as mapping G to a new collection of random variables $\mathcal{T}^\pi G$.

The random-variable operator is appealing because it is concise and easily understood. In many circumstances, this makes it the tool of choice for reasoning about distributional reinforcement learning problems. One issue, however, is that its definition above is mathematically incomplete. This is because it specifies the probability distribution of $(\mathcal{T}^\pi G)(x)$, but not its identity as a mapping from some sample space to the real numbers. As discussed in Section 2.7, without care we may produce random variables that exhibit undesirable behavior: for example, because rewards at different points in time are improperly correlated. More immediately, the theory of contraction mappings needs a clear definition of the space on which an operator is defined – in the case of the random-variable operator, this requires us to specify a space of random variables to operate

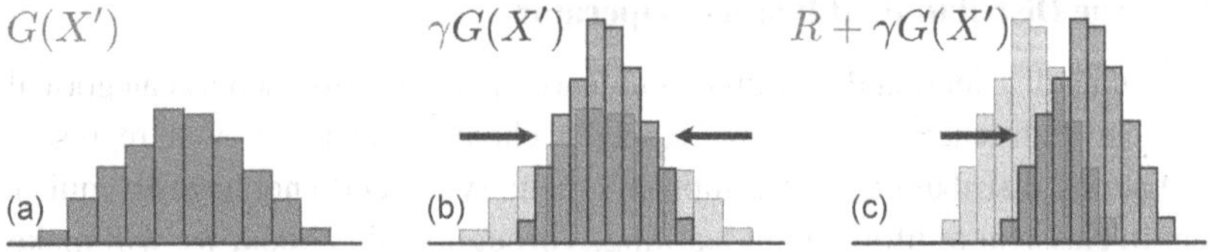

Figure 4.1
The random-variable Bellman operator is composed of three operations: **(a)** indexing into a collection of random variables, **(b)** multiplication by the discount factor, and **(c)** addition of two random variables. Here, we assume that R and X' take on a single value for clarity.

in. Properly defining such a space is possible but requires some technical subtlety and measure-theoretic considerations; we refer the interested reader to Section 4.9.

A more direct solution is to consider the distributional Bellman operator as a mapping on the space of return-distribution functions. Starting with the distributional Bellman equation

$$\eta^\pi(x) = \mathbb{E}_\pi[(\mathrm{b}_{R,\gamma})_\# \eta^\pi(X') \mid X = x],$$

we again view the right-hand side as the result of applying a series of transformations, in this case to probability distributions.

Definition 4.8. The *distributional Bellman operator* $\mathcal{T}^\pi : \mathscr{P}(\mathbb{R})^{\mathcal{X}} \to \mathscr{P}(\mathbb{R})^{\mathcal{X}}$ is the mapping defined by

$$(\mathcal{T}^\pi \eta)(x) = \mathbb{E}_\pi[(\mathrm{b}_{R,\gamma})_\# \eta(X') \mid X = x]. \tag{4.8}$$

△

Here, the operations on probability distributions are expressed (rather compactly) by the expectation in Equation 4.8 and the use of the pushforward distribution derived from the bootstrap function b; these are the operations of mixing, scaling, and translation previously described in Section 2.8.

We can gain additional insight into how the operator transforms a return function η by considering the situation in which the random reward R and the return distributions $\eta(x)$ admit respective probability densities $p_{\mathcal{R}}(r \mid x, a)$ and $p_x(z)$. In this case, the probability density of $(\mathcal{T}^\pi \eta)(x)$, denoted p'_x, is

$$p'_x(z) = \gamma^{-1} \sum_{a \in \mathcal{A}} \pi(a \mid x) \int_{r \in \mathbb{R}} p_{\mathcal{R}}(r \mid x, a) \sum_{x' \in \mathcal{X}} P_{\mathcal{X}}(x' \mid x, a) p_{x'}\Big(\frac{z - r}{\gamma}\Big) \mathrm{d}r. \tag{4.9}$$

Expressed in terms of probability densities, the indexing of a collection of random variables becomes a mixture of densities, while their addition becomes a convolution; this is in fact what is depicted in Figure 4.1. In terms of cumulative distribution functions, we have

$$F_{(\mathcal{T}^\pi\eta)(x)}(z) = \mathbb{E}_\pi\Big[F_{\eta(X')}\Big(\frac{z-R}{\gamma}\Big) \mid X = x\Big].$$

However, we prefer the operator that deals directly with probability distributions (Equation 4.8) as it can be used to concisely express more complex operations on distributions. One such operation is the projection of a probability distribution onto a finitely parameterized set, which we will use in Chapter 5 to construct algorithms for approximating η^π.

Using Definition 4.8, we can formally express the fact that the return-distribution function η^π is the only solution to the equation

$$\eta = \mathcal{T}^\pi\eta.$$

The proof is relatively technical and will be given in Section 4.8.

Proposition 4.9. The return-distribution function η^π satisfies

$$\eta^\pi = \mathcal{T}^\pi\eta^\pi$$

and is the unique fixed point of the distributional Bellman operator $\mathcal{T}^\pi$. △

When working with the distributional Bellman operator, one should be mindful that the random reward R and next state X' are generally not independent, because they both depend on the chosen action A (we briefly mentioned this concern in Section 2.8). In Equation 4.9, this is explicitly handled by the outer sum over actions. Analogously, we can make explicit the dependency on A by introducing a second expectation in Equation 4.8:

$$(\mathcal{T}^\pi\eta)(x) = \mathbb{E}_\pi[\mathbb{E}_\pi[(\mathrm{b}_{R,\gamma})_\#\eta(X') \mid X = x, A] \mid X = x].$$

By conditioning the inner expectation on the action A, we make the random variables R and $\gamma G(X')$ conditionally independent in the inner expectation. We will make use of this technique in proving Theorem 4.25, the main theoretical result of this chapter.

In some circumstances, it is useful to translate between operations on probability distributions and those on random variables. We do this by means of a representative set of random variables called an *instantiation*.

Definition 4.10. Given a probability distribution $\nu \in \mathscr{P}(\mathbb{R})$, we say that a random variable Z is an instantiation of ν if its distribution is ν, written $Z \sim \nu$. Similarly, we say that a collection of random variables $G = (G(x) : x \in \mathcal{X})$ is an instantiation of a return-distribution function $\eta \in \mathscr{P}(\mathbb{R})^{\mathcal{X}}$ if for every $x \in \mathcal{X}$, we have $G(x) \sim \eta(x)$. △

Given a return-distribution function $\eta \in \mathscr{P}(\mathbb{R})^{\mathcal{X}}$, the new return-distribution function $\mathcal{T}^\pi \eta$ can be obtained by constructing an instantiation G of η, performing the transformation on the collection of random variables G as described at the beginning of this section, and then extracting the distributions of the resulting random variables. This is made formal as follows.

Proposition 4.11. Let $\eta \in \mathscr{P}(\mathbb{R})^{\mathcal{X}}$, and let $G = (G(x) : x \in \mathcal{X})$ be an instantiation of η. For each $x \in \mathcal{X}$, let $(X = x, A, R, X')$ be a sample transition independent of G. Then $R + \gamma G(X')$ has the distribution $(\mathcal{T}^\pi \eta)(x)$:

$$\mathcal{D}_\pi(R + \gamma G(X') \mid X = x) = (\mathcal{T}^\pi \eta)(x)\,. \qquad \triangle$$

Proof. The result follows immediately from the definition of the distributional Bellman operator. For clarity, we step through the argument again, mirroring the transformations set out at the beginning of the section. First, the indexing transformation gives

$$\begin{aligned}\mathcal{D}_\pi(G(X') \mid X = x) &= \sum_{x' \in \mathcal{X}} \mathbb{P}_\pi(X' = x' \mid X = x)\eta(x') \\ &= \mathbb{E}_\pi[\eta(X') \mid X = x]\,.\end{aligned}$$

Next, scaling by γ yields

$$\mathcal{D}_\pi(\gamma G(X') \mid X = x) = \mathbb{E}_\pi[(\mathrm{b}_{0,\gamma})_\# \eta(X') \mid X = x]\,,$$

and finally adding the immediate reward R gives the result

$$\mathcal{D}_\pi(R + \gamma G(X') \mid X = x) = \mathbb{E}_\pi[(\mathrm{b}_{R,\gamma})_\# \eta(X') \mid X = x]\,. \qquad \square$$

Proposition 4.11 is an instance of a recurring principle in distributional reinforcement learning that "different routes lead to the same answer." Throughout this book, we will illustrate this point as it arises with a commutative diagram; the particular case under consideration is depicted in Figure 4.2.

4.4 Wasserstein Distances for Return Functions

Many desirable properties of reinforcement learning algorithms (for example, the fact that they produce a good approximation of the value function) are due to the contractive nature of the Bellman operator T^π. In this section, we will establish that the distributional Bellman operator $\mathcal{T}^\pi$, too, is a contraction mapping – analogous to the value-based operator, the application of $\mathcal{T}^\pi$ brings return functions closer together.

One difference between expected-value and distributional reinforcement learning is that the space of return-distribution functions $\mathscr{P}(\mathbb{R})^{\mathcal{X}}$ is substantially

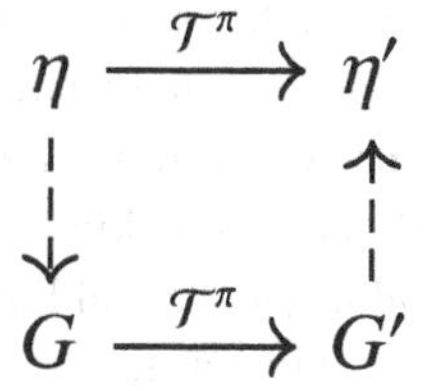

Figure 4.2
A commutative diagram illustrating two perspectives on the application of the distributional Bellman operator. The top horizontal line represents the direct application to the return-distribution function η, yielding η'. The alternative path first instantiates the return-distribution function η as a collection of random variables $G = (G(x) : (x) \in \mathcal{X})$, transforms G to obtain another collection of random variables G', and then extracts the distributions of these random variables to obtain η'.

different from the space of value functions. To measure distances between value functions, we can simply treat them as finite-dimensional vectors, taking the absolute difference of value estimates at individual states. By contrast, it is somewhat less intuitive to see what "close" means when comparing probability distributions. Throughout this chapter, we will consider a number of *probability metrics* that measure distances between distributions, each presenting different mathematical and computational properties. We begin with the family of Wasserstein distances.

Definition 4.12. Let $\nu \in \mathscr{P}(\mathbb{R})$ be a probability distribution with cumulative distribution function F_ν. Let Z be an instantiation of ν (in particular, $F_Z = F_\nu$). The generalized inverse F_ν^{-1} is given by

$$F_\nu^{-1}(\tau) = \inf_{z \in \mathbb{R}} \{z : F_\nu(z) \geq \tau\} .$$

We additionally write $F_Z^{-1} = F_\nu^{-1}$. △

Definition 4.13. Let $p \in [1, \infty)$. The *p-Wasserstein distance* is a function w_p : $\mathscr{P}(\mathbb{R}) \times \mathscr{P}(\mathbb{R}) \to [0, \infty]$ given by

$$w_p(\nu, \nu') = \left(\int_0^1 \left|F_\nu^{-1}(\tau) - F_{\nu'}^{-1}(\tau)\right|^p \mathrm{d}\tau \right)^{1/p} .$$

The ∞-Wasserstein distance $w_\infty : \mathscr{P}(\mathbb{R}) \times \mathscr{P}(\mathbb{R}) \to [0, \infty]$ is

$$w_\infty(\nu, \nu') = \sup_{\tau \in (0,1)} \left|F_\nu^{-1}(\tau) - F_{\nu'}^{-1}(\tau)\right| .$$ △

Graphically, the Wasserstein distances between two probability distributions measure the area between their cumulative distribution functions, with values along the abscissa taken to the pth power; see Figure 4.3. When $p = \infty$, this becomes the largest horizontal difference between the inverse cumulative distribution functions. The p-Wasserstein distances satisfy the definition of a metric, except that they may not be finite for arbitrary pairs of distributions in $\mathscr{P}(\mathbb{R})$; see Exercise 4.6. Properly speaking, they are said to be *extended metrics*, since they may take values on the real line extended to include infinity. Most probability metrics that we will consider are extended metrics rather than metrics in the sense of Definition 4.2. We measure distances between return-distribution functions in terms of the largest Wasserstein distance between probability distributions at individual states.

Definition 4.14. Let $p \in [1, \infty]$. The supremum p-Wasserstein distance $\overline{w}_p$ between two return-distribution functions $\eta, \eta' \in \mathscr{P}(\mathbb{R})^{\mathcal{X}}$ is defined by[27]

$$\overline{w}_p(\eta, \eta') = \sup_{x \in \mathcal{X}} w_p(\eta(x), \eta'(x)) \,. \qquad \triangle$$

The supremum p-Wasserstein distances fulfill all requirements of an extended metric on the space of return-distribution functions $\mathscr{P}(\mathbb{R})^{\mathcal{X}}$; see Exercise 4.7. Based on these distances, we give our first contractivity result regarding the distributional Bellman operator; its proof is given at the end of the section.

Proposition 4.15. The distributional Bellman operator is a contraction mapping on $\mathscr{P}(\mathbb{R})^{\mathcal{X}}$ in the supremum p-Wasserstein distance, for all $p \in [1, \infty]$. More precisely,

$$\overline{w}_p(\mathcal{T}^\pi \eta, \mathcal{T}^\pi \eta') \leq \gamma \overline{w}_p(\eta, \eta') \,,$$

for all $\eta, \eta' \in \mathscr{P}(\mathbb{R})^{\mathcal{X}}$. $\triangle$

Proposition 4.15 is significant in that it establishes a close parallel between the expected-value and distributional operators. Following the line of reasoning given in Section 4.2, it provides the mathematical justification for the development and analysis of computational approaches for finding the return function η^π. More immediately, it also enables us to characterize the convergence of the sequence

$$\eta_{k+1} = \mathcal{T}^\pi \eta_k \qquad (4.10)$$

to the return function η^π.

27. Because we assume that there are finitely many states, we can equivalently write max in the definition of supremum distance. However, we prefer the more generally applicable sup.

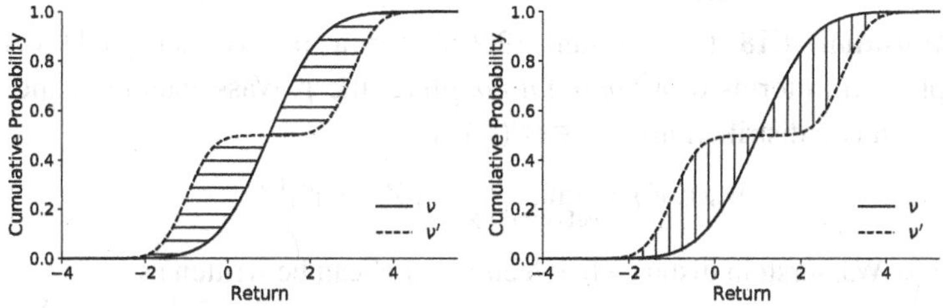

Figure 4.3
Left: Illustration of the p-Wasserstein distance between a normal distribution $\nu = \mathcal{N}(1,1)$ and a mixture of two normal distributions $\nu' = \frac{1}{2}\mathcal{N}(-1, 0.5) + \frac{1}{2}\mathcal{N}(3, 0.5)$. **Right**: Illustration of the ℓ_p metric for the same distributions (see Section 4.5). In both cases, the shading indicates the axis along which the differences are taken to the pth exponent.

Proposition 4.16. Suppose that for each $(x, a) \in \mathcal{X} \times \mathcal{A}$, the reward distribution $P_{\mathcal{R}}(\cdot \mid x, a)$ is supported on the interval $[R_{\text{MIN}}, R_{\text{MAX}}]$. Then for any initial return function η_0 whose distributions are bounded on the interval $\left[\frac{R_{\text{MIN}}}{1-\gamma}, \frac{R_{\text{MAX}}}{1-\gamma}\right]$, the sequence

$$\eta_{k+1} = \mathcal{T}^\pi \eta_k$$

converges to η^π in the supremum p-Wasserstein distance (for all $p \in [1, \infty]$). △

The restriction to bounded rewards in Proposition 4.16 is necessary to make use of the tools developed in Section 4.2, at least without further qualification. This is because Proposition 4.7 requires all distances to be finite, which is not guaranteed under our definition of a probability metric. If, for example, the initial condition η_0 is such that

$$\overline{w}_p(\eta_0, \eta^\pi) = \infty\,,$$

then Proposition 4.15 is not of much use. A less restrictive but more technically elaborate set of assumptions will be presented later in the chapter. For now, we provide the proof of the two preceding results. First, we obtain a reasonably simple proof of Proposition 4.15 by considering an alternative formulation of the p-Wasserstein distances in terms of *couplings*.

Definition 4.17. Let $\nu, \nu' \in \mathscr{P}(\mathbb{R})$ be two probability distributions. A coupling between ν and ν' is a joint distribution $\upsilon \in \mathscr{P}(\mathbb{R}^2)$ such that if (Z, Z') is an instantiation of υ, then also Z has distribution ν and Z' has distribution ν'. We write $\Gamma(\nu, \nu') \subseteq \mathscr{P}(\mathbb{R}^2)$ for the set of all couplings of ν and ν'. △

Proposition 4.18 (see Villani (2008) for a proof)**.** Let $p \in [1, \infty)$. Expressed in terms of an *optimal coupling*, the p-Wasserstein distance between two distributions $\nu, \nu' \in \mathscr{P}(\mathbb{R})$ is

$$w_p(\nu, \nu') = \min_{\upsilon \in \Gamma(\nu,\nu')} \mathop{\mathbb{E}}_{(Z,Z')\sim\upsilon} [|Z - Z'|^p]^{1/p} .$$

The ∞-Wasserstein distance between ν and ν' can be written as

$$w_\infty(\nu, \nu') = \min_{\upsilon \in \Gamma(\nu,\nu')} \inf \Big\{ z \in \mathbb{R} : \mathop{\mathbb{P}}_{(Z,Z')\sim\upsilon} (|Z - Z'| > z) = 0 \Big\} . \qquad \triangle$$

Informally, the optimal coupling finds an arrangement of the two probability distributions that maximizes "agreement": it produces outcomes that are as close as possible. In Proposition 4.18, the optimal coupling takes on a very simple form given by inverse cumulative distribution functions. For $\nu, \nu' \in \mathscr{P}(\mathbb{R})$, an optimal coupling is the probability distribution of the random variable

$$(F_\nu^{-1}(\tau), F_{\nu'}^{-1}(\tau)), \quad \tau \sim \mathcal{U}([0, 1]). \tag{4.11}$$

This can be understood by noting how the 1-Wasserstein distance between ν and ν' is obtained by measuring the horizontal distance between the two cumulative distribution functions, at each level $\tau \in [0, 1]$ (Figure 4.3).

Proof of Proposition 4.15. Let $p \in [1, \infty)$ be fixed. For each $x \in \mathcal{X}$, consider the optimal coupling between $\eta(x)$ and $\eta'(x)$ and instantiate it as the pair of random variables $(G(x), G'(x))$. Next, denote by (x, A, R, X') the random transition beginning in $x \in \mathcal{X}$, constructed to be independent from $G(y)$ and $G'(y)$, for all $y \in \mathcal{X}$. With these variables, write

$$\tilde{G}(x) = R + \gamma G(X') , \quad \tilde{G}'(x) = R + \gamma G'(X') .$$

By Proposition 4.11, $\tilde{G}(x)$ has distribution $(\mathcal{T}^\pi \eta)(x)$ and $\tilde{G}'(x)$ has distribution $(\mathcal{T}^\pi \eta')(x)$. The pair $(\tilde{G}(x), \tilde{G}'(x))$ therefore forms a valid coupling of these distributions. Now

$$\begin{aligned} w_p^p\big((\mathcal{T}^\pi \eta)(x), (\mathcal{T}^\pi \eta')(x)\big) &\overset{(a)}{\leq} \mathbb{E}_\pi \Big[\big|(R + \gamma G(X')) - (R + \gamma G'(X'))\big|^p \mid X = x \Big] \\ &\overset{(b)}{=} \gamma^p \, \mathbb{E} \Big[\big|G(X') - G'(X')\big|^p \mid X = x \Big] \\ &\overset{(c)}{\leq} \gamma^p \sum_{x' \in \mathcal{X}} \mathbb{P}_\pi(X' = x' \mid X = x) \, \mathbb{E} \Big[\big|G(x') - G'(x')\big|^p \Big] \\ &\overset{(d)}{\leq} \gamma^p \sup_{x' \in \mathcal{X}} \mathbb{E} \Big[\big|G(x') - G'(x')\big|^p \Big] \\ &\overset{(e)}{=} \gamma^p \overline{w}_p^p(\eta, \eta') . \end{aligned}$$

Taking a supremum over $x \in \mathcal{X}$ on the left-hand side and the pth root of both sides yields the result. Here, (a) follows since the Wasserstein distance is defined as a minimum over couplings, (b) follows from algebraic manipulation of the expectation, (c) follows from independence of the sample transition $(X = x, A, R, X')$ and the random variables $(G(x), G'(x) : x \in \mathcal{X})$, (d) because the maximum of nonnegative quantities is at least as great as their weighted average, and (e) follows since $(G(x'), G'(x'))$ was defined as an optimal coupling of $\eta(x')$ and $\eta'(x')$. The proof for $p = \infty$ is similar (see Exercise 4.8). □

Proof of Proposition 4.16. Let us denote by $\mathscr{P}_B(\mathbb{R})$ the space of distributions bounded on $[V_{\text{MIN}}, V_{\text{MAX}}]$, where as usual

$$V_{\text{MIN}} = \frac{R_{\text{MIN}}}{1-\gamma}, \quad V_{\text{MAX}} = \frac{R_{\text{MAX}}}{1-\gamma}.$$

We will show that under the assumption of rewards bounded on $[R_{\text{MIN}}, R_{\text{MAX}}]$,

(a) the return function η^π is in $\mathscr{P}_B(\mathbb{R})$, and
(b) the distributional Bellman operator maps $\mathscr{P}_B(\mathbb{R})$ to itself.

Consequently, we can invoke Proposition 4.7 with $O = \mathcal{T}^\pi$ and $M = \mathscr{P}_B(\mathbb{R})$ to conclude that for any initial $\eta_0 \in \mathscr{P}_B(\mathbb{R})^{\mathcal{X}}$, the sequence of iterates $(\eta_k)_{k\geq 0}$ converges to η^π with respect to $d = \overline{w}_p$, for any $p \in [1, \infty]$.

To prove (a), note that for any state $x \in \mathcal{X}$,

$$G^\pi(x) = \sum_{t=0}^{\infty} \gamma^t R_t, \quad X_0 = x,$$

and since $R_t \in [R_{\text{MIN}}, R_{\text{MAX}}]$ for all t, then also $G^\pi(x) \in [V_{\text{MIN}}, V_{\text{MAX}}]$. For (b), let $\eta \in \mathscr{P}_B(\mathbb{R})^{\mathcal{X}}$ and denote by G an instantiation of this return-distribution function. For any $x \in \mathcal{X}$,

$$\begin{aligned}
\mathbb{P}_\pi(R + \gamma G(X') \leq V_{\text{MAX}} \mid X = x) &= \mathbb{P}_\pi(\gamma G(X') \leq V_{\text{MAX}} - R \mid X = x) \\
&\geq \mathbb{P}_\pi(\gamma G(X') \leq V_{\text{MAX}} - R_{\text{MAX}} \mid X = x) \\
&= \mathbb{P}_\pi(G(X') \leq V_{\text{MAX}} \mid X = x) \\
&= 1.
\end{aligned}$$

By the same reasoning,

$$\mathbb{P}_\pi(R + \gamma G(X') \geq V_{\text{MIN}} \mid X = x) = 1.$$

Since $R + \gamma G(X'), X = x$ is an instantiation of $(\mathcal{T}^\pi \eta)(x)$ for each x, we conclude that if $\eta \in \mathscr{P}_B(\mathbb{R})^{\mathcal{X}}$, then also $\mathcal{T}^\pi \eta \in \mathscr{P}_B(\mathbb{R})^{\mathcal{X}}$. □

4.5 ℓ_p Probability Metrics and the Cramér Distance

The previous section established that the distributional Bellman operator is well behaved with respect to the family of Wasserstein distances. However, these are but a few among many standard probability metrics. We will see in Chapter 5 that theoretical analysis sometimes requires us to study the behavior of the distributional operator with respect to other metrics. In addition, many practical algorithms directly optimize a metric (typically expressed as a loss function) as part of their operation (see Chapter 10). The *Cramér distance*, a member of the broader family of ℓ_p metrics, is of particular interest to us.

Definition 4.19. Let $p \in [1, \infty)$. The distance $\ell_p : \mathscr{P}(\mathbb{R}) \times \mathscr{P}(\mathbb{R}) \to [0, \infty]$ is a probability metric defined by

$$\ell_p(\nu, \nu') = \left(\int_{\mathbb{R}} |F_\nu(z) - F_{\nu'}(z)|^p \mathrm{d}z \right)^{1/p} . \tag{4.12}$$

For $p = 2$, this is the *Cramér distance*.[28] The ℓ_∞ or *Kolmogorov–Smirnov distance* is given by

$$\ell_\infty(\nu, \nu') = \sup_{z \in \mathbb{R}} \left| F_\nu(z) - F_{\nu'}(z) \right| .$$

The respective supremum ℓ_p distances are given by ($\eta, \eta' \in \mathscr{P}(\mathbb{R})^{\mathcal{X}}$)

$$\overline{\ell}_p(\eta, \eta') = \sup_{x \in \mathcal{X}} \ell_p(\eta(x), \eta(x')) .$$

These are extended metrics on $\mathscr{P}(\mathbb{R})^{\mathcal{X}}$. △

Where the p-Wasserstein distances measure differences in outcomes, the ℓ_p distances measure differences in the probabilities associated with these outcomes. This is because the exponent p is applied to cumulative probabilities (this is illustrated in Figure 4.3). The distributional Bellman operator is also a contraction mapping under the ℓ_p distances for $p \in [1, \infty)$, albeit with a larger contraction modulus.[29]

28. Historically, the Cramér distance has been defined as the square of ℓ_2. In our context, it seems unambiguous to use the word for ℓ_2 itself.

29. For $p = 1$, the distributional Bellman operator has contraction modulus γ; this is sensible given that $\ell_1 = w_1$.

Proposition 4.20. For $p \in [1, \infty)$, the distributional Bellman operator $\mathcal{T}^\pi$ is a contraction mapping on $\mathscr{P}(\mathbb{R})^{\mathcal{X}}$ with respect to $\bar{\ell}_p$, with contraction modulus $\gamma^{1/p}$. That is,

$$\bar{\ell}_p(\mathcal{T}^\pi \eta, \mathcal{T}^\pi \eta') \le \gamma^{1/p} \bar{\ell}_p(\eta, \eta')$$

for all $\eta, \eta' \in \mathscr{P}(\mathbb{R})^{\mathcal{X}}$. △

The proof of Proposition 4.20 will follow as a corollary of a more general result given in Section 4.6. One way to relate it to our earlier result is to consider the behavior of the sequence defined by

$$\eta_{k+1} = \mathcal{T}^\pi \eta_k. \tag{4.13}$$

As measured in the p-Wasserstein distance, the sequence $(\eta_k)_{k \ge 0}$ approaches η^π at a rate of γ; but if we instead measure distances using the ℓ_p metric, this rate is slower – only $\gamma^{1/p}$. Measured in terms of ℓ_∞ (the Kolmogorov–Smirnov distance), the sequence of iterates may in fact not seem to approach η^π at all. To see this, it suffices to consider a single-state process with zero reward (that is, $P_{\mathcal{X}}(X' = x \mid X = x) = 1$ and $R = 0$) and a discount factor $\gamma = 0.9$. In this case, $\eta^\pi(x) = \delta_0$. For the initial condition $\eta_0(x) = \delta_1$, we obtain

$$\eta_1(x) = (\mathcal{T}^\pi \eta_0)(x) = \delta_\gamma.$$

Now, the (supremum) ℓ_∞ distance between η^π and η_0 is 1, because for any $z \in (0, 1)$,

$$F_{\eta_0(x)}(z) = 0 \quad F_{\eta^\pi(x)}(z) = 1.$$

However, the ℓ_∞ distance between η^π and η_1 is also 1, by the same argument (but now restricted to $z \in (0, \gamma)$). Hence, there is no $\beta \in [0, 1)$ for which

$$\bar{\ell}_\infty(\eta_1, \eta^\pi) < \beta \bar{\ell}_\infty(\eta_0, \eta^\pi).$$

Exercise 4.16 asks you to prove a similar result for a probability metric called the total variation distance (see also Figure 4.4).

The more general point is that different probability metrics are sensitive to different characteristics of probability distributions, and to varying degrees. At one extreme, the ∞-Wasserstein distance is effectively insensitive to the probability associated with different outcomes, while at the other extreme, the Kolmogorov–Smirnov distance is insensitive to the scale of the difference between outcomes. In Section 4.6, we will show that a metric's sensitivity to differences in outcomes determines the contraction modulus of the distributional Bellman operator under that metric; informally speaking, this explains the "nice" behavior of the distributional Bellman operator under the Wasserstein distances.

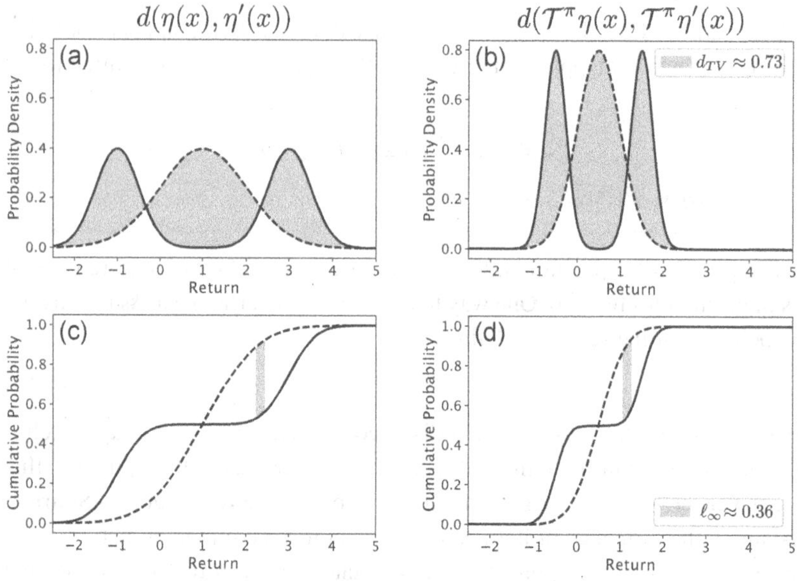

Figure 4.4
The distributional Bellman operator is not a contraction mapping in either the supremum form of **(a, b)** total variation distance (d_{TV}, shaded in the top panels; see Exercise 4.16 for a definition) or **(c, d)** Kolmogorov–Smirnov distance ℓ_∞ (vertical distance in the bottom panels). The left panels show the density **(a)** and cumulative distribution function **(c)** of two distributions $\eta(x)$ (dashed line) and $\eta'(x)$ (solid line). The right panels show the same after applying the distributional Bellman operator **(b, d)**, specifically considering the transformation induced by the discount factor γ. The lack of contractivity can be explained by the fact that neither d_{TV} nor ℓ_∞ is a homogeneous probability metric (Section 4.6).

Before moving on, let us summarize the results established thus far. By combining the theory of contraction mappings with suitable probability metrics, we were able to characterize the behavior of the iterates

$$\eta_{k+1} = \mathcal{T}^\pi \eta_k \,. \tag{4.14}$$

In the following chapters, we will use this as the basis for the design of implementable algorithms that approximate the return distribution η^π and will appeal to contraction mapping theory to provide theoretical guarantees for these algorithms. In particular, in Chapter 6, we will analyze the categorical temporal-difference learning under the lens of the Cramér distance. While the results presented until now suffice for most practical purposes, the following sections deal with some of the more technical considerations that arise from studying Equation 4.14 under general conditions, particularly the issue of infinite distances between distributions.

4.6 Sufficient Conditions for Contractivity

In the remainder of this chapter, we characterize in greater generality the behavior of the sequence of return function estimates described by Equation 4.14, viewed under the lens of different probability metrics. We begin with a formal definition of what it means for a function d to be a probability metric.

Definition 4.21. A probability metric is an extended metric on the space of probability distributions, written

$$d : \mathscr{P}(\mathbb{R}) \times \mathscr{P}(\mathbb{R}) \to [0, \infty] .$$

Its supremum extension is the function $\overline{d} : \mathscr{P}(\mathbb{R})^{\mathcal{X}} \times \mathscr{P}(\mathbb{R})^{\mathcal{X}} \to \mathbb{R}$ defined as

$$\overline{d}(\eta, \eta') = \sup_{x \in \mathcal{X}} d(\eta(x), \eta'(x)) .$$

We refer to $\overline{d}$ as a *return-function metric*; it is an extended metric on $\mathscr{P}(\mathbb{R})^{\mathcal{X}}$. △

Our analysis is based on three properties that a probability metric should possess in order to guarantee contractivity. These three properties relate closely to the three fundamental operations that make up the distributional Bellman operator: scaling, convolution, and mixture of distributions (equivalently: scaling, addition, and indexing of random variables). In this analysis, we will find that some properties are more easily stated in terms of random variables, others in terms of probability distributions. Accordingly, given two probability distributions ν, ν' with instantiations Z, Z', let us overload notation and write

$$d(Z, Z') = d(\nu, \nu').$$

Definition 4.22. Let $c > 0$. The probability metric d is *c-homogeneous* if for any scalar $\gamma \in [0, 1)$ and any two distributions $\nu, \nu' \in \mathscr{P}(\mathbb{R})$ with associated random variables Z, Z', we have

$$d(\gamma Z, \gamma Z') = \gamma^c d(Z, Z') .$$

In terms of probability distributions, this is equivalently given by the condition

$$d((\mathrm{b}_{0,\gamma})_{\#}\nu, (\mathrm{b}_{0,\gamma})_{\#}\nu') = \gamma^c d(\nu, \nu') .$$

If no such c exists, we say that d is not homogeneous. △

Definition 4.23. The probability metric d is *regular* if for any two distributions $\nu, \nu' \in \mathscr{P}(\mathbb{R})$ with associated random variables Z, Z', and an independent random variable W, we have

$$d(W+Z, W+Z') \le d(Z, Z'). \tag{4.15}$$

In terms of distributions, this is

$$d(\mathbb{E}_W[(\mathrm{b}_{W,1})_\# \nu], \mathbb{E}_W[(\mathrm{b}_{W,1})_\# \nu']) \le d(\nu, \nu'). \quad \triangle$$

Definition 4.24. Given $p \in [1, \infty)$, the probability metric d is *p-convex*[30] if for any $\alpha \in (0,1)$ and distributions $\nu_1, \nu_2, \nu_1', \nu_2' \in \mathscr{P}(\mathbb{R})$, we have

$$d^p(\alpha\nu_1 + (1-\alpha)\nu_2, \alpha\nu_1' + (1-\alpha)\nu_2') \le \alpha d^p(\nu_1, \nu_1') + (1-\alpha) d^p(\nu_2, \nu_2'). \quad \triangle$$

Although this p-convexity property is given for a mixture of two distributions, it implies an analogous property for mixtures of finitely many distributions.

Theorem 4.25. Consider a probability metric d. Suppose that d is regular, c-homogeneous for some $c > 0$, and that there exists $p \in [1, \infty)$ such that d is p-convex. Then for all return-distribution functions $\eta, \eta' \in \mathscr{P}(\mathbb{R})^{\mathcal{X}}$, we have

$$\overline{d}(\mathcal{T}^\pi \eta, \mathcal{T}^\pi \eta') \le \gamma^c \overline{d}(\eta, \eta'). \quad \triangle$$

Proof. Fix a state $x \in \mathcal{X}$ and action $a \in \mathcal{A}$. For this state, consider the sample transition $(X = x, A = a, R, X')$ (Equation 2.12), and recall that R and X' are independent given X and A, since

$$R \sim P_{\mathcal{R}}(\cdot \mid X, A) \quad X' \sim P_{\mathcal{X}}(\cdot \mid X, A).$$

Let G and G' be instantiations of η and η', respectively.[31] We introduce a state-action variant of the distributional Bellman operator, $\mathcal{T} : \mathscr{P}(\mathbb{R})^{\mathcal{X}} \to \mathscr{P}(\mathbb{R})^{\mathcal{X} \times \mathcal{A}}$, given by

$$(\mathcal{T}\eta)(x, a) = \mathbb{E}[(\mathrm{b}_{R,\gamma})_\# \eta(X') \mid X = x, A = a].$$

Note that this operator is defined independently of the policy π, since the action a is specified as an argument. We then calculate directly, indicating where each hypothesis of the result is used:

$$\begin{aligned} d^p((\mathcal{T}\eta)(x,a), (\mathcal{T}\eta')(x,a)) &= d^p(R + \gamma G(X'), R + \gamma G'(X')) \\ &\overset{(a)}{\le} d^p(\gamma G(X'), \gamma G'(X')) \\ &\overset{(b)}{=} \gamma^{pc} d^p(G(X'), G'(X')) \end{aligned}$$

30. This matches the usual definition of convexity (for d^p) if one treats the pair (ν_1, ν_2) as a single argument from $\mathscr{P}(\mathbb{R}) \times \mathscr{P}(\mathbb{R})$.

31. Note: the proof does not assume the independence of $G(x)$ and $G(y)$, $x \ne y$. See Exercise 4.12.

$$
\overset{(c)}{\leq} \gamma^{pc} \sum_{x' \in \mathcal{X}} \mathbb{P}_\pi(X' = x' \mid X = x, A = a) d^p(G(x'), G'(x'))
$$

$$
\leq \gamma^{pc} \sup_{x' \in \mathcal{X}} d^p(G(x'), G'(x'))
$$

$$
= \gamma^{pc} \overline{d}^p(\eta, \eta') .
$$

Here, (a) follows from regularity of d, (b) follows from the c-homogeneous property of d, and (c) follows from p-convexity, where the mixture is over the values taken by the random variable X'. We also note that

$$
(\mathcal{T}^\pi \eta)(x) = \sum_{a \in \mathcal{A}} \pi(a \mid x)(\mathcal{T}\eta)(x, a),
$$

and hence by p-convexity of d, we have

$$
d^p((\mathcal{T}^\pi \eta)(x), (\mathcal{T}^\pi \eta')(x)) \leq \sum_{a \in \mathcal{A}} \pi(a \mid x) d^p((\mathcal{T}^\pi \eta)(x, a), (\mathcal{T}^\pi \eta')(x, a))
$$

$$
\leq \gamma^{pc} \overline{d}^p(\eta, \eta') .
$$

Taking the supremum over $x \in \mathcal{X}$ on the left-hand side and taking pth roots then yields the result. □

Theorem 4.25 illustrates how the contractivity of the distributional Bellman operator in the probability metric d (specifically, in its supremum extension) follows from natural properties of d. We see that the contraction modulus is closely tied to the homogeneity of d, which informally characterizes the extent to which scaling random variables by a factor γ brings them "closer together." The theorem provides an alternative to our earlier result regarding Wasserstein distances and enables us to establish the contractivity under the ℓ_p distances but also under other probability metrics. Exercise 4.19 explores contractivity under the so-called maximum mean discrepancy (MMD) family of distances.

Proof of Proposition 4.20 (contractivity in ℓ_p distances). We will apply Theorem 4.25 to the probability metric ℓ_p, for $p \in [1, \infty)$. It is therefore sufficient to demonstrate that ℓ_p is $1/p$-homogeneous, regular, and p-convex.

$1/p$-homogeneity. Let $\nu, \nu' \in \mathscr{P}(\mathbb{R})$ with associated random variables Z, Z'. We make use of the fact that for $\gamma \in [0, 1)$,

$$
F_{\gamma Z}(z) = F_Z\left(\frac{z}{\gamma}\right) .
$$

Writing ℓ_p^p for the pth power of ℓ_p, we have

$$
\ell_p^p(\gamma Z, \gamma Z') = \int_{\mathbb{R}} \left| F_{\gamma Z}(z) - F_{\gamma Z'}(z) \right|^p \mathrm{d}z
$$

$$
\begin{aligned}
&= \int_{\mathbb{R}} \left|F_Z(\tfrac{z}{\gamma}) - F_{Z'}(\tfrac{z}{\gamma})\right|^p \mathrm{d}z \\
&\overset{(a)}{=} \int_{\mathbb{R}} \gamma \left|F_Z(z) - F_{Z'}(z)\right|^p \mathrm{d}z \\
&= \gamma \ell_p^p(Z, Z'),
\end{aligned}
\tag{4.16}
$$

where (a) follows from a change of variables $z/\gamma \mapsto z$ in the integral. Therefore, we deduce that $\ell_p(\gamma Z, \gamma Z') = \gamma^{1/p} \ell_p(Z, Z')$.

Regularity. Let $\nu, \nu' \in \mathscr{P}(\mathbb{R})$, with Z, Z' independent instantiations of ν and ν', respectively, and let W be a random variable independent of Z, Z'. Then,

$$
\begin{aligned}
\ell_p^p(W + Z, W + Z') &= \int_{\mathbb{R}} |F_{W+Z}(z) - F_{W+Z'}(z)|^p \mathrm{d}z \\
&= \int_{\mathbb{R}} |\mathbb{E}_W[F_Z(z - W)] - \mathbb{E}_W[F_{Z'}(z - W)]|^p \mathrm{d}z \\
&\overset{(a)}{\leq} \int_{\mathbb{R}} \mathbb{E}_W[|F_Z(z - W) - F_{Z'}(z - W)|^p] \mathrm{d}z \\
&\overset{(b)}{=} \mathbb{E}_W\Big[\int_{\mathbb{R}} |F_Z(z - W) - F_{Z'}(z - W)|^p] \mathrm{d}z\Big] \\
&= \ell_p^p(Z, Z'),
\end{aligned}
$$

where (a) follows from Jensen's inequality, and (b) follows by swapping the integral and expectation (more formally justified by Tonelli's theorem).

***p*-convexity.** Let $\alpha \in (0, 1)$, and $\nu_1, \nu_1', \nu_2, \nu_2' \in \mathscr{P}(\mathbb{R})$. Note that

$$
F_{\alpha\nu_1 + (1-\alpha)\nu_2}(z) = \alpha F_{\nu_1}(z) + (1 - \alpha) F_{\nu_2}(z),
$$

and similarly for the primed distributions ν_1', ν_2'. By convexity of the function $z \mapsto |z|^p$ on the real numbers and Jensen's inequality,

$$
\begin{aligned}
&|F_{\alpha\nu_1 + (1-\alpha)\nu_2}(z) - F_{\alpha\nu_1' + (1-\alpha)\nu_2'}(z)|^p \\
&\qquad \leq \alpha |F_{\nu_1}(z) - F_{\nu_1'}(z)|^p + (1 - \alpha)|F_{\nu_2}(z) - F_{\nu_2'}(z)|^p,
\end{aligned}
$$

for all $z \in \mathbb{R}$. Hence,

$$
\ell_p^p(\alpha\nu_1 + (1 - \alpha)\nu_2, \alpha\nu_1' + (1 - \alpha)\nu_2') \leq \alpha \ell_p^p(\nu_1, \nu_1') + (1 - \alpha)\ell_p^p(\nu_2, \nu_2'),
$$

and ℓ_p is p-convex. □

4.7 A Matter of Domain

Suppose that we have demonstrated, by means of Theorem 4.25, that the distributional Bellman operator is a contraction mapping in the supremum extension of some probability metric d. Is this sufficient to guarantee that the sequence

$$\eta_{k+1} = \mathcal{T}^\pi \eta_k$$

converges to the return function η^π, by means of Proposition 4.7? In general, no, because d may assign infinite distances to certain pairs of distributions. To invoke Proposition 4.7, we identify a subset of probability distributions $\mathscr{P}_d(\mathbb{R})$ that are all within finite d-distance of each other and then ensure that the distributional Bellman operator is well behaved on this subset. Specifically, we identify a set of conditions under which

(a) the distributional Bellman operator $\mathcal{T}^\pi$ maps $\mathscr{P}_d(\mathbb{R})^\mathcal{X}$ to itself, and
(b) the return function η^π (the fixed point of $\mathcal{T}^\pi$) lies in $\mathscr{P}_d(\mathbb{R})^\mathcal{X}$.

For most common probability metrics and natural problem settings, these requirements are easily verified. In Proposition 4.16, for example, we demonstrated that under the assumption that the reward distributions are bounded, then Proposition 4.7 can be applied with the Wasserstein distances. The aim of this section is to extend the analysis to a broader set of probability metrics but also to a greater number of problem settings, including those where the reward distributions are not bounded.

Definition 4.26. Let d be a probability metric. Its *finite domain* $\mathscr{P}_d(\mathbb{R}) \subseteq \mathscr{P}(\mathbb{R})$ is the set of probability distributions with finite first moment and finite d-distance to the distribution that puts all of its mass on zero:

$$\mathscr{P}_d(\mathbb{R}) = \{\nu \in \mathscr{P}(\mathbb{R}) : d(\nu, \delta_0) < \infty\, ,\ \mathop{\mathbb{E}}_{Z\sim\nu}[|Z|] < \infty\}\,. \tag{4.17}$$

△

By the triangle inequality, for any two distributions $\nu, \nu' \in \mathscr{P}_d(\mathbb{R})$, we are guaranteed $d(\nu, \nu') < \infty$. Although the choice of δ_0 as the reference point is somewhat arbitrary, it is sensible given that many reinforcement learning problems include the possibility of receiving no reward at all (e.g., $G^\pi(x) = 0$). The finite first-moment assumption is made in light of Assumption 2.5, which guarantees that return distributions have well-defined expectations.

Proposition 4.27. Let d be a probability metric satisfying the conditions of Theorem 4.25, with finite domain $\mathscr{P}_d(\mathbb{R})$. Let $\mathcal{T}^\pi$ be the distributional Bellman operator corresponding to a given Markov decision process $(\mathcal{X}, \mathcal{A}, \xi_0, P_\mathcal{X}, P_\mathcal{R})$. Suppose that

(a) $\eta^\pi \in \mathscr{P}_d(\mathbb{R})$, and

(b) $\mathscr{P}_d(\mathbb{R})$ is *closed* under $\mathcal{T}^\pi$: for any $\eta \in \mathscr{P}_d(\mathbb{R})^\mathcal{X}$, we have that $\mathcal{T}^\pi\eta \in \mathscr{P}_d(\mathbb{R})^\mathcal{X}$.

Then for any initial condition $\eta_0 \in \mathscr{P}_d(\mathbb{R})$, the sequence of iterates defined by

$$\eta_{k+1} = \mathcal{T}^\pi \eta_k$$

converges to η^π with respect to $\overline{d}$. △

Proposition 4.27 is a specialization of Proposition 4.7 to the distributional setting and generalizes our earlier result regarding bounded reward distributions. Effectively, it allows us to prove the convergence of the sequence $(\eta_k)_{k\geq 0}$ for a family of Markov decision processes satisfying the two conditions above. The condition $\eta^\pi \in \mathscr{P}_d(\mathbb{R})$, while seemingly benign, does not automatically hold; Exercise 4.20 illustrates the issue using a modified p-Wasserstein distance.

Example 4.28 (ℓ_p **metrics).** For a given $p \in [1, \infty)$, the finite domain of the probability metric ℓ_p is

$$\mathscr{P}_{\ell_p}(\mathbb{R}) = \{\nu \in \mathscr{P}(\mathbb{R}) : \mathop{\mathbb{E}}_{Z\sim\nu}[|Z|] < \infty\},$$

the set of distributions with finite first moment. This follows because

$$\begin{aligned}
\ell_p^p(\nu, \delta_0) &= \int_{-\infty}^0 F_\nu(z)^p \mathrm{d}z + \int_0^\infty (1 - F_\nu(z))^p \mathrm{d}z \\
&\leq \int_{-\infty}^0 F_\nu(z) \mathrm{d}z + \int_0^\infty (1 - F_\nu(z)) \mathrm{d}z \\
&\overset{(a)}{=} \mathbb{E}[\max(0, -Z)] + \mathbb{E}[\max(0, Z)] \\
&= \mathbb{E}[|Z|],
\end{aligned}$$

where (a) follows from expressing $F_\nu(z)$ and $1 - F_\nu(z)$ as integrals and using Tonelli's theorem:

$$\begin{aligned}
\int_0^\infty (1 - F_\nu(z)) \mathrm{d}z &= \int_0^\infty \mathop{\mathbb{E}}_{Z\sim\nu}[\mathbb{1}\{Z > z\}] \mathrm{d}z \\
&= \mathop{\mathbb{E}}_{Z\sim\nu}\Big[\int_0^\infty \mathbb{1}\{Z > z\} \mathrm{d}z\Big]
\end{aligned}$$

$$= \mathop{\mathbb{E}}_{Z \sim \nu} [\max(0, Z)] ,$$

and similarly for the integral from $-\infty$ to 0.

The conditions of Proposition 4.27 are guaranteed by Assumption 2.5 (rewards have finite first moments). From Chapter 2, we know that under this assumption, the random return has finite expected value and hence

$$\eta^\pi(x) \in \mathscr{P}_{\ell_p}(\mathbb{R}), \quad \text{for all } x \in \mathcal{X}.$$

Similarly, we can show from elementary operations that if R and $G(x)$ satisfy

$$\mathbb{E}_\pi [|R| \mid X = x] < \infty , \quad \mathbb{E}[|G(x)|] < \infty, \quad \text{for all } x \in \mathcal{X} ,$$

then also

$$\mathbb{E}_\pi [|R + \gamma G(X')| \mid X = x] < \infty, \quad \text{for all } x \in \mathcal{X} .$$

Following Proposition 4.27, provided that $\eta_0 \in \mathscr{P}_{\ell_p}(\mathbb{R})^{\mathcal{X}}$, then the sequence $(\eta_k)_{k \geq 0}$ converges to η^π with respect to $\overline{\ell}_p$. △

When d is the p-Wasserstein distance ($p \in [1, \infty]$), the finite domain takes on a particularly useful form that we denote by $\mathscr{P}_p(\mathbb{R})$:

$$\mathscr{P}_p(\mathbb{R}) = \{\nu \in \mathscr{P}(\mathbb{R}) : \mathop{\mathbb{E}}_{Z \sim \nu} [|Z|^p] < \infty\} , \quad p \in [0, \infty) ,$$

$$\mathscr{P}_\infty(\mathbb{R}) = \{\nu \in \mathscr{P}(\mathbb{R}) : \exists C > 0 \text{ s.t. } \nu([-C, C]) = 1\} .$$

For $p < \infty$, this is the set of distributions with bounded pth moments; for $p = \infty$, the set of distributions with bounded support. In particular, observe that the finite domains of ℓ_1 and w_1 coincide: $\mathscr{P}_{\ell_p}(\mathbb{R}) = \mathscr{P}_1(\mathbb{R})$.

As with the ℓ_p distances, we can satisfy the conditions of Proposition 4.27 for $d = w_p$ by introducing an assumption on the reward distributions. In this case, we simply require that these be in $\mathscr{P}_p(\mathbb{R})$. As this assumption will recur throughout the book, we state it here in full; Exercise 4.11 goes through the steps of the corresponding result.

Assumption 4.29(p). For each state-action pair $(x, a) \in \mathcal{X} \times \mathcal{A}$, the reward distribution $P_{\mathcal{R}}(\cdot \mid x, a)$ is in $\mathscr{P}_p(\mathbb{R})$. △

Proposition 4.30. Let $p \in [1, \infty]$. Under Assumption 4.29(p), the return function η^π has finite pth moments ($p = \infty$: is bounded). In addition, for any initial return function $\eta_0 \in \mathscr{P}_p(\mathbb{R})$, the sequence defined by

$$\eta_{k+1} = \mathcal{T}^\pi \eta_k$$

converges to η^π with respect to the supremum p-Wasserstein metric. △

4.8 Weak Convergence of Return Functions*

Proposition 4.27 implies that if each distribution $\eta^\pi(x)$ lies in the finite domain $\mathscr{P}_d(\mathbb{R})$ of a given probability metric d that is regular, c-homogeneous, and p-convex, then η^π is the unique solution to the equation

$$\eta = \mathcal{T}^\pi \eta \tag{4.18}$$

in the space $\mathscr{P}_d(\mathbb{R})^{\mathcal{X}}$. It does not, however, rule out the existence of solutions outside this space. This concern can be addressed by showing that for any $\eta_0 \in \mathscr{P}(\mathbb{R})^{\mathcal{X}}$, the sequence of probability distributions $(\eta_k(x))_{k\geq 0}$ defined by

$$\eta_{k+1} = \mathcal{T}^\pi \eta_k$$

converges *weakly* to the return distribution $\eta^\pi(x)$, for each state $x \in \mathcal{X}$. In addition to giving an alternative perspective on the quantitative convergence results of these iterates, the uniqueness of η^π as a solution to Equation 4.18 (stated as Proposition 4.9) follows immediately from Proposition 4.34 below.

Definition 4.31. Let $(\nu_k)_{k\geq 0}$ be a sequence of distributions in $\mathscr{P}(\mathbb{R})$, and let $\nu \in \mathscr{P}(\mathbb{R})$ be another probability distribution. We say that $(\nu_k)_{k\geq 0}$ *converges weakly* to ν if for every $z \in \mathbb{R}$ at which F_ν is continuous, we have $F_{\nu_k}(z) \to F_\nu(z)$. △

We will show that for each $x \in \mathcal{X}$, the sequence $(\eta_k(x))_{k\geq 0}$ converges weakly to $\eta^\pi(x)$. A simple approach is to consider the relationships between well-chosen instantiations of η_k (for each $k \in \mathbb{N}$) and η^π, by means of the following classical result (see e.g., Billingsley 2012). Recall that a sequence of random variables $(Z_k)_{k\geq 0}$ converges to Z with probability 1 if

$$\mathbb{P}(\lim_{k\to\infty} Z_k = Z) = 1 .$$

Lemma 4.32. Let $(\nu_k)_{k\geq 0}$ be a sequence in $\mathscr{P}(\mathbb{R})$ and $\nu \in \mathscr{P}(\mathbb{R})$ be another probability distribution. Let $(Z_k)_{k\geq 0}$ and Z be instantiations of these distributions all defined on the same probability space. If $Z_k \to Z$ with probability 1, then $\nu_k \to \nu$ weakly. △

Lemma 4.32 is not a uniformly applicable approach to demonstrating weak convergence; there always exists such instantiations by Skorokhod's representation theorem (Section 25), but finding such instantiations is not always straightforward. However, in our case, there are a very natural set of instantiations that work, constructed by the following result.

Lemma 4.33. Let $\eta \in \mathscr{P}(\mathbb{R})^{\mathcal{X}}$, and let G be an instantiation of η. For $x \in \mathcal{X}$, if $(X_t, A_t, R_t)_{t\geq 0}$ is a random trajectory with initial state $X_0 = x$ and generated by following π, independent of G, then $\sum_{t=0}^{k-1} \gamma^t R_t + \gamma^k G(X_k)$ is an instantiation of $((\mathcal{T}^\pi)^k \eta)(x)$. △

Proof. This follows by inductively applying Proposition 4.11. □

Proposition 4.34. Let $\eta_0 \in \mathscr{P}(\mathbb{R})^{\mathcal{X}}$, and for $k \geq 0$ define

$$\eta_{k+1} = \mathcal{T}^{\pi}\eta_k .$$

Then we have $\eta_k(x) \to \eta^{\pi}(x)$ weakly for each $x \in \mathcal{X}$, and consequently η^{π} is the unique fixed point of $\mathcal{T}^{\pi}$ in $\mathscr{P}(\mathbb{R})^{\mathcal{X}}$. △

Proof. Fix $x \in \mathcal{X}$, let G_0 be an instantiation of η_0, and on the same probability space, let $(X_t, A_t, R_t)_{t\geq 0}$ be a trajectory generated by π with initial state $X_0 = x$, independent of G_0 (the existence of such a probability space is guaranteed by the Ionescu–Tulcea theorem, as described in Remark 2.1). Then

$$G_k(x) = \sum_{t=0}^{k-1} \gamma^t R_t + \gamma^k G_0(X_k)$$

is an instantiation of $\eta_k(x)$ by Lemma 4.33. Furthermore, $\sum_{t=0}^{\infty} \gamma^t R_t$ is an instantiation of $\eta^{\pi}(x)$. We have

$$\Big|G_k(x) - \sum_{t=0}^{\infty} \gamma^t R_t\Big| = \Big|\gamma^k G_0(X_k) - \sum_{t=k}^{\infty} \gamma^t R_t\Big| \leq \gamma^k \Big|G_0(X_k)\Big| + \Big|\sum_{t=k}^{\infty} \gamma^t R_t\Big| \to 0 ,$$

with probability 1. The convergence of the first term follows since $|G_0(X_k)| \leq \max_{x\in\mathcal{X}} |G_0(x)|$ is bounded with probability 1 (w.p. 1) and $\gamma^k \to 0$, and the convergence of the second term follows from convergence of $\sum_{t=0}^{k-1} \gamma^t R_t$ to $\sum_{t=0}^{\infty} \gamma^t R_t$ w.p. 1. Therefore, we have $G_k(x) \to \sum_{t=0}^{\infty} \gamma^t R_t$ w.p. 1, and so $\eta_k(x) \to \eta^{\pi}(x)$ weakly. Finally, this implies that there can be no other fixed point of $\mathcal{T}^{\pi}$ in $\mathscr{P}(\mathbb{R})^{\mathcal{X}}$; if η_0 were such a fixed point, then we would have $\eta_k = \eta_0$ for all $k \geq 0$ and would simultaneously have $\eta_k(x) \to \eta^{\pi}(x)$ weakly for all $x \in \mathcal{X}$, a contradiction unless $\eta_0 = \eta^{\pi}$. □

As the name indicates, the notion of weak convergence is not as strong as many other notions of convergence of probability distributions. In general, it does not even guarantee convergence of the mean of the sequence of distributions; see Exercise 4.21. In addition, we lose any notion of convergence rate provided by the contractive nature of the distributional operator under specific metrics. The need for stronger guarantees, such as those offered by Proposition 4.27, motivates the contraction mapping theory developed in this chapter.

4.9 Random-Variable Bellman Operators*

In this chapter, we defined the distributional Bellman operator $\mathcal{T}^\pi$ as a mapping on the space of return-distribution functions $\mathscr{P}(\mathbb{R})^\mathcal{X}$. We also saw that the action of the operator on a return function $\eta \in \mathscr{P}(\mathbb{R})^\mathcal{X}$ can be understood both through direct manipulation of the probability distributions or through manipulation of a collection of random variables instantiating these distributions.

Viewing the operator through its effect on the distribution of a collection of representative random variables is a useful tool for understanding distributional reinforcement learning and may prompt the reader to ask whether it is possible to avoid referring to probability distributions at all, working instead directly with random variables. We describe one approach to this below using the tools of probability theory and then discuss some of its shortcomings.

Let $G_0 = (G_0(x) : x \in \mathcal{X})$ be an initial collection of real-valued random variables, indexed by state, supported on a probability space $(\Omega_0, \mathscr{F}_0, \mathbb{P}_0)$. For each $k \in \mathbb{N}^+$, let $(\Omega_k, \mathscr{F}_k, \mathbb{P}_k)$ be another probability space, supporting a collection of random variables $((A_k(x), R_k(x, a), X'_k(x, a)) : x \in \mathcal{X}, a \in \mathcal{A})$, with $A_k(x) \sim \pi(\cdot \mid x)$, and independently $R_k(x, a) \sim P_\mathcal{R}(\cdot \mid x, a)$, $X_k(x, a) \sim P_\mathcal{X}(\cdot \mid x, a)$. We then consider the product probability space on $\Omega = \prod_{k \in \mathbb{N}} \Omega_k$. All random variables defined above can naturally be viewed as functions on this joint probability space which depend on $\omega = (\omega_0, \omega_1, \omega_2, \dots) \in \Omega$ only through the coordinate ω_k that matches the index k on the random variable. Note that under this construction, all random variables with distinct indices are independent.

Now define $\mathscr{X}_\mathbb{N}$ as the set of real-valued random variables on $(\Omega, \mathscr{F}, \mathbb{P})$ (where $\mathscr{F}$ is the product σ-algebra) that depend on only finitely many coordinates of $\omega \in \Omega$. We can define a Bellman operator $\mathcal{T}^\pi : \mathscr{X}_\mathbb{N} \to \mathscr{X}_\mathbb{N}$ as follows. Given $G = (G(x) : x \in \mathcal{X}) \in \mathscr{X}_\mathbb{N}^\mathcal{X}$, let $K \in \mathbb{N}$ be the smallest integer such that the random variables $(G(x) : x \in \mathcal{X})$ depend on $\omega = (\omega_0, \omega_1, \omega_2, \dots) \in \Omega$ only through $\omega_0, \dots, \omega_{K-1}$; such an integer exists due to the definition of $\mathscr{X}_\mathbb{N}$ and the finiteness of $\mathcal{X}$. We then define $\mathcal{T}^\pi G \in \mathscr{X}_\mathbb{N}$ by

$$(\mathcal{T}^\pi G)(x) = R_K(x, A_K(x)) + \gamma G(X'_K(x, A_K(x))) .$$

With this definition, we can obtain a sequence of collections of random variables $(G_k)_{k \geq 0}$, defined iteratively by $G_{k+1} = \mathcal{T}^\pi G_k$, for $k \geq 0$.[32] We have therefore formalized an operator entirely within the realm of random variables, without reference to the distributions of the iterates $(G_k)_{k \geq 0}$. By construction, the distribution of the random variables $(G_k)_{k \geq 0}$ matches the sequence of distributions that would be obtained by working directly on the space of probability distributions with the usual distributional Bellman operator. More concretely, if

32. This is real equality between random variables, rather than in distribution.

$\eta_0 \in \mathscr{P}(\mathbb{R})^{\mathcal{X}}$ is such that $\eta_0(x)$ is the distribution of $G_0(x)$ for each $x \in \mathcal{X}$, then we have that η_k, defined by $\eta_k = (\mathcal{T}^\pi)^k \eta_0$, is such that $\eta_k(x)$ is the distribution of $G_k(x)$, for each $x \in \mathcal{X}$. Thus, the random-variable Bellman operator constructed above is consistent with the distributional Bellman operator that is the main focus of this chapter.

One difficulty with this random variable operator is that it does not have a fixed point; while the *distribution* of the random variables $G_k(\cdot)$ converges to that of $G^\pi(\cdot)$, the random variables themselves, as functions on the probability space, do not converge.[33] Thus, while it is one way to view distributional reinforcement learning purely in terms of random variables, it is much less natural to analyze algorithms from this perspective, rather than through probability distributions as described in this chapter.

4.10 Technical Remarks

Remark 4.1. Our exposition in this chapter has focused on the standard distributional Bellman equation

$$G^\pi(x) \stackrel{\mathcal{D}}{=} R + \gamma G^\pi(X'),\ X = x\,.$$

A similar development is possible with the alternative notions of random return mentioned in Section 2.9, including the random-horizon return. In general, the distributional operators that arise from these alternative notions of the random return are still contraction mappings, although the metrics and contraction moduli involved in these statements differ. For example, while the standard distributional Bellman operator is not a contraction in total variation distance, the distributional Bellman operator associated with the random-horizon return is; Exercise 4.17 explores this point in greater detail. △

Remark 4.2. The ideas developed in this chapter, and indeed the other chapters of the book, can also be applied to learning other properties related to the return. Achab (2020) explores several methods for learning the distribution of the random variables $R + \gamma V^\pi(X')$, under the different possible initial conditions $X = x$; these objects interpolate between the expected return and the full distribution of the return. Exercise 4.22 explores the development of contractive operators for the distributions of these objects. △

Remark 4.3. As well as allowing us to quantify the contractivity of the distributional Bellman operator, the probability metrics described in this chapter can be used to measure the accuracy of algorithms that aim to approximate the return

33. This is a fairly subtle point – the reader is invited to consider what happens to $G_k(x)$ and $G^\pi(x)$ as mappings from Ω to $\mathbb{R}$.

distribution. In particular, they can be used to give quantitative guarantees on the accuracy of the nonparametric distributional Monte Carlo algorithm described in Remark 3.1. These guarantees can then be used to determine how many sample trajectories are required to approximate the true return distribution at a required level of accuracy: for example, when evaluating the performance of the algorithms in Chapters 5 and 6. We assume that K returns $(G_k)_{k=1}^K$ beginning at state x_0 have been generated independently via the policy π and consider the accuracy of the nonparametric distributional Monte Carlo estimator

$$\hat{\eta}^\pi(x_0) = \frac{1}{K}\sum_{k=1}^{K} \delta_{G_k} .$$

Different realizations of the sampled returns will lead to different estimates; the following result provides a guarantee in the Kolmogorov–Smirnov distance ℓ_∞ that holds with high probability over the possible realizations of the returns. For any $\varepsilon > 0$, we have

$$\ell_\infty(\hat{\eta}^\pi(x_0), \eta^\pi(x_0)) \leq \varepsilon\,, \text{ with probability at least } 1 - 2\exp(-2K\varepsilon^2)\,.$$

Thus, for a desired level of accuracy ε and a desired confidence $1-\delta$, K can be selected so that $2\exp(-2K\varepsilon^2) < \delta$, which then yields the guarantee that with probability at least $1-\delta$, we have $\ell_\infty(\hat{\eta}^\pi(x_0), \eta^\pi(x_0)) \leq \varepsilon$. This is in fact a key result in empirical process theory known as the Dvoretzky–Kiefer–Wolfowitz inequality (Dvoretzky et al. 1956; Massart 1990). Its uses specifically in reinforcement learning include the works of Keramati et al. (2020) and Chandak et al. (2021). Similar concentration bounds are possible under other probability metrics (such as the Wasserstein distances; see, e.g., Weed and Bach 2019; Bobkov and Ledoux 2019), though typically some form of a priori information about the return distribution, such as bounds on minimum/maximum returns, is required to establish such bounds. △

4.11 Bibliographical Remarks

4.1–4.2. The use of operator theory to understand reinforcement learning algorithms is standard to most textbooks in the field (Bertsekas and Tsitsiklis 1996; Szepesvári 2010; Puterman 2014; Sutton and Barto 2018), and is commonly used in theoretical reinforcement learning (Munos 2003; Bertsekas 2011; Scherrer 2014). Puterman (2014) provides a thorough introduction to vector notation for Markov decision processes. Improved convergence results can be obtained by studying the eigenspectrum of the transition kernel, as shown by, for example, Morton (1971) and Bertsekas (1994, 2012). The elementary contraction mapping theory described in Section 4.2 goes back to Banach (1922). Our reference on metric spaces is the definitive textbook by Rudin (1976).

Not all reinforcement learning algorithms are readily analyzed using the theory of contraction mappings. This is the case for policy-gradient algorithms (Sutton et al. 2000; but see Ghosh et al. 2020; Bhandari and Russo 2021), but also value-based algorithms such as advantage learning (Baird 1999; Bellemare et al. 2016).

4.3. The random-variable Bellman operator presented here is from our earlier work (Bellemare et al. 2017a), which provided an analysis in the p-Wasserstein distances. There, the technical issues discussed in Section 4.9 were obviated by declaring R, X' and the collection $(G(x) : x \in \mathcal{X})$ to be independent. Those issues were raised in a later paper (Rowland et al. 2018), which also introduced the distributional Bellman equation in terms of probability measures and the pushforward notation. The probability density equation (Equation 4.9) can be found in Morimura et al. (2010b). Earlier instances of distributional operators are given by Chung and Sobel (1987) and Morimura et al. (2010a), who provide an operator on cumulative distribution functions and Jaquette (1976), who provides an operator on Laplace transforms.

4.4. The Wasserstein distance can be traced back to Leonid Kantorovich (1942) and has been rediscovered (and renamed) multiple times in its history. The name we use here is common but a misnomer as Leonid Vaserstein (after whom the distance is named) did not himself do any substantial work on the topic. Among other names, we note the Mallows metric (Bickel and Freedman 1981) and the Earth–Mover distance (Rubner et al. 1998). Much earlier, Monge (1781) was the first to study the problem of optimal transportation from a transport theory perspective. See Vershik (2013) and Panaretos and Zemel (2020) for further historical comments and Villani (2008) for a survey of theoretical properties. A version of the contraction analysis in p-Wasserstein distances was given by Bellemare et al. (2017a); we owe the proof of Proposition 4.15 in terms of optimal couplings to Philip Amortila.

The use of contraction mapping theory to analyze stochastic fixed-point equations was introduced by Rösler (1991), who analyzes the Quicksort algorithm by characterizing the distributional fixed points of contraction mappings in 2-Wasserstein distance. Applications and generalization of this technique include the analysis of further recursive algorithms, models in stochastic geometry, and branching processes (Rösler 1992; Rachev and Rüschendorf 1995; Neininger 1999; Rösler and Rüschendorf 2001; Rösler 2001; Neininger 2001; Neininger and Rüschendorf 2004; Rüschendorf 2006; Rüschendorf and Neininger 2006; Alsmeyer 2012). Although the random-variable Bellman equation (Equation 2.15) can be viewed as a system of recursive distributional equations, the emphasis on a collection of effectively independent random variables (G^π) differs from the usual treatment of such equations.

4.5. The family of ℓ_p distances described in this chapter is covered at length in the work of Rachev et al. (2013), which studies an impressive variety of probability metrics. A version of the contraction analysis in Cramér distance was originally given by Rowland et al. (2018). In two and more dimensions, the Cramér distance is generalized by the *energy distance* (Székely 2002; Székely and Rizzo 2013; Rizzo and Székely 2016), itself a member of the maximum mean discrepancy (MMD) family (Gretton et al. 2012); contraction analysis in terms of MMD metrics was undertaken by Nguyen et al. (2021) (see Exercise 4.19 for further details). Another special case of the ℓ_p metrics considered in this chapter is the Kolmogorov–Smirnov distance (ℓ_∞), which features in results in empirical process theory, such as the Glivenko–Cantelli theorem. Many of these metrics are *integral probability metrics* (Müller 1997), which allows for a dual formulation with appealing algorithmic consequences. Chung and Sobel (1987) provide a nonexpansion result in total variation distance (without naming it as such; the proof uses an integral probability metric formulation).

4.6. The properties of regularity, convexity, and c-homogeneity were introduced by Zolotarev (1976) in a slightly more general setting. Our earlier work presented these in a modern context (Bellemare et al. 2017b), albeit with only a mention of their potential use in reinforcement learning. Although that work proposed the term "sum-invariant" as mnemonically simpler, this is only technically correct when Equation 4.15 holds with equality; we have thus chosen to keep the original name. Theorem 4.25 is new to this book.

The characterization of the Wasserstein distance as an optimal transport problem in Proposition 4.18 is the standard presentation of the Wasserstein distance in more abstract settings, which allows it to be applied to probability distributions over reasonably general metric spaces (Villani 2003, 2008; Ambrosio et al. 2005; Santambrogio 2015). Optimal transport has also increasingly found application within machine learning in recent years, particularly in generative modeling (Arjovsky et al. 2017). Optimal transport and couplings also arise in the study of bisimulation metrics for Markov decision processes (Ferns et al. 2004; Ferns and Precup 2014; Amortila et al. 2019) as well as analytical tools for sample-based algorithms (Amortila et al. 2020). Peyré and Cuturi (2019) provide an overview of algorithms, analysis, and applications associated with optimal transport in machine learning and related disciplines.

4.7–4.8. Villani (2008) gives further discussion on the domain of Wasserstein distances and on their relationship to weak convergence.

4.9. The usefulness of a random-variable operator has been a source of intense debate between the authors of this book. The form we present here is inspired by the "stack of rewards" model from Lattimore and Szepesvári (2020).

4.12 Exercises

Exercise 4.1. Show that the no-loop operator defined in Example 4.6 is a contraction mapping with modulus

$$\beta = \gamma \max_{x \in \mathcal{X}} \mathbb{P}_\pi(X' \neq x \mid X = x). \qquad \triangle$$

Exercise 4.2. For $p \in [1, \infty)$, let $\|\cdot\|_p$ be the L^p norm over $\mathbb{R}^{\mathcal{X}}$, defined as

$$\|V\|_p = \Big(\sum_{x \in \mathcal{X}} |V(x)|^p\Big)^{1/p}.$$

Show that T^π is not a contraction mapping in the metric induced by the L^p norm unless $p = \infty$ (see Equation 4.4 for a definition of the L^∞ metric). *Hint.* A two-state example suffices. △

Exercise 4.3. In this exercise, you will use the ideas of Section 4.1 to study several operators associated with expected-value reinforcement learning.

(i) For $n \in \mathbb{N}^+$, consider the *n-step evaluation operator* $T^\pi_n : \mathbb{R}^{\mathcal{X}} \to \mathbb{R}^{\mathcal{X}}$ defined by

$$(T^\pi_n V)(x) = \mathbb{E}_\pi\Big[\sum_{t=0}^{n-1} \gamma^t R_t + \gamma^n V(X_n) \,\Big|\, X_0 = x\Big].$$

Show that T^π_n has V^π as a fixed point and is a contraction mapping with respect to the L^∞ metric with contraction modulus γ^n. Hence, deduce that repeated application of T^π_n to any initial value function estimate converges to V^π. Show that in fact, $T^\pi_n = (T^\pi)^n$.

(ii) Consider the λ-return operator $T^\pi_\lambda : \mathbb{R}^{\mathcal{X}} \to \mathbb{R}^{\mathcal{X}}$ defined by

$$(T^\pi_\lambda V)(x) = (1-\lambda) \sum_{n=1}^{\infty} \lambda^{n-1} \mathbb{E}_\pi\Big[\sum_{t=0}^{n-1} \gamma^t R_t + \gamma^n V(X_n) \,\Big|\, X_0 = x\Big].$$

Show that T^π_λ has V^π as a fixed point and is a contraction mapping with respect to the L^∞ metric with contraction modulus

$$\gamma\left(\frac{1-\lambda}{1-\gamma\lambda}\right).$$

Hence, deduce that repeated application of T^π_λ to any initial value function estimate converges to V^π. △

Exercise 4.4. Consider the random-variable operator (Equation 4.7). Given a return-variable function G_0, write $G_0(x, \omega) = G_0(x)(\omega)$ for the realization of the random return corresponding to $\omega \in \Omega$. Additionally, for each $x \in \mathcal{X}$, let $(x, A(x), R(x), X'(x))$ be an independent random transition defined on the same probability space, and write $(A(x, \omega), R(x, \omega), X'(x, \omega))$ to denote the dependence of these random variables on $\omega \in \Omega$. Suppose that for each $k \geq 0$, we define the return-variable function G_{k+1} as

$$G_{k+1}(x, \omega) = R(x, \omega) + \gamma G_k(X'(x, \omega), \omega).$$

For a given x, characterize the function

$$G^*(x, \omega) = \lim_{k \to \infty} G_k(x, \omega).$$ △

Exercise 4.5. Suppose that you are given a description of a Markov decision process along with a policy π, with the property that the policy π reaches a terminal state (i.e., one for which the return is zero) in at most $T \in \mathbb{N}$ steps. Describe a recursive procedure that takes in a scalar ω on $[0, 1]$ and outputs a return $z \in \mathbb{R}$ such that $\mathbb{P}(G^\pi(X_0) \leq z) = \omega$. *Hint.* You may want to use the fact that sampling a random variable Z can be emulated by drawing τ uniformly from $[0, 1]$, and returning $F_Z^{-1}(\tau)$. △

Exercise 4.6. Let $p \in [1, \infty]$.

(i) Show that for any $\nu, \nu' \in \mathscr{P}(\mathbb{R})$ with finite pth moments, we have $w_p(\nu, \nu) < \infty$. *Hint.* Use the triangle inequality with intermediate distribution δ_0. Hence, prove that the p-Wasserstein metric is indeed a metric on $\mathscr{P}_p(\mathbb{R})$, for $p \in [1, \infty]$.

(ii) Show that on the space $\mathscr{P}(\mathbb{R})$, the p-Wasserstein metric satisfies all requirements of a metric except finiteness. For each $p \in [1, \infty]$, exhibit a pair of distributions $\nu, \nu' \in \mathscr{P}(\mathbb{R})$ such that $w_p(\nu, \nu') = \infty$.

(iii) Show that if $1 \leq q < p < \infty$, we have $\mathbb{E}[|Z|^p] < \infty \implies \mathbb{E}[|Z|^q] < \infty$ for any random variable Z. Deduce that $\mathscr{P}_p(\mathbb{R}) \subseteq \mathscr{P}_q(\mathbb{R})$. *Hint.* Consider applying Jensen's inequality with the function $z \mapsto |z|^{p/q}$. △

Exercise 4.7. Let $d : \mathscr{P}(\mathbb{R}) \times \mathscr{P}(\mathbb{R}) \to [0, \infty]$ be a probability metric with finite domain $\mathscr{P}_d(\mathbb{R})$.

(i) Prove that the supremum distance $\overline{d}$ is an extended metric on $\mathscr{P}(\mathbb{R})^{\mathcal{X}}$.

(ii) Prove that it is a metric on $\mathscr{P}_d(\mathbb{R})^{\mathcal{X}}$. △

Exercise 4.8. Prove Proposition 4.15 for $p = \infty$. △

Exercise 4.9. Consider two normally distributed random variables with mean μ and μ' and common variance σ^2. Derive an expression for the ∞-Wasserstein distance between these two distributions. Conclude that Assumption 4.29(∞) is sufficient, but not necessary, for two distributions to have finite ∞-Wasserstein distance. How does the situation change if the two normal distributions in question have unequal variances? △

Exercise 4.10. Explain, in words, why Assumption 4.29(p) is needed for Proposition 4.30. By considering the definition of the p-Wasserstein distance, explain why for $p > 1$ and $1 \leq q < p$, Assumption 4.29(q) is not sufficient to guarantee convergence in the p-Wasserstein distance. △

Exercise 4.11. This exercise guides you through the proof of Proposition 4.30.

(i) First, show that under Assumption 4.29(p), the return distributions $\eta^\pi(x)$ have finite pth moments, for all $x \in \mathcal{X}$. You may find it useful to deal with $p = \infty$ separately, and in the case $p \in [1, \infty)$, you may find it useful to rewrite

$$\mathbb{E}_\pi\Big[\big|\sum_{t=0}^{\infty} \gamma^t R_t\big|^p \mid X = x\Big] = (1-\gamma)^{-p}\mathbb{E}_\pi\Big[\big|\sum_{t=0}^{\infty}(1-\gamma)\gamma^t R_t\big|^p \mid X = x\Big]$$

and use Jensen's inequality on the function $z \mapsto |z|^p$.

(ii) Let $\eta \in \mathscr{P}_p(\mathbb{R})^{\mathcal{X}}$, and let G be an instantiation of η. First letting $p \in [1, \infty)$, use the inequality $|z_1 + z_2|^p \leq 2^{p-1}(|z_1|^p + |z_2|^p)$ to argue that if $(X = x, A, R, X')$ is a sample transition independent of G, then

$$\mathbb{E}_\pi[|R + \gamma G(X')|^p \mid X = x] < \infty\,.$$

Hence, argue that under Assumption 4.29(p), $\mathscr{P}_p(\mathbb{R})^{\mathcal{X}}$ is closed under $\mathcal{T}^\pi$. Argue separately that this holds for $p = \infty$ too.

Hence, argue that Proposition 4.27 applies, and hence conclude that Proposition 4.30 holds. △

Exercise 4.12. In the proof of Theorem 4.25, we did not need to assume that for the two distinct states $x, y \in \mathcal{X}$, their associated returns $G(x)$ and $G(y)$ are independent. Explain why. △

Exercise 4.13. The (1,1)-Pareto distribution ν_{PAR} has cumulative distribution

$$F_{\nu_{\text{PAR}}}(z) = \begin{matrix} 0 & \text{if } z < 1, \\ 1 - \frac{1}{z} & \text{if } z \geq 1\,. \end{matrix}$$

Justify the necessity of including Assumption 2.5 (finite-mean rewards) in Definition 4.26 by demonstrating that

$$\ell_2(\nu_{\text{PAR}}, \delta_0) < \infty \quad \text{yet} \quad \mathop{\mathbb{E}}_{Z\sim\nu}[Z] = \infty\,.$$ △

Exercise 4.14. The purpose of this exercise is to contrast the p-Wasserstein and ℓ_p distances. For each of the following, find a pair of probability distributions $\nu, \nu' \in \mathscr{P}(\mathbb{R})$ such that, for a given $\varepsilon > 0$,

(i) $w_2(\nu, \nu')$ smaller than $\ell_2(\nu, \nu')$;
(ii) $w_2(\nu, \nu')$ larger than $\ell_2(\nu, \nu')$;
(iii) $w_\infty(\nu, \nu') = \varepsilon$ and $\ell_\infty(\nu, \nu') = 1$;
(iv) $w_\infty(\nu, \nu') = 1$ and $\ell_\infty(\nu, \nu') = \varepsilon$. △

Exercise 4.15. Show that the dependence on $\gamma^{1/p}$ is tight in Proposition 4.20. △

Exercise 4.16. The *total variation distance* $d_{\text{TV}} : \mathscr{P}(\mathbb{R}) \times \mathscr{P}(\mathbb{R}) \to \mathbb{R}$ is defined by

$$d_{\text{TV}}(\nu, \nu') = \sup_{U \subseteq \mathbb{R}} |\nu(U) - \nu'(U)| \,, \tag{4.19}$$

for all $\nu, \nu' \in \mathscr{P}(\mathbb{R})$.[34] Show, by means of a counterexample, that the distributional Bellman operator is not a contraction mapping in the supremum extension of this distance. △

Exercise 4.17. Consider the alternative notion of the return introduced in Section 2.9, the random-horizon return, for which γ is treated as the probability of continuing. Write down the distributional Bellman operator that corresponds to this random-horizon return. For which metrics considered in this chapter is this distributional Bellman operator a contraction mapping? Show in particular that this distributional Bellman operator is a contraction with respect to the supremum version of the total variation distance over return-distribution functions, introduced in Exercise 4.16. What is its contraction modulus? △

Exercise 4.18. Remark 2.3 describes some differences between Markov decision processes with finite state spaces (as we consider throughout the book) and generalizations with infinite state spaces. The contraction mapping theory in this chapter is one case where stronger assumptions are required when moving to larger state spaces. Using the example described in Remark 2.3, show that Assumption 4.29(w_1) is insufficient to make $\mathscr{P}_1(\mathbb{R})$ closed under the distributional Bellman operator $\mathcal{T}^\pi$ when the state space is countably infinite. How could this assumption be strengthened to guarantee closedness? △

Exercise 4.19 (*). The goal of this exercise is to explore the contractivity of the distributional Bellman operator with respect to a class of metrics known as *maximum mean discrepancies*; this analysis was undertaken by Nguyen et al. (2021). A *kernel* on $\mathbb{R}$ is a function $K : \mathbb{R} \times \mathbb{R} \to \mathbb{R}$ with the property that for any finite set $\{z_1, \ldots, z_n\} \subseteq \mathbb{R}$, the $m \times m$ matrix with (i, j)th entry $K(z_i, z_j)$ is positive semi-definite. A function $K : \mathbb{R} \times \mathbb{R} \to \mathbb{R}$ satisfying the weaker condition that

$$\sum_{i,j=1}^{n} c_i c_j K(z_i, z_j) \geq 0$$

34. For the reader with a measure-theoretic background, the supremum here is over measurable subsets of $\mathbb{R}$.

whenever $\sum_{i=1}^{n} c_i = 0$ is called a *conditionally positive-definite kernel*. Conditionally positive-definite kernels form a measure of similarity between pairs of points in $\mathbb{R}$ and can also be used to define notions of distance over probability distributions. The maximum mean discrepancy (MMD) associated with the conditionally positive-definite kernel K is defined by

$$\mathrm{MMD}_K(\nu, \nu') = \Big[\mathbb{E}_{\substack{X\sim\nu \\ X'\sim\nu}}[K(X, X')] + \mathbb{E}_{\substack{Y\sim\nu' \\ Y'\sim\nu'}}[K(Y, Y')] - 2\mathbb{E}_{\substack{X\sim\nu \\ Y\sim\nu'}}[K(X, Y)]\Big]^{1/2},$$

where each pair of random variables in the expectations above is taken to be independent.

(i) Consider the function $K_\alpha(z_1, z_2) = -|z_1 - z_2|^\alpha$, with $\alpha \in (0, 2)$. Székely and Rizzo (2013, Proposition 2) show that this defines a conditionally positive-definite kernel. Show that MMD_{K_α} is regular, c-homogeneous (for some $c > 0$), and p-convex (for some $p \in [1, \infty)$). Hence, use Theorem 4.25 to establish that the distributional Bellman operator is a contraction with respect to MMD_{K_α}, under suitable assumptions.

(ii) The Gaussian kernel, or squared exponential kernel, with variance $\sigma^2 > 0$ and length scale $\lambda > 0$ is defined by $K(z_1, z_2) = \sigma^2 \exp(-(z_1 - z_2)^2/(2\lambda^2))$. Show, through the use of a counterexample, that the MMD corresponding to the Gaussian kernel is not c-homogeneous for any $c > 0$, and so Theorem 4.25 cannot be applied. Further, find an MDP and policy π that serve as a counterexample to the contractivity of the distributional Bellman operator with respect to this MMD metric. △

Exercise 4.20. Let $p \in [1, \infty]$, and consider a modification of the p-Wasserstein distance, $\tilde{w}_p$, such that $\tilde{w}_p(\nu, \nu') = w_p(\nu, \nu')$ if both $\nu, \nu' \in \mathscr{P}(\mathbb{R})$ are expressible as finite mixtures of Dirac deltas, and $\tilde{w}_p(\nu, \nu') = \infty$ otherwise.

(i) Show that $\mathscr{P}_{\tilde{w}_p}(\mathbb{R})$ is the set of distributions expressible as finite mixtures of Dirac deltas.

(ii) Exhibit a Markov decision process and policy π for which all conditions of Proposition 4.27 hold except for the condition $\eta^\pi \in \mathscr{P}_{\tilde{w}_p}(\mathbb{R})^{\mathcal{X}}$.

(iii) Show that the sequence of iterates $(\eta_k)_{k\geq 0}$ does not converge to η^π under $\tilde{w}_p$ in this case. △

Exercise 4.21. Consider the sequence of distributions $(\nu_k)_{k=1}^{\infty}$ defined by

$$\nu_k = \frac{k-1}{k}\delta_0 + \frac{1}{k}\delta_k .$$

Show that this sequence converges weakly to another distribution ν. From this, deduce that weak convergence does not imply convergence of expectations. △

Exercise 4.22. Achab (2020) considers the random variables

$$\tilde{G}^\pi(x) = R + \gamma V^\pi(X'), \quad X = x.$$

What does the distribution of this random variable capture about the underlying MDP and policy π? Using the tools of this chapter, derive an operator over $\mathscr{P}(\mathbb{R})^{\mathcal{X}}$ that has the collection of distributions corresponding to this random variable under each of the initial conditions in $\mathcal{X}$ as a fixed point, and analyze the properties of this operator. △

5 Distributional Dynamic Programming

Markov decision processes model the dynamics of an agent exerting control over its environment. Once the agent's policy is selected, a Markov decision process gives rise to a sequential system whose behavior we would like to characterize. In particular, *policy evaluation* describes the process of determining the returns obtained from following a policy π. Algorithmically, this translates into the problem of *computing* the value or return-distribution function given the parameters of the Markov decision process and the agent's policy.

Computing the return-distribution function requires being able to describe the output of the algorithm in terms of atomic objects (depending on the programming language, these may be bits, floating point numbers, vectors, or even functions). This is challenging because in general, return distributions take on a continuum of values (i.e., they are infinite-dimensional objects). By contrast, the expected return from a state x is described by a single real number. Defining an algorithm that computes return-distribution functions first requires us to decide how we represent probability distributions in memory, knowing that some approximation error must be incurred if we want to keep things finite.

This chapter takes a look at different representations of probability distributions as they relate to the problem of computing return-distribution functions. We will see that, unlike the relatively straightforward problem of computing value functions, there is no obviously best representation for return-distribution functions and that different finite-memory representations offer different advantages. We will also see that making effective use of different representations requires different algorithms.

5.1 Computational Model

As before, we assume that the environment is described as a finite-state, finite-action Markov decision process. We write $N_{\mathcal{X}}$ and $N_{\mathcal{A}}$ for the size of the state and action spaces $\mathcal{X}$ and $\mathcal{A}$. When describing algorithms in this chapter, we will further assume that the reward distributions $P_{\mathcal{R}}(\cdot \mid x, a)$ are supported on a finite set $\mathcal{R}$ of size $N_{\mathcal{R}}$; we discuss a way of lifting this assumption in Remark 5.1. Of note, having finitely many rewards guarantees the existence of an interval $[V_{\text{MIN}}, V_{\text{MAX}}]$ within which the returns lie.[35] We measure the complexity of a particular algorithm in terms of the number of atomic instructions or memory words it requires, assuming that these can reasonably be implemented in a physical computer, as described by the random-access machine (RAM) model of computation (Cormen et al. 2001).

In classical reinforcement learning, linear algebra provides a simple algorithm for computing the value function of a policy π. In vector notation, the Bellman equation is

$$V^\pi = r^\pi + \gamma P^\pi V^\pi, \tag{5.1}$$

where the transition function P^π is represented as an $N_{\mathcal{X}}$-dimensional square stochastic matrix, and r^π is an $N_{\mathcal{X}}$-dimensional vector. With some matrix algebra, we deduce that

$$\begin{aligned} V^\pi &= r^\pi + \gamma P^\pi V^\pi \\ \iff (I - \gamma P^\pi) V^\pi &= r^\pi \\ \iff V^\pi &= (I - \gamma P^\pi)^{-1} r^\pi . \end{aligned} \tag{5.2}$$

The computational cost of determining V^π is dominated by the matrix inversion, requiring $O(N_{\mathcal{X}}^3)$ operations. The result is exact. The matrix P^π and the vector r^π are constructed entry-wise by writing expectations as sums:

$$P^\pi(x' \mid x) = \sum_{a \in \mathcal{A}} \pi(a \mid x) P_{\mathcal{X}}(x' \mid x, a)$$

$$r^\pi(x) = \sum_{a \in \mathcal{A}} \sum_{r \in \mathcal{R}} \pi(a \mid x) P_{\mathcal{R}}(r \mid x, a) \times r .$$

When the matrix inversion is undesirable, the value function can instead be found by *dynamic programming*. Dynamic programming describes a wide variety of computational methods that find the solution to a given problem by caching intermediate results. In reinforcement learning, the dynamic programming approach for finding the value function V^π, also called *iterative policy evaluation*, begins with an initial estimate $V_0 \in \mathbb{R}^{\mathcal{X}}$ and successively computes

$$V_{k+1} = T^\pi V_k$$

for $k = 1, 2, \ldots$, until some desired number of iterations K have been performed or some stopping criterion is reached. This is possible because, when $\mathcal{X}$, $\mathcal{A}$, and

35. We can always take R_{MIN} and R_{MAX} to be the smallest and largest possible rewards, respectively.

$\mathcal{R}$ are all finite, the Bellman operator can be written in terms of sums:

$$(T^\pi V_k)(x) = \sum_{a \in \mathcal{A}} \sum_{r \in \mathcal{R}} \sum_{x' \in \mathcal{X}} \underbrace{\mathbb{P}_\pi(A = a, R = r, X' = x' \mid X = x)}_{\pi(a \mid x) P_{\mathcal{X}}(x' \mid x, a) P_{\mathcal{R}}(r \mid x, a)} [r + \gamma V_k(x')] \,. \quad (5.3)$$

A naive implementation expresses these sums as nested FOR loops. Since the new value function must be computed at all states, this naive implementation requires on the order of $N = N_{\mathcal{X}}^2 N_{\mathcal{A}} N_{\mathcal{R}}$ operations. We can do better by implementing it in terms of vector operations:

$$V_{k+1} = r^\pi + \gamma P^\pi V_k \,, \quad (5.4)$$

where V_k is stored in memory as an $N_{\mathcal{X}}$-dimensional vector. A single application of the Bellman operator with vectors and matrices requires $O(N_{\mathcal{X}}^2 N_{\mathcal{A}} + N_{\mathcal{X}} N_{\mathcal{A}} N_{\mathcal{R}})$ operations for computing r^π and P^π, and $O(N_{\mathcal{X}}^2)$ operations for the matrix-vector multiplication. As r^π and P^π do not need to be recomputed between iterations, the dominant cost of this process comes from the successive matrix multiplications, requiring $O(KN_{\mathcal{X}}^2)$ operations.[36]

In general, the iterates $(V_k)_{k \geq 0}$ will not reach V^π after any finite number of iterations. However, the contractive nature of the Bellman operator allows us to bound the distance from any iterate to the fixed point V^π.

Proposition 5.1. Let $V_0 \in \mathbb{R}^{\mathcal{X}}$ be an initial value function and consider the iterates

$$V_{k+1} = T^\pi V_k \,. \quad (5.5)$$

For any $\varepsilon > 0$, if we take

$$K_\varepsilon \geq \frac{\log\left(\frac{1}{\varepsilon}\right) + \log(\|V_0 - V^\pi\|_\infty)}{\log\left(\frac{1}{\gamma}\right)} \,,$$

then for all $k \geq K_\varepsilon$, we have that

$$\|V_k - V^\pi\|_\infty \leq \varepsilon \,.$$

For $V_0 = 0$, the dependency on V^π can be simplified by noting that

$$\log(\|V_0 - V^\pi\|_\infty) = \log(\|V^\pi\|_\infty) \leq \log\left(\max(|V_{\text{MIN}}|, |V_{\text{MAX}}|)\right) . \qquad \triangle$$

Proof. Since T^π is a contraction mapping with respect to the L^∞ metric with contraction modulus γ (Proposition 4.4), and V^π is its fixed point (Proposition 2.12),

36. Assuming that the number of states $N_{\mathcal{X}}$ is large compared to the number of actions $N_{\mathcal{A}}$ and rewards $N_{\mathcal{R}}$. For transition functions with special structure (sparsity, low rank, etc.), one can hope to do even better.

we have for any $k \geq 1$

$$\|V_k - V^\pi\|_\infty = \|T^\pi V_{k-1} - T^\pi V^\pi\|_\infty \leq \gamma \|V_{k-1} - V^\pi\|_\infty ,$$

and so by induction we have

$$\|V_k - V^\pi\|_\infty \leq \gamma^k \|V_0 - V^\pi\|_\infty .$$

Setting the right-hand side to be less than or equal to ε and rearranging gives the required inequality for K_ε. □

From Proposition 5.1, we conclude that we can obtain an ε-approximation to V^π in $O(K_\varepsilon N_\mathcal{X}^2)$ operations, by applying the Bellman operator K_ε times to an initial value function $V_0 = 0$. Since the iterate V_k can be represented as an $N_\mathcal{X}$-dimensional vector and is the only object that the algorithm needs to store in memory (other than the description of the MDP itself), this shows that iterative policy evaluation can approximate V^π efficiently.

5.2 Representing Return-Distribution Functions

Now, let us consider what happens in distributional reinforcement learning. As with any computational problem, we first must decide on a data structure that our algorithms operate on. The heart of our data structure is a scheme for representing return-distribution functions in memory. We call such a scheme a *probability distribution representation.*

Definition 5.2. A *probability distribution representation* $\mathscr{F}$, or simply *representation*, is a collection of probability distributions indexed by a parameter θ from some set of allowed parameters Θ:

$$\mathscr{F} = \{P_\theta \in \mathscr{P}(\mathbb{R}) : \theta \in \Theta\} . \qquad \triangle$$

Example 5.3. The *Bernoulli representation* is the set of all Bernoulli distributions:

$$\mathscr{F}_\text{B} = \{(1-p)\delta_0 + p\delta_1 : p \in [0,1]\} . \qquad \triangle$$

Example 5.4. The *uniform representation* is the set of all uniform distributions on finite-length intervals:

$$\mathscr{F}_\text{U} = \{\mathcal{U}([a,b]) : a, b \in \mathbb{R}, a < b\} . \qquad \triangle$$

We represent return functions using a table of probability distributions, each associated with a given state and described in our chosen representation. For example, a uniform return function is described in memory by a table of $2N_\mathcal{X}$ numbers, corresponding to the upper and lower ends of the distribution at each state. By extension, we call such a table a representation of return-distribution functions. Formally, for a representation $\mathscr{F}$, the space of representable return functions is $\mathscr{F}^{\mathcal{X}}$.

With this data structure in mind, let us consider the procedure (introduced by Equation 4.10) that approximates the return function η^π by repeatedly applying the distributional Bellman operator:

$$\eta_{k+1} = \mathcal{T}^\pi \eta_k \,. \tag{5.6}$$

Because an operator is an abstract object, Equation 5.6 describes a mathematical procedure, rather than a computer program. To obtain the latter, we begin by expressing the distributional Bellman operator as a sum, analogous to Equation 5.3. Recall that the distributional operator is defined by an expectation over the random variables R and X':

$$(\mathcal{T}^\pi \eta)(x) = \mathbb{E}_\pi[(\mathrm{b}_{R,\gamma})_\# \eta(X') \mid X = x] \,. \tag{5.7}$$

Here, the expectation describes a mixture of pushforward distributions. By writing this expectation in full, we find that this mixture is given by

$$(\mathcal{T}^\pi \eta)(x) = \sum_{a \in \mathcal{A}} \sum_{r \in \mathcal{R}} \sum_{x' \in \mathcal{X}} \mathbb{P}_\pi(A = a, R = r, X' = x' \mid X = x)(\mathrm{b}_{r,\gamma})_\# \eta(x') \,. \tag{5.8}$$

The pushforward operation scales and then shifts (by γ and r, respectively) the support of the probability distribution $\eta(x')$, as depicted in Figure 2.5. Implementing the distributional Bellman operator therefore requires being able to efficiently perform the shift-and-scale operation and compute mixtures of probability distributions; we caught a glimpse of what that might entail when we derived categorical temporal-difference learning in Chapter 3.

Cumulative distribution functions allow us to rewrite Equations 5.7–5.8 in terms of vector-like objects, providing a nice parallel with the usual vector notation for the expected-value setting. Let us write $F_\eta(x, z) = F_{\eta(x)}(z)$ to denote the cumulative distribution function of $\eta(x)$. We can equally express Equation 5.7 as[37]

$$(\mathcal{T}^\pi F_\eta)(x, z) = \mathbb{E}_\pi \Big[F_\eta\big(X', \tfrac{z-R}{\gamma}\big) \mid X = x\Big]. \tag{5.9}$$

As a weighted sum of cumulative distribution functions, Equation 5.9 is

$$(\mathcal{T}^\pi F_\eta)(x, z) = \sum_{a \in \mathcal{A}} \sum_{r \in \mathcal{R}} \sum_{x' \in \mathcal{X}} \mathbb{P}_\pi(A = a, R = r, X' = x' \mid X = x) F_\eta(x', \tfrac{z-r}{\gamma}). \tag{5.10}$$

Similar to Equation 5.1, we can set $F_\eta = F_{\eta^\pi}$ on both sides of Equation 5.10 to obtain the linear system:

$$F_{\eta^\pi}(x, z) = \sum_{a \in \mathcal{A}} \sum_{r \in \mathcal{R}} \sum_{x' \in \mathcal{X}} \mathbb{P}_\pi(A = a, R = r, X' = x' \mid X = x) F_{\eta^\pi}(x', \tfrac{z-r}{\gamma}) \,.$$

37. With Chapter 4 in mind, note that we are overloading operator notation when we apply $\mathcal{T}^\pi$ to collections of cumulative distribution functions, rather than return-distribution functions.

However, this particular set of equations is infinite-dimensional. This is because cumulative distribution functions are themselves infinite-dimensional objects, more specifically elements of the space of monotonically increasing functions mapping $\mathbb{R}$ to $[0, 1]$. Because of this, we cannot describe them in physical memory, at least not on a modern-day computer. This gives a concrete argument as to why we cannot simply "store" a probability distribution but must instead use a probability distribution representation as our data structure. For the same reason, a direct algebraic solution to Equation 5.10 is not possible, in contrast to the expected-value setting (Equation 5.2). This justifies the need for a dynamic programming method to approximate η^π.

Creating an algorithm for computing the return-distribution function requires us to implement the distributional Bellman operator in terms of our chosen representation. Conversely, we should choose a representation that supports an efficient implementation. Unlike the value function setting, however, there is no single best representation – making this choice requires balancing available memory, accuracy, and the downstream uses of the return-distribution function. The rest of this chapter is dedicated to studying these trade-offs and developing a theory of what makes for a good representation. Like Goldilocks faced with her choices, we first consider the situation where memory and computation are plentiful, then the use of normal distributions to construct a minimally viable return-distribution function, before finally introducing fixed-size empirical representations as a sensible and practical middle ground.

5.3 The Empirical Representation

Simple representations like the Bernoulli representation are ill-suited to describe, say, the different outcomes in blackjack (Example 2.7) or the variations in the return distributions from different policies (Example 2.9). Although there are scenarios in which a representation with few parameters gives a reasonable approximation, the most general-purpose algorithms for computing return distributions should be based on representations that are sufficiently expressive.

To understand what "sufficiently expressive" might mean, let us consider what it means to implement the iterative procedure

$$\eta_{k+1} = \mathcal{T}^\pi \eta_k . \tag{5.11}$$

The most direct implementation is a FOR loop over the iteration number $k = 0, 1, \ldots,$ interleaving

(a) determining $(\mathcal{T}^\pi \eta_k)(x)$ for each $x \in \mathcal{X}$, and
(b) expressing the outcome as a return-distribution function $\eta_{k+1} \in \mathscr{F}$.

The first step of this procedure is an algorithm that emulates the operator $\mathcal{T}^\pi$, which we may call the *operator-algorithm.* When used as part of the for loop, the output of this operator-algorithm at iteration k becomes the input at iteration $k+1$. As such, it is desirable for the inputs and outputs of the operator-algorithm to have the same type: given as input a return function represented by $\mathscr{F}$, the operator-algorithm should produce a new return function that is also represented by $\mathscr{F}$. A prerequisite is that the representation $\mathscr{F}$ be *closed* under the operator $\mathcal{T}^\pi$, in the sense that

$$\eta \in \mathscr{F}^{\mathcal{X}} \implies \mathcal{T}^\pi \eta \in \mathscr{F}^{\mathcal{X}} . \tag{5.12}$$

The *empirical representation* satisfies this desideratum.

Definition 5.5. The *empirical representation* is the set $\mathscr{F}_{\mathrm{E}}$ of *empirical distributions*

$$\mathscr{F}_{\mathrm{E}} = \left\{ \sum_{i=1}^{m} p_i \delta_{\theta_i} : m \in \mathbb{N}^+ , \theta_i \in \mathbb{R}, p_i \geq 0, \sum_{i=1}^{m} p_i = 1 \right\} . \qquad \triangle$$

An empirical distribution $\nu \in \mathscr{F}_{\mathrm{E}}$ can be stored in memory as a finite list of pairs $(\theta_i, p_i)_{i=1}^m$. We call individual elements of such a distribution *particles*, each consisting of a probability and a location. Notationally, we extend the empirical representation to return distributions by writing

$$\eta(x) = \sum_{i=1}^{m(x)} p_i(x) \delta_{\theta_i(x)} \tag{5.13}$$

for the return distribution corresponding to state x.

The application of the distributional Bellman operator to empirical probability distributions has a particular convenient form that we formalize with the following lemma and proposition.

Lemma 5.6. Let $\nu \in \mathscr{F}_{\mathrm{E}}$ be an empirical distribution with parameters m and $(\theta_i, p_i)_{i=1}^m$. For $r \in \mathbb{R}$ and $\gamma \in \mathbb{R}$, we have

$$(\mathrm{b}_{r,\gamma})_\# \nu = \sum_{i=1}^{m} p_i \delta_{r + \gamma \theta_i} . \qquad \triangle$$

In words, the application of the bootstrap function to an empirical distribution ν shifts and scales the locations of that distribution (see Exercise 5.7). This property was implicit in our description of the pushforward operation in Chapter 2, and we made use of it (also implicitly) to derive the categorical temporal-difference learning algorithm in Chapter 3.

Proposition 5.7. Provided that the set of possible rewards $\mathcal{R}$ is finite, the empirical representation $\mathscr{F}_{\mathrm{E}}$ is closed under $\mathcal{T}^\pi$. In particular, if $\eta \in \mathscr{F}_{\mathrm{E}}^{\mathcal{X}}$ is a return-distribution with parameters $\big((p_i(x), \theta_i(x))_{i=1}^{m(x)} : x \in \mathcal{X}\big)$ then

$$(\mathcal{T}^\pi \eta)(x) = \sum_{a\in\mathcal{A}} \sum_{r\in\mathcal{R}} \sum_{x'\in\mathcal{X}} \mathbb{P}_\pi(A = a, R = r, X' = x' \mid X = x) \sum_{i=1}^{m(x')} p_i(x')\delta_{r+\gamma\theta_i(x')}. \tag{5.14}$$

△

Proof. Pick a state $x \in \mathcal{X}$. For a triple $(a, r, x') \in \mathcal{A} \times \mathcal{R} \times \mathcal{X}$, write $P_{a,r,x'} = \mathbb{P}_\pi(A = a, R = r, X' = x' \mid X = x)$. Then,

$$\begin{aligned}
(\mathcal{T}^\pi \eta)(x) &\overset{(a)}{=} \sum_{a\in\mathcal{A}} \sum_{r\in\mathcal{R}} \sum_{x'\in\mathcal{X}} P_{a,r,x'} (\mathrm{b}_{r,\gamma})_\# \eta(x') \\
&= \sum_{a\in\mathcal{A}} \sum_{r\in\mathcal{R}} \sum_{x'\in\mathcal{X}} P_{a,r,x'} (\mathrm{b}_{r,\gamma})_\# \Big(\sum_{i=1}^{m(x')} p_i(x')\delta_{\theta_i(x')} \Big) \\
&\overset{(b)}{=} \sum_{a\in\mathcal{A}} \sum_{r\in\mathcal{R}} \sum_{x'\in\mathcal{X}} P_{a,r,x'} \sum_{i=1}^{m(x')} p_i(x')\delta_{r+\gamma\theta_i(x')} \\
&\equiv \sum_{j=1}^{m'} p'_j \delta_{\theta'_j}
\end{aligned}$$

for some collection $(\theta'_j, p'_j)_{j=1}^{m'}$. Line (a) is Equation 5.8 and (b) follows from Lemma 5.6. We conclude that $(\mathcal{T}^\pi \eta)(x) \in \mathscr{F}_{\mathrm{E}}$, and hence $\mathscr{F}_{\mathrm{E}}$ is closed under $\mathcal{T}^\pi$. □

Algorithm 5.1 uses Proposition 5.7 to compute the application of the distributional Bellman operator to any η represented by $\mathscr{F}_{\mathrm{E}}$. It implements Equation 5.14 almost verbatim, with two simplifications. First, it uses the fact that the particle locations for a distribution $(\mathcal{T}^\pi \eta)(x)$ only depend on r and x' but not on a. This allows us to produce a single particle for each reward-next-state pair. Second, it also encodes the fact that the return is 0 from the terminal state, making dynamic programming more effective from this state (Exercise 5.3 asks you to justify this claim). Since the output of Algorithm 5.1 is also a return function from $\mathscr{F}_{\mathrm{E}}$, we can use it to produce the iterates $\eta_1, \eta_2, \ldots$ from an initial return function $\eta_0 \in \mathscr{F}_{\mathrm{E}}^{\mathcal{X}}$.

Because $\mathscr{F}_{\mathrm{E}}$ is closed under $\mathcal{T}^\pi$, we can analyze the behavior of this procedure using the theory of contraction mappings (Chapter 4). Here we bound the number of iterations needed to obtain an ε-approximation to the

Algorithm 5.1: Empirical representation distributional Bellman operator

Algorithm parameters: η, expressed as $\theta = ((\theta_i(x), p_i(x))_{i=1}^{m(x)} : x \in \mathcal{X})$

```
foreach x ∈ X do
    θ'(x) ← EMPTY LIST
    foreach x' ∈ X do
        foreach r ∈ R do
            α_{r,x'} ← Σ_{a∈A} π(a | x) P_R(r | x, a) P_X(x' | x, a)
            if x' is terminal then
                APPEND (r, α_{r,x'}) to θ'(x)
            else
                for i = 1, ..., m(x') do
                    APPEND (r + γθ_i(x'), α_{r,x'} p_i(x')) to θ'(x)
                end for
        end foreach
    end foreach
end foreach
return θ'
```

return-distribution function η^π, as measured by a supremum p-Wasserstein metric.

Proposition 5.8. Consider an initial return function $\eta_0(x) = \delta_0$ for all $x \in \mathcal{X}$ and the dynamic programming approach that iteratively computes

$$\eta_{k+1} = \mathcal{T}^\pi \eta_k$$

by means of Algorithm 5.1. Let $\varepsilon > 0$ and let

$$K_\varepsilon \geq \frac{\log\left(\frac{1}{\varepsilon}\right) + \log\left(\max(|V_{\text{MIN}}|, |V_{\text{MAX}}|)\right)}{\log\left(\frac{1}{\gamma}\right)}.$$

Then, for all $k \geq K_\varepsilon$, we have that

$$\overline{w}_p(\eta_k, \eta^\pi) \leq \varepsilon \quad \forall p \in [1, \infty],$$

where $\overline{w}_p$ is the supremum p-Wasserstein distance. △

Proof. Similar to the proof of Proposition 5.1, since $\mathcal{T}^\pi$ is a contraction mapping with respect to the $\overline{w}_p$ metric with contraction modulus γ (Proposition 4.15), and η^π is its fixed point (Propositions 2.17 and 4.9), we can deduce

$$\overline{w}_p(\eta_k, \eta^\pi) \leq \gamma^k \overline{w}_p(\eta_0, \eta^\pi) .$$

Since $\eta_0 = \delta_0$ and η^π is supported on $[V_{\text{MIN}}, V_{\text{MAX}}]$, we can upper-bound $\overline{w}_p(\eta_0, \eta^\pi)$ by $\max(|V_{\text{MIN}}|, |V_{\text{MAX}}|)$. Setting the right-hand side to be less than or equal to ε and rearranging gives the required inequality for K_ε. □

Although we state Proposition 5.8 in terms of p-Wasserstein distance for concreteness, we can also obtain similar results more generally for probability metrics under which the distributional Bellman operator is contractive.

The analysis of this section shows that the empirical representation is sufficiently expressive to support an iterative procedure for approximating the return function η^π to an arbitrary precision, due to being closed under the distributional Bellman operator. This result is perhaps somewhat surprising: even if $\eta^\pi \notin \mathscr{F}_{\text{E}}$, we are able to obtain an arbitrarily accurate approximation within $\mathscr{F}_{\text{E}}$.

Example 5.9. Consider the single-state Markov decision process of Example 2.10, with Bernoulli reward distribution $\mathcal{U}(\{0, 1\})$ and discount factor $\gamma = 1/2$. Beginning with $\eta_0(x)$, the return distributions of the iterates

$$\eta_{k+1} = \mathcal{T}^\pi \eta_k$$

are a collection of uniformly weighted, evenly spaced Dirac deltas:

$$\eta_k(x) = \frac{1}{2^k} \sum_{i=0}^{2^k-1} \delta_{\theta_i}, \quad \theta_i = \frac{i}{2^{k-1}}. \tag{5.15}$$

As suggested by Figure 2.3, the sequence of distributions $(\eta_k(x))_{k \geq 0}$ converges to the uniform distribution $\mathcal{U}([0, 2])$ in the p-Wasserstein distances, for all $p \in [1, \infty]$. However, this limit is not itself an empirical distribution. △

The downside is that the algorithm is typically intractable for anything but a small number of iterations K. This is because the lists that describe η_k may grow exponentially in length with K, as shown in the example above. Even when η_0 is initialized to be δ_0 at all states (as in Proposition 5.8), representing the kth iterate requires $O(N_{\mathcal{X}}^k N_{\mathcal{R}}^k)$ particles per state, corresponding to all achievable discounted returns of length k.[38] This is somehow unavoidable: in a certain sense, the problem of computing return functions is NP-hard (see Remark 5.2). This motivates the need for a more complex procedure that forgoes closedness in favor of tractability.

38. A smarter implementation only requires $O(N_{\mathcal{R}}^k)$ particles per state. See Exercise 5.8.

5.4 The Normal Representation

To avoid the computational costs associated with unbounded memory usage, we may restrict ourselves to probability distributions described by a fixed number of parameters. A simple choice is to model the return with a normal distribution, which requires only two parameters per state: a mean μ and variance σ^2.

Definition 5.10. The *normal representation* is the set of normal distributions

$$\mathscr{F}_{\mathrm{N}} = \{\mathcal{N}(\mu, \sigma^2) : \mu \in \mathbb{R}, \sigma^2 \geq 0\}. \qquad \triangle$$

With this representation, a return function is described by a total of $2N_{\mathcal{X}}$ parameters.

More often than not, however, the random returns are not normally distributed. This may be because the rewards themselves are not normally distributed, or because the transition kernel is stochastic. Figure 5.1 illustrates the effect of the distributional Bellman operator on a normal return-distribution function η: mixing the return distributions at successor states results in a mixture of normal distributions, which is not normally distributed except in trivial situations. In other words, the normal representation $\mathscr{F}_{\mathrm{N}}$ is generally not closed under the distributional Bellman operator.

Rather than represent the return distribution with high accuracy, as with the empirical distribution, let us consider the more modest goal of determining the best normal approximation to the return-distribution function η^π. We define "best normal approximation" as

$$\hat{\eta}^\pi(x) = \mathcal{N}\big(V^\pi(x), \mathrm{Var}(G^\pi(x))\big). \tag{5.16}$$

Given that a normal distribution is parameterized by its mean and variance, this is an obviously sensible choice. In some cases, this choice can also be justified by arguing that $\hat{\eta}^\pi(x)$ is the normal distribution closest to $\eta^\pi(x)$ in terms of what is called the Kullback–Leibler divergence.[39] As we now show, this choice also leads to a particularly efficient algorithm for computing $\hat{\eta}^\pi$.

We will construct an iterative procedure that operates on return functions from a normal representation and converges to $\hat{\eta}^\pi$. We start with the random variable operator

$$(\mathcal{T}^\pi G)(x) \overset{\mathcal{D}}{=} R + \gamma G(X'), \quad X = x, \tag{5.17}$$

39. Technically, this is only true when the return distribution $\eta^\pi(x)$ has a probability density function. When $\eta^\pi(x)$ does not have a density, a similar argument can be made in terms with the cross-entropy loss; see Exercises 5.5 and 5.6.

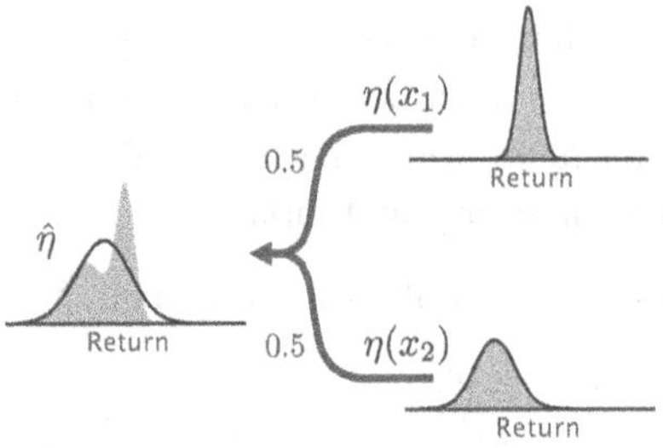

Figure 5.1
Applying the distributional Bellman operator to a return function η described by the normal representation generally produces return distributions that are not normally distributed. Here, a uniform mixture of two normal distributions (shown as probability densities) results in a bimodal distribution (in light gray). The best normal approximation $\hat{\eta}$ to that distribution is depicted by the solid curve.

and take expectations on both sides of the equation:[40]

$$\begin{aligned}\mathbb{E}\left[(\mathcal{T}^\pi G)(x)\right] &= \mathbb{E}_\pi\left[R + \gamma \mathbb{E}[G(X') \mid X'] \mid X = x\right] \\ &= \mathbb{E}_\pi[R \mid X = x] + \gamma\, \mathbb{E}_\pi[\mathbb{E}[G(X') \mid X'] \mid X = x] \\ &= \mathbb{E}_\pi[R \mid X = x] + \gamma \sum_{x' \in \mathcal{X}} \mathbb{P}_\pi(X' = x' \mid X = x)\, \mathbb{E}[G(x')]\,, \end{aligned} \tag{5.18}$$

where the last step follows from the assumption that the random variables G are independent of the next-state X'. When applied to the return-variable function G^π, Equation 5.18 is none other than the classical Bellman equation.

The same technique allows us to relate the variance of $(\mathcal{T}^\pi G)(x)$ to the next-state variances. For a random variable Z, recall that

$$\mathrm{Var}(Z) = \mathbb{E}\left[(Z - \mathbb{E}[Z])^2\right].$$

The random reward R and next-state X' are by definition conditionally independent given X and A. However, to simplify the exposition, we assume that they are also conditionally independent given only X (in Chapter 8, we will see how this assumption can be avoided).

Let us use the notation Var_π to make explicit the dependency of certain random variables on the sample transition model, analogous to our use of $\mathbb{E}_\pi$. Denote the value function corresponding to G by $V(x) = \mathbb{E}[G(x)]$. We have

$$\begin{aligned}\mathrm{Var}((\mathcal{T}^\pi G)(x)) &= \mathrm{Var}_\pi(R + \gamma G(X') \mid X = x) \\ &= \mathrm{Var}_\pi(R \mid X = x) + \mathrm{Var}_\pi(\gamma G(X') \mid X = x) \\ &= \mathrm{Var}_\pi(R \mid X = x) + \gamma^2 \mathrm{Var}_\pi(G(X') \mid X = x) \\ &= \mathrm{Var}_\pi(R \mid X = x) + \gamma^2 \mathrm{Var}_\pi(V(X') \mid X = x)\end{aligned}$$

40. In Equation 5.18, some of the expectations are taken solely with respect to the sample transition model, which we denote by $\mathbb{E}_\pi$ as usual. The expectation with respect to the random variable G, on the other hand, is not part of the model; we use the unsubscripted $\mathbb{E}$ to emphasize this point.

$$+\gamma^2 \sum_{x' \in \mathcal{X}} \mathbb{P}_\pi(X' = x' \mid X = x) \mathrm{Var}(G(x')), \tag{5.19}$$

where the last line follows by the law of total variance:

$$\mathrm{Var}(A) = \mathrm{Var}[\,\mathbb{E}[A \mid B]] + \mathbb{E}\,[\mathrm{Var}[A \mid B]].$$

Equation 5.19 shows how to compute the variances of the return function $\mathcal{T}^\pi G$. Specifically, these variances depend on the reward variance, the variance in next-state values, and the expected next-state variances. When $G = G^\pi$, this is the Bellman equation for the *variance* of the random return. Writing $\sigma^2_\pi(x) = \mathrm{Var}(G^\pi(x))$, we obtain

$$\begin{aligned}\sigma^2_\pi(x) &= \mathrm{Var}_\pi(R \mid X = x) + \gamma^2 \mathrm{Var}_\pi(V^\pi(X') \mid X = x) \\ &\quad + \gamma^2 \sum_{x' \in \mathcal{X}} \mathbb{P}_\pi(X' = x' \mid X = x) \sigma^2_\pi(x).\end{aligned} \tag{5.20}$$

We now combine the results above to obtain a dynamic programming procedure for finding $\hat{\eta}^\pi$. For each state $x \in \mathcal{X}$, let us denote by $\mu_0(x)$ and $\sigma^2_0(x)$ the parameters of a return-distribution distribution η_0 with $\eta_0(x) = \mathcal{N}(\mu_0(x), \sigma^2_0(x))$. For all x, we simultaneously compute

$$\mu_{k+1}(x) = \mathbb{E}_\pi[R + \gamma\mu_k(X') \mid X = x] \tag{5.21}$$

$$\sigma^2_{k+1}(x) = \mathrm{Var}_\pi(R \mid X = x) + \gamma^2 \mathrm{Var}_\pi(\mu_k(X') \mid X = x) + \gamma^2\, \mathbb{E}_\pi[\sigma^2_k(X') \mid X = x]. \tag{5.22}$$

We can view these two updates as the implementation of a bona fide operator over the space of normal return-distribution functions. Indeed, we can associate to each iteration the return function

$$\eta_k(x) = \mathcal{N}(\mu_k(x), \sigma^2_k(x)) \in \mathscr{F}_{\mathrm{N}}.$$

Analyzing the behavior of the sequence $(\eta_k)_{k \geq 0}$ require some care, but we shall see in Chapter 8 that the iterates converge to the best approximation $\hat{\eta}^\pi$ (Equation 5.16), in the sense that for all $x \in \mathcal{X}$,

$$\mu_k(x) \to V^\pi(x) \qquad \sigma^2_k(x) \to \mathrm{Var}(G^\pi(x)).$$

This derivation shows that the normal representation can be used to create a tractable algorithm for approximating the return distribution. However, in our work, we have found that the normal representation rarely gives a satisfying depiction of the agent's interactions with its environment; it is not sufficiently expressive. In many problems, outcomes are discrete in nature: success or failure, food or hunger, forward motion or fall. This arises in video games in which the game ends once the player's last life is spent. Timing is also critical: to catch the diamond, key, or mushroom, the agent must press "jump" at just the right moment, again leading to discrete outcomes. Even in relatively continuous systems, such as the application of reinforcement learning to stratospheric balloon flight, the return distributions tend to be skewed or multimodal. In short, normal distributions are a poor fit for the wide gamut of problems found in reinforcement learning.

5.5 Fixed-Size Empirical Representations

The empirical representation is expressive because it can use more particles to describe more complex probability distributions. This "blank check" approach to memory and computation, however, results in an intractable algorithm. On the other hand, the simple normal distribution is rarely sufficient to give a good approximation of the return distribution. A good middle ground is to preserve the form of the empirical representation while imposing a limit on its expressivity. Our approach is to fix the number and type of particles used to represent probability distributions.

Definition 5.11. The *m-quantile representation* parameterizes the location of m equally weighted particles. That is,

$$\mathscr{F}_{\mathrm{Q},m} = \left\{ \frac{1}{m} \sum_{i=1}^{m} \delta_{\theta_i} : \theta_i \in \mathbb{R} \right\}. \qquad \triangle$$

Definition 5.12. Given a collection of m evenly spaced locations $\theta_1 < \cdots < \theta_m$, the *m-categorical representation* parameterizes the probability of m particles at these fixed locations:

$$\mathscr{F}_{\mathrm{C},m} = \left\{ \sum_{i=1}^{m} p_i \delta_{\theta_i} : p_i \geq 0, \sum_{i=1}^{m} p_i = 1 \right\}.$$

We denote the stride between successive particles by $\varsigma_m = \frac{\theta_m - \theta_1}{m-1}$. $\triangle$

This definition corresponds to the categorical representation used in Chapter 3. Note that because of the constraint that probabilities should sum to 1, a m-categorical distribution is described by $m - 1$ parameters. In addition, although the representation depends on the choice of locations $\theta_1, \ldots, \theta_m$, we omit this dependence in the notation $\mathscr{F}_{\mathrm{C},m}$ to keep things concise.

In our definition of the m-categorical representation, we assume that the locations $(\theta_i)_{i=1}^m$ are given a priori and not part of the description of a particular probability distribution. This is sensible when we consider that algorithms such as categorical temporal-difference learning use the same set of locations to describe distributions at different states and keep these locations fixed across the learning process. For example, a common choice is $\theta_1 = V_{\text{MIN}}$ and $\theta_m = V_{\text{MAX}}$.

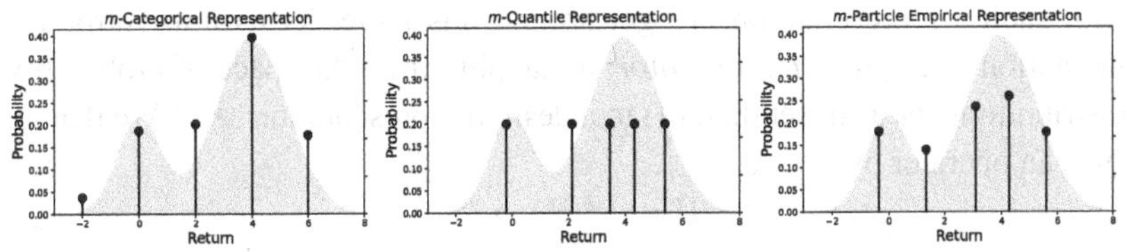

Figure 5.2
A distribution ν (in light gray), as approximated with a m-categorical, m-quantile, or m-particle representation, for $m = 5$.

When it is desirable to adjust both the locations and probabilities of different particles, we instead make use of the *m-particle representation.*

Definition 5.13. The m-particle representation parameterizes both the probability and location of m particles:

$$\mathscr{F}_{\mathrm{E},m} = \Big\{ \sum_{i=1}^{m} p_i \delta_{\theta_i} : \theta_i \in \mathbb{R}, p_i \geq 0, \sum_{i=1}^{m} p_i = 1 \Big\}.$$

The m-particle representation contains both the m-quantile representation and the m-categorical representation; a distribution from $\mathscr{F}_{\mathrm{E},m}$ is defined by $2m - 1$ parameters. △

Mathematically, all three representations described above are subsets of the empirical representation $\mathscr{F}_{\mathrm{E}}$; accordingly, we call them *fixed-size empirical representations*. Fixed-size empirical representations are flexible and can approximate both continuous and discrete outcome distributions (Figure 5.2). The categorical representation is so called because it models the probability of a set of fixed outcomes. This is somewhat of a misnomer: the "categories" are not arbitrary but instead correspond to specific real values. The quantile representation is named for its relationship to the quantiles of the return distribution. Although it might seem like the m-particle representation is a strictly superior choice, we will see that committing to either fixed locations or fixed probabilities simplifies algorithmic design. For an equal number of parameters, it is also not clear whether one should prefer m fully parameterized particles or, say, $2m - 1$ uniformly weighted particles.

Like the normal representation, fixed-size empirical representations are not closed under the distributional Bellman operator $\mathcal{T}^\pi$. As discussed in Section 5.3, the consequence is that we cannot implement the iterative procedure

$$\eta_{k+1} = \mathcal{T}^\pi \eta_k$$

with such a representation. To get around this issue, let us now introduce the notion of a *projection operator*: a mapping from the space of probability distributions (or a subset thereof) to a desired representation $\mathscr{F}$.[41] We denote such an operator by

$$\Pi_{\mathscr{F}} : \mathscr{P}(\mathbb{R}) \to \mathscr{F} .$$

Definitionally, we require that projection operators satisfy, for any $\nu \in \mathscr{F}$,

$$\Pi_{\mathscr{F}} \nu = \nu .$$

We extend the notation $\Pi_{\mathscr{F}}$ to the space of return-distribution functions:

$$(\Pi_{\mathscr{F}} \eta)(x) = \Pi_{\mathscr{F}} (\eta(x)) .$$

The categorical projection Π_{c}, first encountered in Chapter 3, is one such operator; we will study it in greater detail in the remainder of this chapter. We also made implicit use of a projection step in deriving an algorithm for the normal representation: at each iteration, we kept track of the mean and variance of the process but discarded the rest of the distribution, so that the return function iterates could be described with the normal representation.

Algorithmically, we introduce a projection step following the application of $\mathcal{T}^\pi$, leading to a *projected distributional Bellman operator* $\Pi_{\mathscr{F}} \mathcal{T}^\pi$. By definition, this operator maps $\mathscr{F}$ to itself, allowing for the design of distributional algorithms that represent each iterate of the sequence

$$\eta_{k+1} = \Pi_{\mathscr{F}} \mathcal{T}^\pi \eta_k$$

using a bounded amount of memory. We will discuss such algorithmic considerations in Section 5.7, after describing particular projection operators for the categorical and quantile representations. Combined with numerical integration, the use of a projection step also makes it possible to perform dynamic programming with continuous reward distributions (see Exercise 5.9).

5.6 The Projection Step

We now describe projection operators for the categorical and quantile representations, correspondingly called *categorical projection* and *quantile projection*. In both cases, these operators can be seen as finding the best approximation to a given probability distribution, as measured according to a specific probability metric.

To begin, recall that for a probability metric d, $\mathscr{P}_d(\mathbb{R}) \subseteq \mathscr{P}(\mathbb{R})$ is the set of probability distributions with finite mean and finite distance from the reference

41. In Section 4.1, we defined an operator as mapping elements from a space to itself. The term "projection operator" here is reasonable given that $\mathscr{F} \subseteq \mathscr{P}(\mathbb{R})$.

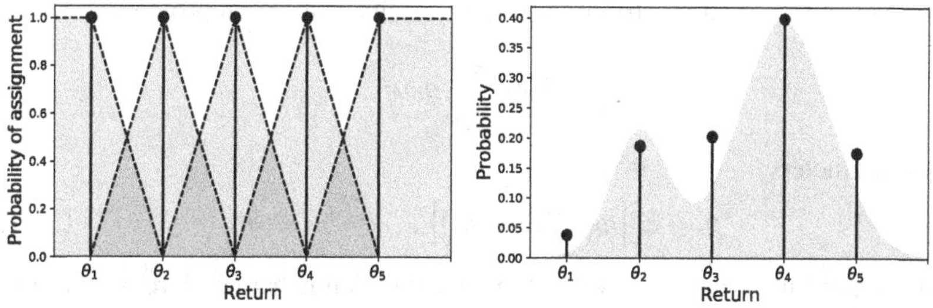

Figure 5.3
Left: The categorical projection assigns probability mass to each location according to a triangular kernel (central locations) and half-triangular kernels (boundary locations). Here, $m = 5$. **Right**: The $m = 5$ categorical projection of a given distribution, shown in gray.

distribution δ_0 (Equation 4.26). For a representation $\mathscr{F} \subseteq \mathscr{P}_d(\mathbb{R})$, a *d-projection* of $\nu \in \mathscr{P}_d(\mathbb{R})$ onto $\mathscr{F}$ is a function $\Pi_{\mathscr{F},d} : \mathscr{P}_d(\mathbb{R}) \to \mathscr{F}$ that finds a distribution $\hat{\nu} \in \mathscr{F}$ that is d-closest to ν:

$$\Pi_{\mathscr{F},d}\nu \in \arg\min_{\hat{\nu} \in \mathscr{F}} d(\nu, \hat{\nu}) . \tag{5.23}$$

Although both the categorical and quantile projections that we present here satisfy this definition, it is worth noting that in the most general setting, neither the existence nor uniqueness of a d-projection $\Pi_{\mathscr{F},d}$ is actually guaranteed (see Remark 5.3). We lift the notion of a d-projection to return-distribution functions in our usual manner; the $\bar{d}$-projection of $\eta \in \mathscr{P}_d(\mathbb{R})^{\mathcal{X}}$ onto $\mathscr{F}^{\mathcal{X}}$ is

$$(\Pi_{\mathscr{F}^{\mathcal{X}},\bar{d}}\eta)(x) = \Pi_{\mathscr{F},d}(\eta(x)) .$$

When unambiguous, we overload notation and write $\Pi_{\mathscr{F},d}\eta$ to denote the projection onto $\mathscr{F}^{\mathcal{X}}$.

It is natural to think of the $\bar{d}$-projection of the return-distribution function η onto $\mathscr{F}^{\mathcal{X}}$ as the best achievable approximation within this representation, measured in terms of d. We thus call $\Pi_{\mathscr{F},d}\nu$ and $\Pi_{\mathscr{F}^{\mathcal{X}},\bar{d}}\eta$ the $(d, \mathscr{F})$-optimal approximations to $\nu \in \mathscr{P}(\mathbb{R})$ and η, respectively.

Categorical projection. In Chapter 3, we defined the categorical projection Π_c as assigning the probability mass q of a particle located at z to the two locations nearest to z in the fixed support $\{\theta_1, \ldots, \theta_m\}$. Specifically, the categorical projection assigns this mass q in (inverse) proportion to the distance to these two neighbors. We now extend this idea to the case where we wish to more generally project a distribution $\nu \in \mathscr{P}_1(\mathbb{R})$ onto the m-categorical representation.

Given a probability distribution $\nu \in \mathscr{P}_1(\mathbb{R})$, its categorical projection

$$\Pi_c \nu = \sum_{i=1}^{m} p_i \delta_{\theta_i}$$

has parameters

$$p_i = \mathop{\mathbb{E}}_{Z \sim \nu}\left[h_i\left(\varsigma_m^{-1}(Z - \theta_i)\right)\right], \quad i = 1, \ldots, m, \tag{5.24}$$

for a set of functions $h_i : \mathbb{R} \to [0, 1]$ that we will define below. Here we write p_i in terms of an expectation rather than a sum, with the idea that this expectation can be efficiently computed (this is the case when ν is itself a m-categorical distribution).

When $i = 2, \ldots, m-1$, the function h_i is the triangular kernel

$$h_i(z) = h(z) = \max(0, 1 - |z|).$$

We use this notation to describe the proportional assignment of probability mass for the inner locations $\theta_2, \ldots, \theta_{m-1}$. One can verify that the triangular kernel assigns probability mass from ν to the location θ_i in proportion to the distance to its neighbors (Figure 5.3).

The parameters of the extreme locations are computed somewhat differently, as these also capture the probability mass associated with values greater than θ_m and smaller than θ_1. For these, we use the half-triangular kernels

$$h_1(z) = \begin{cases} 1 & z \leq 0 \\ \max(0, 1 - |z|) & z > 0 \end{cases} \qquad h_m(z) = \begin{cases} \max(0, 1 - |z|) & z \leq 0 \\ 1 & z > 0. \end{cases}$$

Exercise 5.10 asks you to prove that the projection described here matches to deterministic projection of Section 3.5.

Our derivation gives a mathematical formalization of the idea of assigning proportional probability mass to the locations nearest to a given particle, described in Chapter 3. In fact, it also describes the projection of ν in the Cramér distance (ℓ_2) onto the m-categorical representation. This is stated formally as follows and proven in Remark 5.4.

Proposition 5.14. Let $\nu \in \mathscr{P}_1(\mathbb{R})$. The m-categorical probability distribution whose parameters are given by Equation 5.24 is the (unique) ℓ_2-projection onto $\mathscr{F}_{C,m}$. △

Quantile projection. We call the *quantile projection* of a probability distribution of $\nu \in \mathscr{P}(\mathbb{R})$ a specific projection of ν in the 1-Wasserstein distance (w_1) onto the m-quantile representation ($\Pi_{\mathscr{F}_{Q,m}, w_1}$). With this choice of distance, this projection can be expressed in closed form and is easily implemented. In addition, we will see in Section 5.9 that it leads to a well-behaved dynamic programming algorithm. As with the categorical projection, we introduce the shorthand Π_{Q} for the projection operator $\Pi_{\mathscr{F}_{\text{Q},m},w_1}$.

Consider a probability distribution $\nu \in \mathscr{P}_1(\mathbb{R})$. We are interested in a probability distribution $\Pi_{\text{Q}}\nu \in \mathscr{F}_{\text{Q},m}$ that minimizes the 1-Wasserstein distance from ν:

$$\text{minimize } w_1(\nu, \nu') \text{ subject to } \nu' \in \mathscr{F}_{\text{Q},m}\,.$$

By definition, such a solution must take the form

$$\Pi_{\text{Q}}\nu = \frac{1}{m}\sum_{i=1}^{m} \delta_{\theta_i}\,.$$

The following establishes that choosing $(\theta_i)_{i=1}^{m}$ to be a particular set of quantiles of ν yields a valid w_1-projection of ν.

Proposition 5.15. Let $\nu \in \mathscr{P}_1(\mathbb{R})$. The m-quantile probability distribution whose parameters are given by

$$\theta_i = F_\nu^{-1}\left(\frac{2i-1}{2m}\right) \qquad i = 1, \dots, m \tag{5.25}$$

is a w_1-projection of ν onto $\mathscr{F}_{\text{Q},m}$. △

Equation 5.25 arises because the ith particle of a m-quantile distribution is "responsible" for the portion of the 1-Wasserstein distance measured on the interval $[\frac{i-1}{m}, \frac{i}{m}]$ (Figure 5.4). As formalized by the following lemma, the choice of the midpoint quantile $\frac{2i-1}{2m}$ minimizes the 1-Wasserstein distance to ν on this interval.

Lemma 5.16. Let $\nu \in \mathscr{P}_1(\mathbb{R})$ with cumulative distribution function F_ν. Let $0 \le a < b \le 1$. Then a solution to

$$\min_{\theta \in \mathbb{R}} \int_a^b \left|F_\nu^{-1}(\tau) - \theta\right| \mathrm{d}\tau \tag{5.26}$$

is given by the quantile midpoint

$$\theta = F_\nu^{-1}\left(\frac{a+b}{2}\right).$$ △

The proof is given as Remark 5.5.

Proof of Proposition 5.15. Let $\nu' = \frac{1}{m}\sum_{i=1}^{m} \delta_{\theta_i}$ be a m-quantile distribution. Assume that its locations are sorted: that is, $\theta_1 \le \theta_2 \le \cdots \le \theta_m$. For $\tau \in (0, 1)$, its inverse cumulative distribution function is

$$F_{\nu'}^{-1}(\tau) = \theta_{\lceil \tau m \rceil}\,.$$

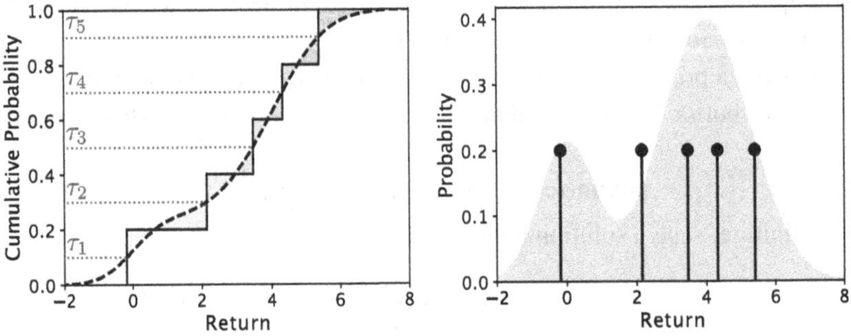

Figure 5.4
Left: The quantile projection finds the quantiles of the distribution ν (the dashed line depicts its cumulative distribution function) for $\tau_i = \frac{2i-1}{2m}, i = 1, \dots m$. The shaded area corresponds to the 1-Wasserstein distance between ν and its quantile projection $\Pi_Q\nu$ (solid line, $m = 5$). **Right**: The optimal $(w_1, \mathscr{F}_{Q,m})$-approximation to the distribution ν, shown in gray.

This function is constant on the intervals $(0, \frac{1}{m}), [\frac{1}{m}, \frac{2}{m}), \dots, [\frac{m-1}{m}, 1)$. The 1-Wasserstein distance between ν and ν' therefore decomposes into a sum of m terms:

$$w_1(\nu, \nu') = \int_0^1 \left|F_\nu^{-1}(\tau) - F_{\nu'}^{-1}(\tau)\right| d\tau$$

$$= \sum_{i=1}^{m} \int_{\frac{(i-1)}{m}}^{\frac{i}{m}} \left|F_\nu^{-1}(\tau) - \theta_i\right| d\tau .$$

By Lemma 5.16, the ith term of the sum is minimized by the quantile midpoint $F_\nu^{-1}(\tau_i)$, where

$$\tau_i = \frac{1}{2}\left[\frac{i-1}{m} + \frac{i}{m}\right] = \frac{2i-1}{2m} .$$

□

Unlike the categorical-Cramér case, in general, there is not a unique m-quantile distribution $\nu' \in \mathscr{F}_{Q,m}$ that is closest in w_1 to a given distribution $\nu \in \mathscr{P}_1(\mathbb{R})$. The following example illustrates how the issue might take shape.

Example 5.17. Consider the set of Dirac distributions

$$\mathscr{F}_{Q,1} = \{\delta_\theta : \theta \in \mathbb{R}\} .$$

Let $\nu = \frac{1}{2}\delta_0 + \frac{1}{2}\delta_1$ be the Bernoulli(1/2) distribution. For any $\theta \in [0, 1]$, $\delta_\theta \in \mathscr{F}_{Q,1}$ is an optimal $(w_1, \mathscr{F}_{Q,1})$-approximation to ν:

$$w_1(\nu, \delta_\theta) = \min_{\nu' \in \mathscr{F}_{Q,1}} w_1(\nu, \nu'),$$

Perhaps surprisingly, this shows that the distribution $\delta_{1/2}$, halfway between the two possible outcomes and an intuitive one-particle approximation to ν, is a no better choice than δ_0 when measured in terms of 1-Wasserstein distance. △

5.7 Distributional Dynamic Programming

We embed the projected Bellman operator in an for loop to obtain an algorithmic template for approximating the return function (Algorithm 5.2). We call this template *distributional dynamic programming* (DDP),[42] as it computes

$$\eta_{k+1} = \Pi_{\mathscr{F}} \mathcal{T}^\pi \eta_k \tag{5.27}$$

by iteratively applying a projected distributional Bellman operator. A special case is when the representation is closed under $\mathcal{T}^\pi$, in which case no projection is needed. However, by contrast with Equation 5.6, the use of a projection allows us to consider algorithms for a greater variety of representations. Summarizing the results of the previous sections, instantiating this template involves three parts:

Choice of representation. We first need a probability distribution representation $\mathscr{F}$. Provided that this representation uses finitely many parameters, this enables us to store return functions in memory, using the implied mapping from parameters to probability distributions.

Update step. We then need a subroutine for computing a single application of the distributional Bellman operator to a return function represented by $\mathscr{F}$ (Equation 5.8).

Projection step. We finally need a subroutine that maps the outputs of the update step to probability distributions in $\mathscr{F}$. In particular, when $\Pi_{\mathscr{F}}$ is a d-projection, this involves finding an optimal $(d, \mathscr{F})$-approximation at each iteration.

For empirical representations, including the categorical and quantile representations, the update step can be implemented by Algorithm 5.1. That is, when there are finitely many rewards, the output of $\mathcal{T}^\pi$ applied to any m-particle representation is a collection of empirical distributions:

$$\eta \in \mathscr{F}_{E,m} \implies \mathcal{T}^\pi \eta \in \mathscr{F}_E .$$

42. More precisely, this is distributional dynamic programming applied to the problem of policy evaluation. A sensible but less memorable alternative is "iterative distributional policy evaluation."

Algorithm 5.2: Distributional dynamic programming

Algorithm parameters: representation $\mathscr{F}$, projection $\Pi_{\mathscr{F}}$
desired number of iterations K,
initial return function $\eta_0 \in \mathscr{F}^{\mathcal{X}}$

```
Initialize η ← η₀
for k = 1, . . . , K do
    η′ ← T^π η                          ▷ Algorithm 5.1
    foreach state x ∈ X do
        η(x) ← (Π_F η′)(x)
    end foreach
end for
return η
```

As a consequence, it is sufficient to have an efficient subroutine for projecting empirical distributions back to $\mathscr{F}_{\text{C},m}$, $\mathscr{F}_{\text{Q},m}$, or $\mathscr{F}_{\text{E},m}$.

For example, consider the projection of the empirical distribution

$$\nu = \sum_{j=1}^{n} q_j \delta_{z_j}$$

onto the m-categorical representation (Definition 5.12). For $i = 1, \ldots, m$, Equation 5.24 becomes

$$p_i = \sum_{j=1}^{n} q_j h_i(\varsigma_m^{-1}(z_j - \theta_i)),$$

which can be implemented in linear time with two FOR loops (as was done in Algorithm 3.4).

Similarly, the projection of ν onto the m-quantile representation (Definition 5.11) is achieved by sorting the locations z_i to construct the cumulative distribution function from their associated probabilities q_i:

$$F_\nu(z) = \sum_{j=1}^{n} q_j \mathbb{1}_{\{z_j \leq z\}},$$

from which quantile midpoints can be extracted.

Because the categorical-Cramér and quantile-w_1 pairs recur so often throughout this book, it is convenient to give a name to the algorithms that iteratively apply their respective projected operators. We call these algorithms *categorical*

Algorithm 5.3: Categorical dynamic programming

Algorithm parameters: representation parameters $\theta_1, \ldots, \theta_m$, m,
initial probabilities $((p_i(x))_{i=1}^m : x \in \mathcal{X})$,
desired number of iterations K

for $k = 1, \ldots, K$ **do**
  $\eta(x) = \sum_{i=1}^{m} p_i(x)\delta_{\theta_i}$ for $x \in \mathcal{X}$
  $((z'_j(x), p'_j(x))_{j=1}^{m(x)} : x \in \mathcal{X}) \leftarrow \mathcal{T}^\pi \eta$ ▷ `Algorithm 5.1`
  foreach state $x \in \mathcal{X}$ **do**
    for $i = 1, \ldots, m$ **do**
      $p_i(x) \leftarrow \sum_{j=1}^{m(x)} p'_j(x) h_i(\varsigma_m^{-1}(z'_j - \theta_i))$
    end for
  end foreach
end for
return $((p_i(x))_{i=1}^m : x \in \mathcal{X})$

and *quantile dynamic programming*, respectively (CDP and QDP; Algorithms 5.3 and 5.4).[43]

For both categorical and quantile dynamic programming, the computational cost is dominated by the number of particles produced by the distributional Bellman operator, prior to projection. Since the number of particles in the representation is a constant m, we have that per state, there may be up to $N = mN_{\mathcal{R}}N_{\mathcal{X}}$ particles in this intermediate step. Thus, the cost of K iterations with the m-categorical representation is $O(KmN_{\mathcal{R}}N_{\mathcal{X}}^2)$, not too dissimilar to the cost of performing iterative policy evaluation with value functions. Due to the sorting operation, the cost of K iterations with the m-quantile representation is larger, at $O(KmN_{\mathcal{R}}N_{\mathcal{X}}^2 \log(mN_{\mathcal{R}}N_{\mathcal{X}}))$. In Chapter 6, we describe an incremental algorithm that avoids the explicit sorting operation and is in some cases more computationally efficient.

In designing distributional dynamic programming algorithms, there is a good deal of flexibility in the choice of representation and in the projection step once a representation has been selected. There are basic properties we would like

43. To be fully accurate, we should call these the categorical-Cramér and quantile-w_1 dynamic programming algorithms, given that they combine particular choices of probability representation and projection. However, brevity has its virtues.

Algorithm 5.4: Quantile dynamic programming

Algorithm parameters: initial locations $((\theta_i(x))_{i=1}^m : x \in \mathcal{X})$,
desired number of iterations K

for $k = 1, \ldots, K$ **do**
  $\eta(x) = \sum_{i=1}^{m} \frac{1}{m} \delta_{\theta_i(x)}$ for $x \in \mathcal{X}$
  $\eta' = ((z'_j(x), p'_j(x))_{j=1}^{m(x)} : x \in \mathcal{X}) \leftarrow \mathcal{T}^\pi \eta$ ▷ `Algorithm 5.1`
  foreach state $x \in \mathcal{X}$ **do**
    $(z'_j(x), p'_j(x))_{i=1}^{m(x)} \leftarrow \texttt{sort}((z'_j(x), p'_j(x))_{i=1}^{m(x)})$
    for $j = 1, \ldots, m$ **do**
      $P_j(x) \leftarrow \sum_{i=1}^{j} p'_i(x)$
    end for
    for $i = 1, \ldots, m$ **do**
      $j \leftarrow \min\{l : P_l(x) \geq \tau_i\}$
      $\theta_i(x) \leftarrow z'_j(x)$
    end for
  end foreach
end for
return $((\theta_i(x))_{i=1}^m : x \in \mathcal{X})$

from the sequence defined by Equation 5.27, such as a guarantee of convergence, and further a limit that does not depend on our choice of initialization. Certain combinations of representation and projection will ensure these properties hold, as we explore in Section 5.9, while others may lead to very poorly behaved algorithms (see Exercise 5.19). In addition, using a representation and projection also necessarily incurs some approximation error relative to the true return function. It is often possible to obtain quantitative bounds on this approximation error, as Section 5.10 describes, but often judgment must be used as to what qualitative types of approximation are the most acceptable for task in hand; we return to this point in Section 5.11.

5.8 Error Due to Diffusion

In Section 5.4, we showed that the distributional algorithm for the normal representation finds the best fit to the return function η^π, as measured in Kullback–Leibler divergence. Implicit in our derivation was the fact that we

$$
\begin{array}{ccccccc}
\eta_0 & \xrightarrow{\mathcal{T}^\pi} & \eta_1 & \xrightarrow{\mathcal{T}^\pi} & \cdots & \xrightarrow{\mathcal{T}^\pi} & \eta_k \\
\big\downarrow \Pi_{\mathscr{F}} & & \big\downarrow \Pi_{\mathscr{F}} & & & & \big\downarrow \Pi_{\mathscr{F}} \\
\hat{\eta}_0 & \xrightarrow{\Pi_{\mathscr{F}}\mathcal{T}^\pi} & \hat{\eta}_1 & \xrightarrow{\Pi_{\mathscr{F}}\mathcal{T}^\pi} & \cdots & \xrightarrow{\Pi_{\mathscr{F}}\mathcal{T}^\pi} & \hat{\eta}_k
\end{array}
$$

Figure 5.5
A diffusion-free projection operator $\Pi_{\mathscr{F}}$ yields a distributional dynamic programming procedure that is equivalent to first computing an exact return function and then projecting it.

could interleave projection and update steps to obtain the same solution as if we had first determined η^π without approximation and then found its best fit in $\mathscr{F}_{\mathrm{N}}$. We call a projection operator with this property *diffusion-free* (Figure 5.5).

Definition 5.18. Consider a representation $\mathscr{F}$ and a projection operator $\Pi_{\mathscr{F}}$ for that representation. The projection operator $\Pi_{\mathscr{F}}$ is said to be *diffusion-free* if, for any return function $\eta \in \mathscr{F}^{\mathcal{X}}$, we have

$$\Pi_{\mathscr{F}}\mathcal{T}^\pi\Pi_{\mathscr{F}}\eta = \Pi_{\mathscr{F}}\mathcal{T}^\pi\eta\,.$$

As a consequence, for any $k \geq 0$ and any $\eta \in \mathscr{F}^{\mathcal{X}}$, a diffusion-free projection operator satisfies

$$(\Pi_{\mathscr{F}}\mathcal{T}^\pi)^k\eta = \Pi_{\mathscr{F}}(\mathcal{T}^\pi)^k\eta\,. \qquad \triangle$$

Algorithms that implement diffusion-free projection operators are quite appealing, because they behave as if no approximation had been made until the final iteration. Unfortunately, such algorithms are the exception, rather than the rule. By contrast, without this guarantee, an algorithm may accumulate excess error from iteration to iteration – we say that the iterates $\eta_0, \eta_1, \eta_2, \ldots$ undergo *diffusion*. Known projection algorithms for m-particle representations suffer from this issue, as the following example illustrates.

Example 5.19. Consider a chain of n states with a deterministic left-to-right transition function (Figure 5.6). The last state of this chain, x_n, is terminal and produces a reward of 1; all other states yield no reward. For $t \geq 0$, the discounted return at state x_{n-t} is deterministic and has value γ^t; let us denote its distribution by $\eta(x_{n-t})$. If we approximate this distribution with $m = 11$ particles uniformly spaced from 0 to 1, the best categorical approximation assigns probability to the two particles closest to γ^t.

If we instead use categorical dynamic programming, we find a different solution. Because each iteration of the projected Bellman operator must produce

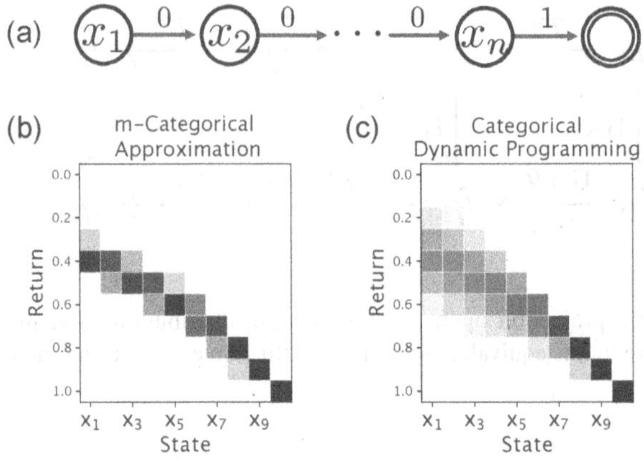

Figure 5.6
(a) An n-state chain with a single nonzero reward at the end. **(b)** The $(\ell_2, \mathscr{F}_{C,m})$-optimal approximation to the return function, for $n = 10$, $m = 11$, and $\theta_1 = 0, \ldots, \theta_{11} = 1$. The probabilities assigned to each location are indicated in grayscale (white = 0, black = 1). **(c)** The approximation found by categorical dynamic programming with the same representation.

a categorical distribution, the iterates undergo diffusion. Far from the terminal state, the return distribution found by Algorithm 5.2 is distributed on a much larger support than the best categorical approximation of $\eta(x_i)$. △

The diffusion in Example 5.19 can be explained analytically. Let us associate each particle with an integer $j = 0, \ldots, 10$, corresponding to the location $\frac{j}{10}$. For concreteness, let $\gamma = 1 - \varepsilon$ for some $0 < \varepsilon \ll 0.1$, and consider the return-distribution function obtained after N iterations of categorical dynamic programming:

$$\hat{\eta} = (\Pi_c \mathcal{T}^\pi)^n \eta_0.$$

Because there are no cycles, one can show that further iterations leave the approximation unchanged.

If we interpret the m particle locations as states in a Markov chain, then we can view the return distribution $\hat{\eta}(x_{n-t})$ as the probability distribution of this Markov chain after t time steps ($t \in \{0, \ldots, n-1\}$). The transition function for states $j = 1 \ldots 10$ is

$$P(\lfloor \gamma j \rfloor \mid j) = \lceil \gamma j \rceil - \gamma j$$
$$P(\lceil \gamma j \rceil \mid j) = 1 - \lceil \gamma j \rceil + \gamma j.$$

For $j = 0$, we simply have $P(0 \mid 0) = 1$ (i.e., state 0 is terminal). When γ is sufficiently small compared to the gap $\varsigma_m = 0.1$ between neighboring particle locations, the process can be approximated with a binomial distribution:

$$\hat{G}(x_{n-t}) \sim 1 - t^{-1}\text{Binom}(1-\gamma, t),$$

where $\hat{G}$ is the return-variable function associated with $\hat{\eta}$. Figure 5.6c gives a rough illustration of this point, with a bell-shaped distribution emerging in the return distribution at state x_1.

5.9 Convergence of Distributional Dynamic Programming

We now use the theory of operators developed in Chapter 4 to characterize the behavior of distributional dynamic programming in the policy evaluation setting: its convergence rate, its point of convergence, and also the approximation error incurred. Specifically, this theory allows us to measure how these quantities are impacted by different choices of representation and projection. Although the algorithmic discussion in previous sections has focused on implementations of distributional dynamic programming in the case of finitely supported reward distributions, the results presented here apply without this assumption (see Exercise 5.9 for indications of how distributional dynamic programming algorithms may be implemented in such cases). As we will see in Chapter 6, the theory developed here also informs incremental algorithms for learning the return-distribution function.

Let us consider a probability metric d, possibly different from the metric under which the projection is performed (when applicable), and which we call the *analysis metric*. We use the analysis metric to characterize instances of Algorithm 5.2 in terms of the *Lipschitz constant* of the projected operator.

Definition 5.20. Let (M, d) be a metric space, and let $\mathcal{O}: M \to M$ be a function on this space. The Lipschitz constant of $\mathcal{O}$ under the metric d is

$$\|\mathcal{O}\|_{\mathrm{d}} = \sup_{\substack{U,U' \in M \\ U \neq U'}} \frac{\mathrm{d}(\mathcal{O}U, \mathcal{O}U')}{\mathrm{d}(U, U')}\,. \qquad \triangle$$

When $\mathcal{O}$ is a contraction mapping, its Lipschitz constant is simply its contraction modulus. That is, under the conditions of Theorem 4.25 applied to a c-homogeneous d, we have

$$\|\mathcal{T}^\pi\|_{\bar{\mathrm{d}}} \leq \gamma^c\,.$$

Definition 5.20 extends the notion of a contraction modulus to operators, such as projections, that are not contraction mappings.

Lemma 5.21. Let (M, d) be a metric space, and let $\mathcal{O}_1, \mathcal{O}_2 : M \to M$ be functions on this space. Write $\mathcal{O}_1\mathcal{O}_2$ for the composition of these mappings. Then,

$$\|\mathcal{O}_1\mathcal{O}_2\|_{\mathrm{d}} \leq \|\mathcal{O}_1\|_{\mathrm{d}}\|\mathcal{O}_2\|_{\mathrm{d}}\,. \qquad \triangle$$

Proof. By applying the definition of the Lipschitz constant twice, we have

$$\mathrm{d}(O_1O_2U, O_1O_2U') \leq \|O_1\|_{\mathrm{d}} d(O_2U, O_2U') \leq \|O_1\|_{\mathrm{d}}\|O_2\|_{\mathrm{d}} \mathrm{d}(U, U'),$$

as required. □

Lemma 5.21 gives a template for validating and understanding different instantiations of Algorithm 5.2. Here, we consider the metric space defined by the finite domain of our analysis metric, $(\mathscr{P}_{\mathrm{d}}(\mathbb{R})^{\mathcal{X}}, \overline{\mathrm{d}})$, and use Lemma 5.21 to characterize $\Pi_{\mathscr{F}}\mathcal{T}^\pi$ in terms of the Lipschitz constants of its parts ($\Pi_{\mathscr{F}}$ and $\mathcal{T}^\pi$). By analogy with the conditions of Proposition 4.27, where needed, we will assume that the environment (more specifically, its random quantities) is reasonably behaved under d.

Assumption 5.22(d). The finite domain $\mathscr{P}_{\mathrm{d}}(\mathbb{R})^{\mathcal{X}}$ is closed under both the projection $\Pi_{\mathscr{F}}$ and the Bellman operator $\mathcal{T}^\pi$, in the sense that

$$\eta \in \mathscr{P}_{\mathrm{d}}(\mathbb{R})^{\mathcal{X}} \implies \mathcal{T}^\pi\eta, \Pi_{\mathscr{F}}\eta \in \mathscr{P}_{\mathrm{d}}(\mathbb{R})^{\mathcal{X}}.$$ △

Assumption 5.22(d) guarantees that distributions produced by the distributional Bellman operator and the projection have finite distances from one another. For many choices of d of interest, Assumption 5.22 can be easily shown to hold when the reward distributions are bounded; contrast with Proposition 4.16.

The Lipschitz constant of projection operators must be at least 1, since for any $\nu \in \mathscr{F}$,

$$\Pi_{\mathscr{F}}\nu = \nu.$$

In the case of the Cramér distance ℓ_2, we can show that it behaves much like a Euclidean metric, from which we obtain the following result on ℓ_2 projections.

Lemma 5.23. Consider a representation $\mathscr{F} \subseteq \mathscr{P}_{\ell_2}(\mathbb{R})$. If $\mathscr{F}$ is complete with respect to ℓ_2 and convex, then the ℓ_2-projection $\Pi_{\mathscr{F},\ell_2} : \mathscr{P}_{\ell_2}(\mathbb{R}) \to \mathscr{F}$ is a *nonexpansion* in that metric:

$$\|\Pi_{\mathscr{F},\ell_2}\|_{\ell_2} = 1.$$

Furthermore, the result extends to return functions and the supremum extension of ℓ_2:

$$\|\Pi_{\mathscr{F},\ell_2}\|_{\overline{\ell}_2} = 1.$$ △

The proof is given in Remark 5.6. Exercise 5.15 asks you show that the m-categorical representation is convex and complete with respect to the Cramér distance. From this, we immediately derive the following.

Lemma 5.24. The projected distributional Bellman operator instantiated with $\mathscr{F} = \mathscr{F}_{\mathrm{C},m}$ and $d = \ell_2$ has contraction modulus $\gamma^{1/2}$ with respect to $\overline{\ell}_2$ on the space $\mathscr{P}_{\ell_2}(\mathbb{R})^{\mathcal{X}}$. That is,

$$\|\Pi_c \mathcal{T}^\pi\|_{\bar{\ell}_2} \le \gamma^{1/2}. \qquad \triangle$$

Lemma 5.24 gives a formal reason for why the use of the m-categorical representation, coupled with a projection based on triangular kernel, produces a convergent distributional dynamic programming algorithm.

The m-quantile representation, however, is not convex. This precludes a direct appeal to an argument based on a quantile analogue to Lemma 5.23. Nor is the issue simply analytical: the w_1 Lipschitz constant of the projected Bellman operator $\Pi_Q \mathcal{T}^\pi$ is greater than 1 (see Exercise 5.16). Instead, we need to use the ∞-Wasserstein distance as our analysis metric. As the w_1-projection onto $\mathscr{F}_{Q,m}$ is a nonexpansion in w_∞, we obtain the following result.

Lemma 5.25. Under Assumption 5.22(w_∞), the projected distributional Bellman operator $\Pi_Q \mathcal{T}^\pi : \mathscr{P}_\infty(\mathbb{R})^{\mathcal{X}} \to \mathscr{P}_\infty(\mathbb{R})^{\mathcal{X}}$ has contraction modulus γ in $\overline{w}_\infty$. That is,

$$\|\Pi_Q \mathcal{T}^\pi\|_{\overline{w}_\infty} \le \gamma. \qquad \triangle$$

Proof. Since $\mathcal{T}^\pi$ is a γ-contraction in $\overline{w}_\infty$ by Proposition 4.15, it is sufficient to prove that $\Pi_Q : \mathscr{P}_\infty(\mathbb{R}) \to \mathscr{P}_\infty(\mathbb{R})$ is a nonexpansion in w_∞; the result then follows from Lemma 5.21. Given two distributions $\nu_1, \nu_2 \in \mathscr{P}_\infty(\mathbb{R})$, we have

$$w_\infty(\nu_1, \nu_2) = \sup_{\tau \in (0,1)} |F_{\nu_1}^{-1}(\tau) - F_{\nu_2}^{-1}(\tau)|.$$

Now note that

$$\Pi_Q \nu_1 = \frac{1}{m} \sum_{i=1}^m \delta_{F_{\nu_1}^{-1}(\tau_i)}, \quad \Pi_Q \nu_2 = \frac{1}{m} \sum_{i=1}^m \delta_{F_{\nu_2}^{-1}(\tau_i)},$$

where $\tau_i = \frac{2i-1}{2m}$. We have

$$\begin{aligned} w_\infty(\Pi_Q \nu_1, \Pi_Q \nu_2) &= \max_{i=1,\dots,m} \left|F_{\nu_1}^{-1}(\tau_i) - F_{\nu_2}^{-1}(\tau_i)\right| \\ &\le \sup_{\tau \in (0,1)} \left|F_{\nu_1}^{-1}(\tau) - F_{\nu_2}^{-1}(\tau)\right| = w_\infty(\nu_1, \nu_2), \end{aligned}$$

as required. □

The derivation of a contraction modulus for a projected Bellman operator provides us with two results. By Proposition 4.7, it establishes that if $\Pi_{\mathscr{F}} \mathcal{T}^\pi$ has a fixed point $\hat{\eta}$, then Algorithm 5.2 converges to this fixed point. Second, it also establishes the existence of such a fixed point when the representation is complete with respect to the analysis metric d, based on *Banach's fixed point theorem*; a proof is provided in Remark 5.7.

Theorem 5.26 (Banach's fixed point theorem)**.** Let (M, d) be a complete metric space and let $\mathcal{O}$ be a contraction mapping on M, with respect to d. Then $\mathcal{O}$ has a unique fixed point $U^* \in M$. △

Proposition 5.27. Let d be a c-homogeneous, regular probability metric that is p-convex for some $p \in [1, \infty)$ and let $\mathscr{F}$ be a representation complete with respect to d. Consider Algorithm 5.2 instantiated with a projection step described by the operator $\Pi_{\mathscr{F}}$, and suppose that Assumption 5.22(d) holds. If $\Pi_{\mathscr{F}}$ is a nonexpansion in $\overline{\mathrm{d}}$, then the corresponding projected Bellman operator $\Pi_{\mathscr{F}}\mathcal{T}^\pi$ has a unique fixed point $\hat{\eta}^\pi$ in $\mathscr{P}_{\mathrm{d}}(\mathbb{R})^{\mathcal{X}}$ satisfying

$$\hat{\eta}^\pi = \Pi_{\mathscr{F}}\mathcal{T}^\pi\hat{\eta}^\pi .$$

Additionally, for any $\varepsilon > 0$, if K the number of iterations is such that

$$K \geq \frac{\log\left(\frac{1}{\varepsilon}\right) + \log \overline{\mathrm{d}}(\eta_0, \hat{\eta}^\pi)}{c \log\left(\frac{1}{\gamma}\right)}$$

with $\eta_0(x) \in \mathscr{P}_{\mathrm{d}}(\mathbb{R})$ for all x, then the output η_K of Algorithm 5.2 satisfies

$$\overline{\mathrm{d}}(\eta_K, \hat{\eta}^\pi) \leq \varepsilon .$$ △

Proof. By the assumptions in the statement, we have that $\mathcal{T}^\pi$ is a γ^c-contraction on $\mathscr{P}_{\mathrm{d}}(\mathbb{R})^{\mathcal{X}}$ by Theorem 4.25, and by Lemma 5.21, $\Pi_{\mathscr{F}}\mathcal{T}^\pi$ is a γ^c-contraction on $\mathscr{P}_{\mathrm{d}}(\mathbb{R})^{\mathcal{X}}$. By Banach's fixed point theorem, there is a unique fixed point $\hat{\eta}^\pi$ for $\Pi_{\mathscr{F}}\mathcal{T}^\pi$ in $\mathscr{P}_{\mathrm{d}}(\mathbb{R})^{\mathcal{X}}$. Now note

$$\overline{\mathrm{d}}(\eta_K, \hat{\eta}^\pi) = \overline{\mathrm{d}}(\Pi_{\mathscr{F}}\mathcal{T}^\pi\eta_{K-1}, \Pi_{\mathscr{F}}\mathcal{T}^\pi\hat{\eta}^\pi) \leq \gamma^c\overline{\mathrm{d}}(\eta_{K-1}, \hat{\eta}^\pi),$$

so by induction

$$\overline{\mathrm{d}}(\eta_K, \hat{\eta}^\pi) \leq \gamma^{cK}\overline{\mathrm{d}}(\eta_0, \hat{\eta}^\pi).$$

Setting the right-hand side to less than ε and rearranging yields the result. □

Although Proposition 5.27 does not allow us to conclude that a particular algorithm will fail, it cautions us against projections in certain probability metrics. For example, because the distributional Bellman operator is only a nonexpansion in the supremum total variation distance, we cannot guarantee a good approximation with respect to this metric after any finite number of iterations. Because we would like the projection step to be computationally efficient, this argument also gives us a criterion with which to choose a representation. For example, although the m-particle representation is clearly more flexible than either the categorical or quantile representations, it is currently not known how to efficiently and usefully project onto it.

5.10 Quality of the Distributional Approximation

Having identified conditions under which the iterates produced by distributional dynamic programming converge, we now ask: to what do they converge? We answer this question by measuring how close the fixed point $\hat{\eta}^\pi$ of the projected Bellman operator (computed by Algorithm 5.2 in the limit of the number of iterations) is to the true return function η^π. The quality of this approximation depends on a number of factors: the choice and size of representation $\mathscr{F}$, which determines the optimal approximation of η^π within $\mathscr{F}$, as well as properties of the projection step.

Proposition 5.28. Let d be a c-homogeneous, regular probability metric that is p-convex for some $p \in [1, \infty)$, and let $\mathscr{F} \subseteq \mathscr{P}_{\text{d}}(\mathbb{R})$ be a representation complete with respect to d. Let $\Pi_{\mathscr{F}}$ be a projection operator that is a nonexpansion in $\overline{\text{d}}$, and suppose Assumption 5.22(d) holds. Consider the projected Bellman operator $\Pi_{\mathscr{F}}\mathcal{T}^\pi$ with fixed point $\hat{\eta}^\pi \in \mathscr{P}_{\text{d}}(\mathbb{R})^{\mathcal{X}}$. Then,

$$\overline{\text{d}}(\hat{\eta}^\pi, \eta^\pi) \leq \frac{\overline{\text{d}}(\Pi_{\mathscr{F}}\eta^\pi, \eta^\pi)}{1 - \gamma^c}.$$

When $\Pi_{\mathscr{F}}$ is a d-projection in some probability metric d, $\Pi_{\mathscr{F}}\eta^\pi$ is a $(d, \mathscr{F})$-optimal approximation of the return function η^π. △

Proof. We have

$$\begin{aligned}\overline{\text{d}}(\eta^\pi, \hat{\eta}^\pi) &\leq \overline{\text{d}}(\eta^\pi, \Pi_{\mathscr{F}}\eta^\pi) + \overline{\text{d}}(\Pi_{\mathscr{F}}\eta^\pi, \hat{\eta}^\pi) \\ &= \overline{\text{d}}(\eta^\pi, \Pi_{\mathscr{F}}\eta^\pi) + \overline{\text{d}}(\Pi_{\mathscr{F}}\mathcal{T}^\pi\eta^\pi, \Pi_{\mathscr{F}}\mathcal{T}^\pi\hat{\eta}^\pi) \\ &\leq \overline{\text{d}}(\eta^\pi, \Pi_{\mathscr{F}}\eta^\pi) + \gamma^c\overline{\text{d}}(\eta^\pi, \hat{\eta}^\pi).\end{aligned}$$

Rearranging then gives the result. □

From this, we immediately derive a result regarding the fixed point $\hat{\eta}^\pi_{\text{c}}$ of the projected operator $\Pi_{\text{c}}\mathcal{T}^\pi$.

Corollary 5.29. The fixed point $\hat{\eta}^\pi_{\text{c}}$ of the categorical-projected Bellman operator $\Pi_{\text{c}}\mathcal{T}^\pi : \mathscr{P}_1(\mathbb{R}) \to \mathscr{P}_1(\mathbb{R})$ satisfies

$$\overline{\ell}_2(\hat{\eta}^\pi_{\text{c}}, \eta^\pi) \leq \frac{\overline{\ell}_2(\Pi_{\text{c}}\eta^\pi, \eta^\pi)}{1 - \gamma^{1/2}}.$$

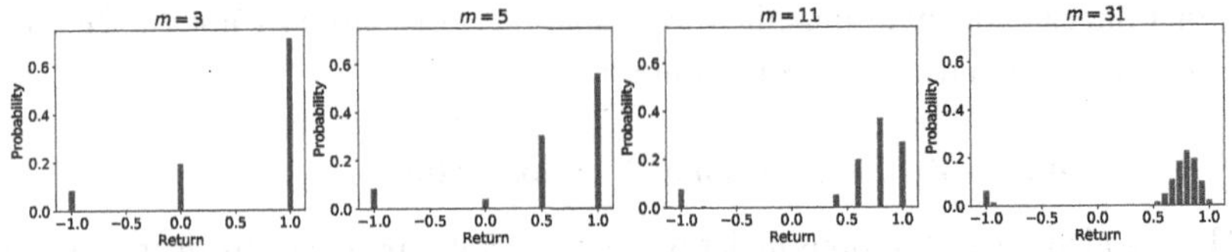

Figure 5.7
The return-distribution function estimates obtained by applying categorical dynamic programming to the Cliffs domain (Example 2.9; here, with the safe policy). Each panel corresponds to a different number of particles.

If the return distributions $(\eta^\pi(x) : x \in \mathcal{X})$ are supported on $[\theta_1, \theta_m]$, we further have that, for each $x \in \mathcal{X}$,

$$\min_{\nu \in \mathscr{F}_{C,m}} \ell_2^2(\nu, \eta^\pi(x)) \le \varsigma_m \sum_{i=1}^{m} \Big(F_{\eta^\pi(x)}(\theta_1 + i\varsigma_m) - F_{\eta^\pi(x)}(\theta_1 + (i-1)\varsigma_m)\Big)^2 \le \varsigma_m\,, \tag{5.28}$$

and hence

$$\bar{\ell}_2(\hat{\eta}^\pi_{C}, \eta^\pi) \le \frac{1}{1-\gamma^{1/2}} \frac{\theta_m - \theta_1}{m-1}\,.$$

△

Proposition 5.28 suggests that the excess error may be greater when the projection is performed in a metric for which c is small. In the particular case of the Cramér distance, the constant in Corollary 5.29 can in fact be strengthened to $\frac{1}{\sqrt{1-\gamma}}$ by arguing about the geometry of the probability space under ℓ_2 under certain conditions; see Remark 5.6 for a discussion of this geometry and Rowland et al. (2018) for details on the strengthened bound.

Varying the number of particles in the categorical representation allows the user to control both the complexity of the associated dynamic programming algorithm and the error of the fixed point $\hat{\eta}^\pi_C$ (Figure 5.7). As discussed in the first part of this chapter, both the memory and time complexity of computing with this representation increase with m. On the other hand, Corollary 5.29 establishes that increasing m also reduces the approximation error incurred from dynamic programming.

Proposition 5.28 can be similarly used to analyze quantile dynamic programming. This result, under the conditions of Lemma 5.25, shows that the fixed point $\hat{\eta}^\pi_Q$ of the projected operator $\Pi_Q \mathcal{T}^\pi$ for the m-quantile representation satisfies

$$\overline{w}_\infty(\hat{\eta}^\pi_Q, \eta^\pi) \le \frac{\overline{w}_\infty(\Pi_Q \eta^\pi, \eta^\pi)}{1-\gamma}\,.$$

Unfortunately, the distance $\overline{w}_\infty(\Pi_Q \eta^\pi, \eta^\pi)$ does not necessarily vanish as the number of particles m in the quantile representation increases. This is because w_∞ is in some sense a very strict distance (Exercise 5.18 makes this point precise). Nevertheless, the dynamic programming algorithm associated with the quantile representation enjoys a similar trade-off between complexity and accuracy as established for the categorical algorithm above. This can be shown by instead analyzing the algorithm via the more lenient w_1 distance; Exercise 5.20 provides a guide to a proof of this result.

Lemma 5.30. Suppose that for each $(x, a) \in \mathcal{X} \times \mathcal{A}$, the reward distribution $P_\mathcal{R}(\cdot \mid x, a)$ is supported on the interval $[R_{\text{MIN}}, R_{\text{MAX}}]$. Then the m-quantile fixed point $\hat{\eta}^\pi_Q$ satisfies

$$\overline{w}_1(\hat{\eta}^\pi_Q, \eta^\pi) \le \frac{3(V_{\text{MAX}} - V_{\text{MIN}})}{2m(1-\gamma)} .$$

In addition, consider an initial return function η_0 with distributions with support bounded in $[V_{\text{MIN}}, V_{\text{MAX}}]$ and the iterates

$$\eta_{k+1} = \Pi_Q \mathcal{T}^\pi \eta_k$$

produced by quantile dynamic programming. Then we have, for all $k \ge 0$,

$$\overline{w}_1(\eta_k, \eta^\pi) \le \gamma^k (V_{\text{MAX}} - V_{\text{MIN}}) + \frac{3(V_{\text{MAX}} - V_{\text{MIN}})}{2m(1-\gamma)} . \qquad \triangle$$

5.11 Designing Distributional Dynamic Programming Algorithms

Although both categorical and quantile dynamic programming algorithms arise from natural choices, there is no reason to believe that they lead to the best approximation of the return-distribution function for a fixed number of parameters. For example, can the error be reduced by using a $\frac{m}{2}$-particle representation instead? Are there measurable characteristics of the environment that could guide the choice of distribution representation?

As a guide to further investigations, Table 5.1 summarizes the desirable properties of representations and projections that were studied in this chapter, as they pertain to the design of distributional dynamic programming algorithms. Because these properties arise from the combination of a representation with a particular projection, every such combination is likely to exhibit a different set of properties. This highlights how new DDP algorithms with desirable characteristics may be produced by simply trying out different combinations. Because these properties arise from the combination of a representation with a particular projection, one can naturally expect new algorithms for example, one can imagine a new algorithm based on the quantile representation that is a nonexpansion in the 1- or 2-Wasserstein distances.

Included in this table is whether a representation is *mean-preserving*. We say that a representation $\mathscr{F}$ and its associated projection operator $\Pi_{\mathscr{F}}$ are mean-preserving if for any distribution ν from a suitable space, the mean of $\Pi_{\mathscr{F}}\nu$ is the same as that of ν. The m-categorical algorithm presented in this chapter is mean-preserving provided the range of considered returns does not exceed the boundaries of its support; the m-quantile algorithm is not. In addition to the mean, we can also consider what happens to other aspects of the approximated distributions. For example, the m-categorical algorithm produces probability distributions that in general have more variance than those of the true return function.

The choice of representation also has consequences beyond distributional dynamic programming. In the next chapter, for example, we will consider the design of incremental algorithms for learning return-distribution functions from samples. There, we will see that the projected operator derived from the categorical representation directly translates into an incremental algorithm, while the operator derived from the quantile representation does not. In Chapter 9, we will also see how the choice of representation interplays with the use of parameterized models such as neural networks to represent the return function compactly. Another consideration, not listed here, concerns the sensitivity of the algorithm to its parameters: for example, for small values of m, the m-quantile representation tends to be a better choice than the m-categorical representation, which suffers from large gaps between its particles' locations (as an extreme example, take $m = 2$).

The design and study of representations remains an active topic in distributional reinforcement learning. The representations we presented here are by no mean an exhaustive portrait of the field. For example, Barth-Maron et al. (2018) considered using mixtures of m normal distributions in domains with vector-valued action spaces. Analyzing representations like these is more challenging because known projection methods suffer from local minima, which in turn implies that dynamic programming may give different (and possibly suboptimal) solutions for different initial conditions.

5.12 Technical Remarks

Remark 5.1 (Finiteness of $\mathcal{R}$). In the algorithms presented in this chapter, we assumed that rewards are distributed on a finite set $\mathcal{R}$. This is not actually needed for most of our analysis but makes it possible to compute expectations and convolutions in finite time and hence devise concrete dynamic programming algorithms. However, there are many problems in which the rewards are better modeled using continuous or unbounded distributions. Rewards generated from

	$\mathcal{T}^\pi$-closed	Tractable	Expressive	Diffusion-free	Mean-preserving
Empirical DP	✓		✓	✓	✓
Normal DP		✓		✓	✓
Categorical DP		✓	✓		*
Quantile DP		✓	✓		

Table 5.1
Desirable characteristics of distributional dynamic programming algorithms. "Empirical DP" and "Normal DP" refer to distributional dynamic programming with the empirical and normal distributions, respectively (Sections 5.3 and 5.4). While no known representation–projection pair satisfies all of these, the categorical-Cramér and quantile-w_1 choices offer a good compromise. *The categorical representation is mean-preserving provided its support spans the range of possible returns.

observations of a physical process are often well modeled by a normal distribution to account for sensor noise. Rewards derived from a queuing process, such as the number of customers who make a purchase at an ice cream shop in a give time interval, can be modeled by a Poisson distribution. △

Remark 5.2 (NP-hardness of computing return distributions). Recall that a problem is said to be *NP-hard* if its solution can also be used to solve all problems in the class NP, by means of a polynomial-time reduction (Cormen et al. 2001). This remark illustrates how the problem of computing certain aspects of the return-distribution function for a given Markov decision process is NP-hard, by reduction from one of Karp's original NP-complete problems, the *Hamiltonian cycle problem*. Reductions from the Hamiltonian cycle problem have previously been used to prove the NP-hardness of a variety of problems relating to constrained Markov decision processes (Feinberg 2000).

Let $G = (V, E)$ be a graph, with $V = \{1, \ldots, n\}$ for some $n \in \mathbb{N}^+$. The Hamiltonian cycle problem asks whether there is a permutation of vertices $\sigma : V \to V$ such that $\{\sigma(i), \sigma(i+1)\} \in E$ for each $i \in [n-1]$, and $\{\sigma(n), \sigma(1)\} \in E$. Now consider an MDP whose state space is the set of integers from 1 to n, denoted $\mathcal{X} = \{1, \ldots, n\}$, and with a singleton action set $\mathcal{A} = \{a\}$, a transition kernel that encodes a random walk over the graph G, and a uniform initial state distribution ξ_0. Further, specify reward distributions as $P_{\mathcal{R}}(\cdot \mid x, a) = \delta_x$ for each $x \in \mathcal{X}$, and set $\gamma < 1/n+2$. For such a discount factor, there is a one-to-one mapping between trajectories and returns. As the action set is a singleton, there is a single policy π. It can be shown that there is a Hamiltonian cycle in G if and only if the support of the return distribution has nonempty intersection with the set

$$\bigcup_{\sigma \in \mathbb{S}_n} \left[\sum_{t=0}^{n-1} \gamma^t \sigma(t+1) + \gamma^n \sigma(1), \sum_{t=0}^{n-1} \gamma^t \sigma(t+1) + \gamma^n (\sigma(1)+1) \right). \tag{5.29}$$

Since the MDP description is clearly computable in polynomial time from the graph specifying the Hamiltonian cycle problem, it must therefore be the case that ascertaining whether the set in Expression 5.29 has nonempty intersection with the return distribution support is NP-hard. The full proof is left as Exercise 5.21. △

Remark 5.3 (Existence of d-projections). As an example of a setting where d-projections do not exist, let $\theta_1, \ldots, \theta_m \in \mathbb{R}$ and consider the *softmax categorical representation*

$$\mathscr{F} = \Big\{ \sum_{i=1}^{m} \frac{e^{\varphi_i}}{\sum_{j=1}^{m} e^{\varphi_j}} \delta_{\theta_i} : \varphi_1, \ldots, \varphi_m \in \mathbb{R} \Big\}.$$

For the distribution δ_{θ_1} and metric $d = w_2$, there is no optimal $(\mathscr{F}, d)$-approximation to δ_{θ_1}, since for any distribution in $\mathscr{F}$, there is another distribution in $\mathscr{F}$ that lies closer to δ_{θ_1} with respect to w_2. The issue is that there are elements of the representation which are arbitrarily close to the distribution to be projected, but there is no distribution in the representation that dominates all others as in Equation 5.23. An example of a sufficient condition for the existence of an optimal $(\mathscr{F}, d)$-approximation to $\hat{\nu} \in \mathscr{P}_d(\mathbb{R})$ is that the metric space $(\mathscr{F}, d)$ has the *Bolzano–Weierstrass property*, also known as *sequential compactness*: for any sequence $(\nu_k)_{k\geq 0}$ in $\mathscr{F}$, there is subsequence that converges to a point in $\mathscr{F}$ with respect to d. When this assumption holds, we may take the sequence $(\nu_k)_{k\geq 0}$ in $\mathscr{F}$ to satisfy

$$d(\nu_k, \hat{\nu}) \leq \inf_{\nu \in \mathscr{F}} d(\nu, \hat{\nu}) + \frac{1}{k+1}.$$

Using the Bolzano–Weierstrass property, we may pass to a subsequence $(\nu_{k_n})_{n\geq 0}$ converging to $\nu^* \in \mathscr{F}$. We then observe

$$d(\nu^*, \hat{\nu}) \leq d(\nu^*, \nu_{k_n}) + d(\nu_{k_n}, \hat{\nu}) \to \inf_{\nu \in \mathscr{F}} d(\nu, \hat{\nu}),$$

showing that ν^* is an optimal $(\mathscr{F}, d)$-approximation to $\hat{\nu}$. The softmax representation (under the w_2 metric) fails to have the Bolzano–Weierstrass property. As an example, the sequence of distributions $(\frac{k}{k+1}\delta_{\theta_1} + \frac{1}{k+1}\delta_{\theta_2})_{k\geq 0}$ in $\mathscr{F}$ has no subsequence that converges to a point in $\mathscr{F}$. △

Remark 5.4 (Proof of Proposition 5.14). We will demonstrate the following Pythagorean identity. For any $\nu \in \mathscr{P}_{\ell_2}(\mathbb{R})$, and $\nu' \in \mathscr{F}_{\text{C},m}$, we have

$$\ell_2^2(\nu, \nu') = \ell_2^2(\nu, \Pi_c \nu) + \ell_2^2(\Pi_c \nu, \nu') .$$

From this, it follows that $\nu' = \Pi_c \nu$ is the unique ℓ_2-projection of ν onto $\mathscr{F}_{\text{C},m}$, since this choice of ν' uniquely minimizes the right-hand side. To show this identity, we first establish an interpretation of the cumulative distribution function (CDF) of $\Pi_c \nu$ as averaging the CDF values of ν on each interval (θ_i, θ_{i+1}) for $i = 1, \ldots, m-1$. First note that for $z \in (\theta_i, \theta_{i+1}]$, we have

$$\begin{aligned} h_i(\varsigma_m^{-1}(z - \theta_i)) &= 1 - \varsigma_m^{-1}|z - \theta_i| \\ &= \frac{1}{\theta_{i+1} - \theta_i}(\theta_{i+1} - \theta_i + \theta_i - z) \\ &= \frac{\theta_{i+1} - z}{\theta_{i+1} - \theta_i} . \end{aligned}$$

Now, for $i = 1, \ldots, m-1$, we have

$$\begin{aligned} F_{\Pi_c \nu}(\theta_i) &= \sum_{j=1}^{i} \mathop{\mathbb{E}}_{Z\sim\nu}[h_j(Z)] \\ &= \mathop{\mathbb{E}}_{Z\sim\nu}\left[\mathbb{1}\{Z \le \theta_i\} + \mathbb{1}\{Z \in (\theta_i, \theta_{i+1}]\}\frac{\theta_{i+1} - Z}{\theta_{i+1} - \theta_i}\right] \\ &= \frac{1}{\theta_{i+1} - \theta_i}\int_{\theta_i}^{\theta_{i+1}} F_\nu(z)\mathrm{d}z . \end{aligned}$$

Now note

$$\begin{aligned} \ell_2^2(\nu, \nu') &= \int_{-\infty}^{\infty} (F_\nu(z) - F_{\nu'}(z))^2 \mathrm{d}z \\ &\overset{(a)}{=} \int_{-\infty}^{\infty} (F_\nu(z) - F_{\Pi_c\nu}(z))^2 \mathrm{d}z + \int_{-\infty}^{\infty} (F_{\Pi_c\nu}(z) - F_{\nu'}(z))^2 \mathrm{d}z \\ &\quad + 2\int_{-\infty}^{\infty} (F_\nu(z) - F_{\Pi_c\nu}(z))(F_{\Pi_c\nu}(z) - F_{\nu'}(z))\mathrm{d}z \\ &= \ell_2^2(\nu, \Pi_c\nu) + \ell_2^2(\Pi_c\nu, \nu') \\ &\quad + 2\Big(\int_{-\infty}^{\theta_1} + \sum_{i=1}^{m-1}\int_{\theta_i}^{\theta_{i+1}} + \int_{\theta_m}^{\infty}\Big)(F_\nu(z) - F_{\Pi_c\nu}(z))(F_{\Pi_c\nu}(z) - F_{\nu'}(z))\mathrm{d}z \\ &\overset{(b)}{=} \ell_2^2(\nu, \Pi_c\nu) + \ell_2^2(\Pi_c\nu, \nu') \end{aligned}$$

establishing the identity as required. Here, (a) follows by adding and subtracting $F_{\Pi_c\nu}$ inside the parentheses and expanding, and (b) follows by noting that on $(-\infty, \theta_1)$ and (θ_m, ∞), $F_{\Pi_c\nu} = F_{\nu'}$, and on each interval (θ_i, θ_{i+1}), $F_{\Pi_c\nu} - F_{\nu'}$ is constant and $F_{\Pi_c \nu}$ is constant and equals the average of F_ν on the interval, meaning that

$$\int_{\theta_i}^{\theta_{i+1}} (F_\nu(z) - F_{\Pi_c \nu}(z))(F_{\Pi_c \nu}(z) - F_{\nu'}(z))dz$$

$$= (F_{\Pi_c \nu}(\theta_i) - F_{\nu'}(\theta_i)) \int_{\theta_i}^{\theta_{i+1}} (F_\nu(z) - F_{\Pi_c \nu}(z))dz = 0\,,$$

as required. △

Remark 5.5 (Proof of Lemma 5.16). Assume that F_ν^{-1} is continuous at $\tau^* = \frac{a+b}{2}$; this is not necessary but simplifies the proof. For any τ,

$$\left|F_\nu^{-1}(\tau) - \theta\right| \tag{5.30}$$

is a convex function, and hence so is Equation 5.26. A subgradient[44] for Equation 5.30 is

$$g_\tau(\theta) = \begin{cases} 1 & \text{if } \theta < F_\nu^{-1}(\tau) \\ -1 & \text{if } \theta > F_\nu^{-1}(\tau) \\ 0 & \text{if } \theta = F_\nu^{-1}(\tau)\,. \end{cases}$$

A subgradient for Equation 5.26 is therefore

$$\begin{aligned} g_{[a,b]}(\theta) &= \int_{\tau=a}^{b} g_\tau(\theta) d\tau \\ &= \int_{\tau=a}^{F_\nu(\theta)} -1 \, \mathrm{d}\tau + \int_{\tau=F_\nu(\theta)}^{b} 1 \, \mathrm{d}\tau \\ &= -(F_\nu(\theta) - a) + (b - F_\nu(\theta)). \end{aligned}$$

Setting the subgradient to zero and solving for θ,

$$\begin{aligned} 0 &= a + b - 2F_\nu(\theta) \\ \implies F_\nu(\theta) &= \frac{a+b}{2} \\ \implies \theta &= F_\nu^{-1}\left(\frac{a+b}{2}\right). \end{aligned}$$

△

Remark 5.6 (Proof of Lemma 5.23). The argument follows the standard proof that finite-dimensional Euclidean projections onto closed convex sets are non-expansions. Throughout, we write Π for $\Pi_{\mathscr{F},\ell_2}$ for conciseness. We begin with

44. For a convex function $f: \mathbb{R} \to \mathbb{R}$, we say that $g: \mathbb{R} \to \mathbb{R}$ is a subgradient for f if for all $z_1, z_2 \in \mathbb{R}$, we have $f(z_2) \geq f(z_1) + g(z_1)(z_2 - z_1)$.

the observation that for any $\nu \in \mathscr{P}_{\ell_2}(\mathbb{R})$ and $\nu' \in \mathscr{F}$, we have

$$\int_{\mathbb{R}} (F_\nu(z) - F_{\Pi\nu}(z))(F_{\nu'}(z) - F_{\Pi\nu}(z))\mathrm{d}z \le 0\,, \tag{5.31}$$

since if not, we have $(1-\varepsilon)\Pi\nu + \varepsilon\nu' \in \mathscr{F}$ for all $\varepsilon \in (0, 1)$ by convexity, and

$$\begin{aligned}
&\ell_2^2(\nu, (1-\varepsilon)\Pi\nu + \varepsilon\nu') \\
&= \int_{\mathbb{R}} (F_\nu(z) - (1-\varepsilon)F_{\Pi\nu}(z) - \varepsilon F_{\nu'}(z))^2 \mathrm{d}z \\
&= \ell_2^2(\nu, \Pi\nu) - 2\varepsilon \int_{\mathbb{R}} (F_\nu(z) - F_{\Pi\nu}(z))(F_{\nu'}(z) - F_{\Pi\nu}(z))\mathrm{d}z + O(\varepsilon^2)\,.
\end{aligned}$$

This final line must be at least as great as $\ell_2^2(\nu, \Pi\nu)$ for all $\varepsilon \in (0, 1)$, by definition of Π. It must therefore be the case that Inequality 5.31 holds, since if not, we could select $\varepsilon > 0$ sufficiently small to make $\ell_2^2(\nu, (1-\varepsilon)\Pi\nu + \varepsilon\nu')$ smaller than $\ell_2^2(\nu, \Pi\nu)$.

Now take $\nu_1, \nu_2 \in \mathscr{P}_{\ell_2}(\mathbb{R})$. Applying the above inequality twice yields

$$\int_{\mathbb{R}} (F_{\nu_1}(z) - F_{\Pi\nu_1}(z))(F_{\Pi\nu_2}(z) - F_{\Pi\nu_1}(z))\mathrm{d}z \le 0\,,$$

$$\int_{\mathbb{R}} (F_{\nu_2}(z) - F_{\Pi\nu_2}(z))(F_{\Pi\nu_1}(z) - F_{\Pi\nu_2}(z))\mathrm{d}z \le 0\,.$$

Adding these inequalities then yields

$$\begin{aligned}
&\int_{\mathbb{R}} (F_{\nu_1}(z) - F_{\nu_2}(z) + F_{\Pi\nu_1}(z) - F_{\Pi\nu_2}(z))(F_{\Pi\nu_1}(z) - F_{\Pi\nu_2}(z))\mathrm{d}z \le 0 \\
&\Longrightarrow \ell_2^2(\Pi\nu_1, \Pi\nu_2) + \int_{\mathbb{R}} (F_{\nu_1}(z) - F_{\nu_2}(z))(F_{\Pi\nu_1}(z) - F_{\Pi\nu_2}(z))\mathrm{d}z \le 0 \\
&\Longrightarrow \ell_2^2(\Pi\nu_1, \Pi\nu_2) \le \int_{\mathbb{R}} (F_{\nu_2}(z) - F_{\nu_1}(z))(F_{\Pi\nu_1}(z) - F_{\Pi\nu_2}(z))\mathrm{d}z\,.
\end{aligned}$$

Applying the Cauchy–Schwarz inequality to the remaining integral then yields

$$\ell_2^2(\Pi\nu_1, \Pi\nu_2) \le \ell_2(\Pi\nu_1, \Pi\nu_2)\ell_2(\nu_1, \nu_2)\,,$$

from which the result follows by rearranging. △

Remark 5.7 (Proof of Banach's fixed point theorem (Theorem 5.26)). The map O has at most one fixed point by Proposition 4.5. It therefore suffices to exhibit a fixed point in M. Let $\beta \in [0, 1)$ be the contraction modulus of O, and let $U_0 \in M$. Consider the sequence $(U_k)_{k\ge 0}$ defined by $U_{k+1} = OU_k$ for $k \ge 0$. For any $l > k$, we have

$$d(U_l, U_k) \le \beta^k d(U_{l-k}, U_0)$$

$$\leq \beta^k \sum_{j=0}^{l-k-1} d(U_j, U_{j+1})$$

$$\leq \beta^k \sum_{j=0}^{l-k-1} \beta^j d(U_1, U_0)$$

$$\leq \frac{\beta^k}{1-\beta} d(U_1, U_0) .$$

Therefore, as $k \to \infty$, we have $d(U_l, U_k) \to 0$, so $(U_k)_{k\geq 0}$ is a Cauchy sequence. By completeness of (M, d), $(U_k)_{k\geq 0}$ has a limit $U^* \in M$. Finally, for any $k > 0$, we have

$$d(U^*, OU^*) \leq d(U^*, U_k) + d(U_k, OU^*) \leq d(U^*, U_k) + \beta d(U_{k-1}, U^*) ,$$

and as $d(U^*, U_k) \to 0$, we deduce that $d(U^*, OU^*) = 0$. Hence, U^* is the unique fixed point of O. △

5.13 Bibliographical Remarks

5.1. The term "dynamic programming" and the Bellman equation are due to Bellman (1957b). The relationship between the Bellman equation, value functions, and linear systems of equations is studied at length by Puterman (2014) and Bertsekas (2012). Bertsekas (2011) provides a treatment of iterative policy evaluation generalized to matrices other than discounted stochastic matrices. The advantages of the iterative process are well documented in the field and play a central role in the work of Sutton and Barto (2018), which is our source for the term "iterative policy evaluation."

5.2. Our notion of a probability distribution representation reflects the common principle in machine learning of modeling distributions with simple parametric families of distributions; see, for example, the books by MacKay (2003), Bishop (2006), Wainwright and Jordan (2008), and Murphy (2012). The Bernoulli representation was introduced in the context of distributional reinforcement learning, mostly as a curio, by Bellemare et al. (2017a). Normal approximations have been extensively used in reinforcement learning, often in Bayesian settings (Dearden et al. 1998; Engel et al. 2003; Morimura et al. 2010b; Lee et al. 2013). Similar to Equation 5.10, Morimura et al. (2010a) present a distributional Bellman operator in terms of cumulative distribution functions; see also Chung and Sobel (1987).

5.3. Li et al. (2022) consider what is effectively distributional dynamic programming with the empirical representation; they show that in the undiscounted, finite-horizon setting with reward distributions supported on finitely many integers, exact computation is tractable. Our notion of the empirical representation has roots in particle filtering (Doucet et al. 2001; Robert and Casella 2004; Brooks et al. 2011; Doucet and Johansen 2011) and is also a representation in modern variational inference algorithms (Liu and Wang 2016). The NP-hardness result (properly discussed in Remark 5.2) is new as given here, but Mannor and Tsitsiklis (2011) give a related result in the context of mean-variance optimization.

5.4. Sobel (1982) is usually noted as the source of the Bellman equation for the variance and its associated operator. The equation plays an important role in theoretical exploration (Lattimore and Hutter 2012; Azar et al. 2013). Tamar et al. (2016) study the variance equation in the context of function approximation.

5.5. The m-categorical representation was used in a distributional setting by Bellemare et al. (2017a), inspired by the success of categorical representations in generative modeling (van den Oord et al. 2016). Dabney et al. (2018b) introduced the quantile representation to avoid the inefficiencies in using a fixed set of evenly spaced locations, as well as deriving an algorithm more closely grounded in the Wasserstein distances.

Morimura et al. (2010a) used the m-particle representation to design a risk-sensitive distributional reinforcement learning algorithm. In a similar vein, Maddison et al. (2017) used the same representation in the context of exponential utility reinforcement learning. Both approaches are closely related to particle filtering and sequential Monte Carlo methods (Gordon et al. 1993; Doucet et al. 2001; Särkkä 2013; Naesseth et al. 2019; Chopin and Papaspiliopoulos 2020), which rely on stochastic sampling and resampling procedures, by contrast to the deterministic dynamic programming methods of this chapter.

5.6, 5.8. The categorical projection was originally proposed as an ad hoc solution to address the need to map the output of the distributional Bellman operator back onto the support of the distribution. Its description as the expectation of a triangular kernel was shown by Rowland et al. (2018), justifying its use from a theoretical perspective and providing the proof of Proposition 5.14. Lemma 5.16 is due to Dabney et al. (2018b).

5.7, 5.9–5.10. The language and analysis of projected operators is inherited from the theoretical analysis of linear function approximation in reinforcement learning; a canonical exposition may be found in Tsitsiklis and Van Roy (1997) and Lagoudakis and Parr (2003). Because the space of probability distributions is not a vector space, the analysis is somewhat different and, among other things, requires more technical care (as discussed in Chapter 4). A version of Theorem 5.28 in the special case of CDP appears in Rowland et al. (2018). Of note, in the linear function approximation setting, the main technical argument revolves around the noncontractive nature of the stochastic matrix P^π in a weighted L_2 norm, whereas here it is due to the c-homogeneity of the analysis metric (and does not involve P^π). See Chapter 9.

5.11. A discussion of the mean-preserving property is given in Lyle et al. (2019).

5.14 Exercises

Exercise 5.1. Suppose that a Markov decision process is acyclic, in the sense that for any policy π and nonterminal state $x \in \mathcal{X}$,

$$\mathbb{P}_\pi(X_t = x \mid X_0 = x) = 0 \text{ for all } t > 0 .$$

Consider applying iterative policy evaluation to this MDP, beginning with the initial condition $V(x_\varnothing) = 0$ for all terminal states $x_\varnothing \in \mathcal{X}$. Show that it converges to V^π in a finite number of iterations K, and give a bound on K. △

Exercise 5.2. Show that the normal representation (Definition 5.10) is closed under the distributional Bellman operator $\mathcal{T}^\pi$ under the following conditions:

(i) the policy is deterministic: $\pi(a \mid x) \in \{0, 1\}$;
(ii) the transition function is deterministic: $P_{\mathcal{X}}(x' \mid x, a) \in \{0, 1\}$; and
(iii) the rewards are normally distributed, $P_{\mathcal{R}}(\cdot \mid x, a) = \mathcal{N}(\mu_{x,a}, \sigma^2_{x,a})$. △

Exercise 5.3. In Algorithm 5.1, we made use of the fact that the return $G^\pi(x_\varnothing)$ from the terminal state $x_\varnothing$ is 0, arguing that this is a more effective procedure.

(i) Explain how this change affects the output of categorical and quantile dynamic programming, compared to the algorithm that explicitly maintains and computes a return-distribution estimate for $x_\varnothing$.
(ii) Explain how this changes affects the analysis given in Section 5.9 onward. △

Exercise 5.4. Provide counterexamples showing that if any of the conditions of the previous exercise do not hold, then the normal representation may not be closed under $\mathcal{T}^\pi$. △

Exercise 5.5. Consider a probability distribution $\nu \in \mathscr{P}_2(\mathbb{R})$ with probability density f_ν. For a normal distribution $\nu' \in \mathscr{F}_{\text{N}}$ with probability density $f_{\nu'}$, define the Kullback–Leibler divergence

$$\text{KL}(\nu \,\|\, \nu') = \int_{\mathbb{R}} f_\nu(z) \log \frac{f_\nu(z)}{f_{\nu'}(z)} \mathrm{d}z .$$

Show that the normal distribution $\hat{\nu} = \mathcal{N}(\mu, \sigma^2)$ minimizing $\text{KL}(\nu \,\|\, \hat{\nu})$ has parameters given by

$$\mu = \mathop{\mathbb{E}}_{Z\sim\nu}[Z] , \qquad \sigma^2 = \mathop{\mathbb{E}}_{Z\sim\nu}[(Z-\mu)^2] . \tag{5.32}$$

△

Exercise 5.6. Consider again a probability distribution $\nu \in \mathscr{P}_2(\mathbb{R})$ with instantiation Z, and for a normal distribution $\nu' \in \mathscr{F}_{\text{N}}$, define the *cross-entropy loss*

$$\text{CE}(\nu, \nu') = \mathop{\mathbb{E}}_{Z \sim \nu} [\log f_{\nu'}(z)].$$

Show that the normal distribution ν' that minimizes this cross-entropy loss has parameters given by Equation 5.32. Contrasting with the preceding exercise, explain why this result applies irrespective of whether ν' has a probability density. △

Exercise 5.7. Prove Lemma 5.6 from the definition of the pushforward distribution. △

Exercise 5.8. The naive implementation of Algorithm 5.1 requires $O(N_{\mathcal{X}}^{k+1} N_{\mathcal{R}}^k)$ memory to perform the kth iteration. Describe an implementation that reduces this cost to $O(N_{\mathcal{X}} N_{\mathcal{R}}^k)$. △

Exercise 5.9. Consider a Markov decision process in which the rewards are normally distributed:

$$P_{\mathcal{R}}(\cdot \mid x, a) = \mathcal{N}(\mu(x,a), \sigma^2(x,a)), \quad \text{for } x \in \mathcal{X}, a \in \mathcal{A}.$$

Suppose that we represent our distributions with the m-categorical representation, with projection in Cramér distance Π_{c}:

$$\eta(x) = \sum_{i=1}^{m} p_i(x) \delta_{\theta_i}.$$

(i) Show that

$$(\Pi_{\text{c}} \mathcal{T}^\pi \eta)(x)$$
$$= \sum_{a \in \mathcal{A}} \sum_{x' \in \mathcal{X}} \pi(a \mid x) P_{\mathcal{X}}(x' \mid x, a) \sum_{i=1}^{m} p_i(x') \mathbb{E}_\pi [\Pi_{\text{c}} \delta_{R + \gamma \theta_i} \mid X = x, A = a].$$

(ii) Construct a numerical integration scheme that approximates the terms

$$\mathbb{E}_\pi [\Pi_{\text{c}} \delta_{R + \gamma \theta_i} \mid X = x, A = a]$$

to ε precision on individual probabilities, for any x, a.

(iii) Use this scheme to construct a distributional dynamic programming algorithm for Markov decision processes with normally distributed rewards. What is its per-iteration computational cost?

(iv) Suppose that the support $\theta_1, \ldots, \theta_m$ is evenly spaced on $[\theta_{\text{MIN}}, \theta_{\text{MAX}}]$, that $\sigma^2(x,a) = 1$ for all x, a, and additionally, $\mu(x,a) \in [(1-\gamma)\theta_{\text{MIN}}, (1-\gamma)\theta_{\text{MAX}}]$. Analogous to the results of Section 5.10, bound the approximation error resulting from this algorithm, as a function of m, ε, the number of iterations K. △

Exercise 5.10. Prove that the deterministic projection of Section 3.5, defined in terms of a map to two neighboring particles, is equivalent to the triangular kernel formulation presented in Section 5.6. △

Exercise 5.11. Show that each of the representations of Definitions 5.11–5.13 is complete with respect to w_p, for $p \in [1, \infty]$. That is, for $\mathscr{F} \in \{\mathscr{F}_{\mathrm{Q},m}, \mathscr{F}_{\mathrm{C},m}, \mathscr{F}_{\mathrm{E},m}\}$, show that if $\nu \in \mathscr{P}_p(\mathbb{R})$ is the limit of a sequence of probability distributions $(\nu_k)_{k\geq 0} \in \mathscr{F}$, then $\nu \in \mathscr{F}$. △

Exercise 5.12. Show that the empirical representation $\mathscr{F}_{\mathrm{E}}$ (Definition 5.5) is not complete with respect to the 1-Wasserstein distance. Explain, concretely, why this causes difficulties in defining a distance-based projection onto $\mathscr{F}_{\mathrm{E}}$. △

Exercise 5.13. Explain what happens if the inverse cumulative distribution function F_ν^{-1} is not continuous in Lemma 5.16. When does that arise? What are the implications for the w_1-projection onto the m-quantile representation? △

Exercise 5.14. Implement distributional dynamic programming with the empirical representation in the programming language of your choice. Apply it to the deterministic Markov decision process depicted in Figure 5.6, for a finite number of iterations K, beginning with $\eta_0(x) = \delta_0$. Plot the supremum 1-Wasserstein distance between the iterates η_k and η^π as a function of k. △

Exercise 5.15. Prove that the m-categorical representation $\mathscr{F}_{\mathrm{C},m}$ is convex and complete with respect to the Cramér distance ℓ_2, for all $m \in \mathbb{N}^+$. △

Exercise 5.16. Show that the projected Bellman operator, instantiated with the m-quantile representation $\mathscr{F}_{\mathrm{Q},m}$ and the w_1-projection, is *not* a contraction in w_1. △

Exercise 5.17. Consider the metric space given by the open interval $(0, 1)$ equipped with the Euclidean metric on the real line, $d(x, y) = |x - y|$.

(i) Show that this metric space is not complete.

(ii) Construct a simple contraction map on this metric space that has no fixed point, illustrating the necessity of the condition of completeness in Banach's fixed-point theorem. △

Exercise 5.18. Let $p = \frac{\sqrt{2}}{2}$, and consider the probability distribution

$$\nu = (1 - p)\delta_0 + p\delta_1 .$$

Show that, for all $m \in \mathbb{N}^+$, the projection of ν onto the m-quantile representation, written $\Pi_Q \nu$, is such that

$$w_\infty(\Pi_Q \nu, \nu) = 1 .$$

Comment on the utility of the supremum-w_∞ metric in analyzing the convergence and the quality of the fixed point of quantile dynamic programming. △

Exercise 5.19. Consider the Bernoulli representation:

$$\{p\delta_0 + (1-p)\delta_1 : p \in [0,1]\} ;$$

this is also the categorical representation with $m = 2$, $\theta_1 = 0$, $\theta_2 = 1$. Consider the distributional dynamic programming obtained by combining this representation with a w_∞-projection.

(i) Describe the projection operator mathematically.

(ii) Consider a two-state, single-action MDP with states x, y with transition dynamics such that each state transitions immediately to the other and rewards that are deterministically 0. Show that with $\gamma > 1/2$, the distributional dynamic programming operator defined above is not a contraction mapping and in particular has multiple fixed points. △

Exercise 5.20. The purpose of this exercise is to develop a guarantee of the approximation error incurred by the fixed point $\hat{\eta}^\pi_Q$ of the projected distributional Bellman operator $\Pi_Q \mathcal{T}^\pi$ and the m-quantile representation; a version of this analysis originally appeared in Rowland et al. (2019). Here, we write $\mathscr{P}_B(\mathbb{R})$ for the space of probability distributions bounded on $[V_{\text{MIN}}, V_{\text{MAX}}]$.

(i) For a distribution $\nu \in \mathscr{P}_B(\mathbb{R})$ and the quantile projection $\Pi_Q : \mathscr{P}_1(\mathbb{R}) \to \mathscr{F}_{Q,m}$, show that we have

$$w_1(\Pi_Q \nu, \nu) \le \frac{V_{\text{MAX}} - V_{\text{MIN}}}{2m} .$$

(ii) Hence, using the triangle inequality, show that for any $\nu, \nu' \in \mathscr{P}_B(\mathbb{R})$, we have

$$w_1(\Pi_Q \nu, \Pi_Q \nu') \le w_1(\nu, \nu') + \frac{V_{\text{MAX}} - V_{\text{MIN}}}{m} .$$

Thus, while Π_Q is not an nonexpansion under w_1, per Exercise 5.16, it is in some sense "not far" from satisfying this condition.

(iii) Hence, using the triangle inequality with the return functions $\hat{\eta}^\pi_Q$, $\Pi_Q \eta^\pi$, and η^π, show that if $\eta^\pi \in \mathscr{P}_B(\mathbb{R})^{\mathcal{X}}$, then we have

$$\overline{w}_1(\hat{\eta}^\pi_Q, \eta^\pi) \le \frac{3(V_{\text{MAX}} - V_{\text{MIN}})}{2m(1-\gamma)} .$$

(iv) Finally, show that $w_1(\nu, \nu') \leq w_\infty(\nu, \nu')$ for any $\nu, \nu' \in \mathscr{P}_B(\mathbb{R})$. Thus, starting from the observation that

$$\overline{w}_1(\eta_k, \eta^\pi) \leq \overline{w}_1(\eta_k, \hat{\eta}^\pi_Q) + \overline{w}_1(\hat{\eta}^\pi_Q, \eta^\pi),$$

deduce that

$$\overline{w}_1(\eta_k, \eta^\pi) \leq \gamma^k (V_{\text{MAX}} - V_{\text{MIN}}) + \frac{3(V_{\text{MAX}} - V_{\text{MIN}})}{2m(1-\gamma)}.$$

△

Exercise 5.21. The aim of this exercise is to fill out the details in the reduction described in Remark 5.2.

(i) Consider an MDP where $r(x, a) > 0$ is the (deterministic) reward received from choosing action a in state x (that is, the distribution $P_{\mathcal{R}}(\cdot \mid x, a)$ is a Dirac delta at $r(x, a)$). Deduce that if the values $r(x, a)$ are distinct across state-action pairs and that the discount factor γ satisfies

$$\gamma < \frac{R_{\text{MIN}}}{R_{\text{MIN}} + R_{\text{MAX}}}$$

then there is an injective mapping from trajectories to returns.

(ii) Hence, show that for the class of MDPs described in Remark 5.2, the trajectories whose returns lie in the set

$$\bigcup_{\sigma \in \mathbb{S}_n} \left[\sum_{t=0}^{n-1} \gamma^t \sigma(t+1) + \gamma^n \sigma(1), \sum_{t=0}^{n-1} \gamma^t \sigma(t+1) + \gamma^n (\sigma(1) + 1) \right)$$

are precisely the trajectories whose initial $n+1$ states correspond to a Hamiltonian cycle. △

6 Incremental Algorithms

The concept of experience is central to reinforcement learning. Methods such as TD and categorical TD learning iteratively update predictions on the basis of transitions experienced by interacting with an environment. Such incremental algorithms are applicable in a wide range of scenarios, including those in which no model of the environment is known, or in which the model is too complex to allow for dynamic programming methods to be applied. Incremental algorithms are also often easier to implement. For these reasons, they are key in the application of reinforcement learning to many real-world domains.

With this ease of use, however, comes an added complication. In contrast to dynamic programming algorithms, which steadily make progress toward the desired goal, there is no guarantee that incremental methods will generate consistently improving estimates of return distributions from iteration to iteration. For example, an unusually high reward in a sampled transition may actually lead to a short-term degrading of the value estimate for the corresponding state. In practice, this requires making sufficiently small steps with each update, to average out such variations. In theory, the stochastic nature of incremental updates makes the analysis substantially more complicated than contraction mapping arguments.

This chapter takes a closer look at the behavior and design of incremental algorithms – distributional and not. Using the language of operators and probability distribution representations, we also formalize what it means for an incremental algorithm to perform well and discuss how to analyze its asymptotic convergence to the desired estimate.

6.1 Computation and Statistical Estimation

Iterative policy evaluation finds an approximation to the value function V^π by successively computing the iterates

$$V_{k+1} = T^\pi V_k\,, \tag{6.1}$$

defined by an arbitrary initial value function estimate $V_0 \in \mathbb{R}^{\mathcal{X}}$. We can also think of temporal-difference learning as computing an approximation to the value function V^π, albeit with a different mode of operation. To begin, recall from Chapter 3 the TD learning update rule:

$$V(x) \leftarrow (1-\alpha)V(x) + \alpha(r + \gamma V(x')) . \tag{6.2}$$

One of the aims of this chapter is to study the long-term behavior of the value function estimate V (and, eventually, of estimates produced by incremental, distributional algorithms).

At the heart of our analysis is the behavior of a single update. That is, for a fixed $V \in \mathbb{R}^{\mathcal{X}}$, we may understand the learning dynamics of temporal-difference learning by considering the random value function estimate V' defined via the sample transition model $(X = x, A, R, X')$:

$$V'(x) = (1-\alpha)V(x) + \alpha(R + \gamma V(X')) , \tag{6.3}$$
$$V'(y) = V(y) , \quad y \neq x .$$

There is a close connection between the expected effect of the update given by Equation 6.3 and iterative policy evaluation. Specifically, the expected value of the quantity $R + \gamma V(X')$ precisely corresponds to the application of the Bellman operator to V, evaluated at the source state x:

$$\mathbb{E}_\pi[R + \gamma V(X') \mid X = x] = (T^\pi V)(x) .$$

Consequently, in expectation TD learning adjusts its value function estimate at x in the direction given by the Bellman operator:

$$\mathbb{E}_\pi[V'(x) \mid X = x] = (1-\alpha)V(x) + \alpha(T^\pi V)(x) . \tag{6.4}$$

To argue that temporal-difference learning correctly finds an approximation to V^π, we must also be able to account for the random nature of TD updates. An effective approach is to rewrite Equation 6.3 as the sum of an expected target and a mean-zero noise term:

$$V'(x) = (1-\alpha)V(x) + \alpha\Big(\underbrace{(T^\pi V)(x)}_{\text{expected target}} + \underbrace{R + \gamma V(X') - (T^\pi V)(x)}_{\text{noise}} \Big) ; \tag{6.5}$$

with this decomposition, we may simultaneously analyze the mean dynamics of TD learning as well as the effect of the noise on the value function estimates. In the second half of the chapter, we will use Equation 6.5 to establish that under appropriate conditions, these dynamics can be controlled so as to guarantee the convergence of temporal-difference learning to V^π and analogously the convergence of categorical temporal-difference learning to the fixed point $\hat{\eta}^\pi_{\mathrm{c}}$.

6.2 From Operators to Incremental Algorithms

As illustrated in the preceding section, we can explain the behavior of temporal-difference learning (an incremental algorithm) by relating it to the Bellman operator. New incremental algorithms can also be obtained by following the reverse process – by deriving an update rule from a given operator. This technique is particularly effective in distributional reinforcement learning, where one often needs to implement incremental counterparts to a variety of dynamic programming algorithms. To describe how one might derive an update rule from an operator, we now introduce an abstract framework based on what is known as stochastic approximation theory.[45]

Let us assume that we are given a contractive operator $\mathcal{O}$ over some state-indexed quantity and that we are interested in determining the fixed point U^* of this operator. With dynamic programming methods, we obtain an approximation of U^* by computing the iterates

$$U_{k+1}(x) = (\mathcal{O}U_k)(x), \quad \text{for all } x \in \mathcal{X}.$$

To construct a corresponding incremental algorithm, we must first identify what information is available at each update; this constitutes our *sampling model.* For example, in the case of temporal-difference learning, this is the sample transition model (X, A, R, X'). For Monte Carlo algorithms, the sampling model is the random trajectory $(X_t, A_t, R_t)_{t \geq 0}$ (see Exercise 6.1). In the context of this chapter, we assume that the sampling model takes the form (X, Y), where X is the *source state* to be updated, and Y comprises all other information in the model, which we term the *sample experience.*

Given a step size α and realizations x and y of the source state variable X and sample experience Y, respectively, we consider incremental algorithms whose update rule can be expressed as

$$U(x) \leftarrow (1 - \alpha)U(x) + \alpha\hat{\mathcal{O}}(U, x, y). \tag{6.6}$$

Here, $\hat{\mathcal{O}}(U, x, y)$ is a *sample target* that may depend on the current estimate U. Typically, the particular setting we are in also imposes some limitation on the form of $\hat{\mathcal{O}}$. For example, when $\mathcal{O}$ is the Bellman operator T^π, although $\hat{\mathcal{O}}(U, x, y) = (\mathcal{O}U)(x)$ is a valid instantiation of Equation 6.6, its implementation might require knowledge of the environment's transition kernel and reward function. Implicit within Equation 6.6 is the notion that the space that the estimate U occupies supports a mixing operation; this will indeed be the case

45. Our treatment of incremental algorithms and their relation to stochastic approximation theory is far from exhaustive; the interested reader is invited to consult the bibliographical remarks.

for the algorithms we consider in this chapter, which work either with finite-dimensional parameter sets or probability distributions themselves.

With this framework in mind, the question is what makes a sensible choice for $\hat{O}$.

Unbiased update. An important case is when the sample target $\hat{O}$ can be chosen so that in expectation, it corresponds to the application of the operator O:

$$\mathbb{E}[\hat{O}(U, X, Y) \mid X = x] = (OU)(x) . \tag{6.7}$$

In general, when Equation 6.7 holds, the resulting incremental algorithm is also well behaved. More formally, we will see that under reasonable conditions, the estimates produced by such an algorithm are guaranteed to converge to U^* – a generalization of our earlier statement that temporal-difference learning converges to V^π.

Conversely, when the operator O can be expressed as an expectation over some function of U, X, and Y, then it is possible to derive a sample target simply by *substituting* the random variables involved with their realizations. In effect, we then use the sample experience to construct an unbiased estimate of $(OU)(x)$. As a concrete example, the TD target, expressed in terms of random variables, is

$$\hat{O}(V, X, Y) = R + \gamma V(X') ;$$

the corresponding update rule is

$$V(x) \leftarrow (1 - \alpha)V(x) + \alpha \underbrace{(r + \gamma V(x'))}_{\text{sample target}} .$$

In the next section, we will show how to use this approach to derive categorical temporal-difference learning (introduced in Chapter 3) from the categorical-projected Bellman operator.

Example 6.1. The *consistent Bellman operator* is an operator over state-action value functions based on the idea of making consistent choices at each state. At a high level, the consistent operator adds the constraint that actions that leave the state unchanged should be repeated. This operator is formalized as

$$T^\pi_{\mathrm{C}} Q(x, a) = \mathbb{E}_\pi \big[R + \gamma \max_{a' \in \mathcal{A}} Q(X', a')\mathbb{1}_{\{X' \neq x\}} + \gamma Q(x, a)\mathbb{1}_{\{X' = x\}} \mid X = x\big] .$$

Let (x, a, r, x') be drawn according to the sample transition model. The update rule derived by substitution is

$$Q(x, a) \leftarrow \begin{cases} (1 - \alpha)Q(x, a) + \alpha(r + \gamma \max_{a' \in \mathcal{A}} Q(x', a')) & \text{if } x' \neq x \\ (1 - \alpha)Q(x, a) + \alpha(r + \gamma Q(x, a)) & \text{otherwise.} \end{cases}$$

Compared to Q-learning (Section 3.7), the consistent update rule increases the *action gap* at each state, in the sense that its operator's fixed point Q^*_c has the property that for all $(x, a) \in \mathcal{X} \times \mathcal{A}$,

$$\max_{a' \in \mathcal{A}} Q^*_c(x, a') - Q^*_c(x, a) \geq \max_{a' \in \mathcal{A}} Q^*(x, a') - Q^*(x, a),$$

with strict inequality whenever $P_{\mathcal{X}}(x \mid x, a) > 0$. △

A general principle. Sometimes, expressing the operator O in the form of Equation 6.7 requires information that is not available to our sampling model. In this case, it is sometimes still possible to construct an update rule whose *repeated* application approximates the operator O. More precisely, given a fixed estimate $\tilde{U}$, with this approach we look for a sample target function $\hat{O}$ such that from a suitable initial condition, repeated updates of the form

$$U(x) \leftarrow (1 - \alpha)U(x) + \alpha\hat{O}(\tilde{U}, x, y)$$

lead to $U \approx O\tilde{U}$. In this case, a necessary condition for $\hat{O}$ to be a suitable sample target is that it should leave the fixed point U^* unchanged, in expectation:

$$\mathbb{E}_\pi[\hat{O}(U^*, X, Y) \mid X = x] = U^*(x).$$

In Section 6.4, we will introduce *quantile temporal-difference learning*, an algorithm that applies this principle to find the fixed point of the quantile-projected Bellman operator.

6.3 Categorical Temporal-Difference Learning

Categorical dynamic programming (CDP) computes a sequence $(\eta_k)_{k\geq 0}$ of return-distribution functions, defined by iteratively applying the projected distributional Bellman operator $\Pi_c\mathcal{T}^\pi$ to an initial return-distribution function η_0:

$$\eta_{k+1} = \Pi_c\mathcal{T}^\pi\eta_k.$$

As we established in Section 5.9, the sequence generated by CDP converges to the fixed point $\hat{\eta}^\pi_c$. Let us express this fixed point in terms of a collection of probabilities $((p^\pi_i(x))_{i=1}^m : x \in \mathcal{X})$ associated with m particles located at $\theta_1, \ldots, \theta_m$:

$$\hat{\eta}^\pi_c(x) = \sum_{i=1}^m p^\pi_i(x)\delta_{\theta_i}.$$

To derive an incremental algorithm from the categorical-projection Bellman operator, let us begin by expressing the projected distributional operator $\Pi_c\mathcal{T}^\pi$ in terms of an expectation over the sample transition $(X = x, A, R, X')$:

$$(\Pi_c\mathcal{T}^\pi\eta)(x) = \Pi_c\,\mathbb{E}_\pi\Big[(\mathrm{b}_{R,\gamma})_\#\eta^\pi(X') \mid X = x\Big]. \tag{6.8}$$

Following the line of reasoning from Section 6.2, in order to construct an unbiased sample target by substituting R and X' with their realizations, we need to rewrite Equation 6.8 with the expectation outside of the projection Π_c. The following establishes the validity of exchanging the order of these two operations.

Proposition 6.2. Let $\eta \in \mathscr{F}_{C,m}^{\mathcal{X}}$ be a return function based on the m-categorical representation. Then for each state $x \in \mathcal{X}$,

$$(\Pi_c \mathcal{T}^\pi \eta)(x) = \mathbb{E}_\pi \left[\Pi_c((\mathrm{b}_{R,\gamma})_\# \eta(X')) \mid X = x \right]. \qquad \triangle$$

Proposition 6.2 establishes that the projected operator $\Pi_c \mathcal{T}^\pi$ can be written in such a way that the substitution of random variables with their realizations can be performed. Consequently, we deduce that the random sample target

$$\hat{O}(\eta, x, (R, X')) = \Pi_c(\mathrm{b}_{R,\gamma})_\# \eta(X')$$

provides an unbiased estimate of $(\Pi_c \mathcal{T}^\pi \eta)(x)$. For a given realization (x, a, r, x') of the sample transition, this leads to the update rule[46]

$$\eta(x) \leftarrow (1-\alpha)\eta(x) + \alpha \underbrace{\Pi_c \left((\mathrm{b}_{r,\gamma})_\# \eta(x') \right)}_{\text{sample target}} . \tag{6.9}$$

The last part of the CTD derivation is to express Equation 6.9 in terms of the actual parameters being updated. These parameters are the probabilities $((p_i(x))_{i=1}^m : x \in \mathcal{X})$ of the return-distribution function estimate η:

$$\eta(x) = \sum_{i=1}^m p_i(x) \delta_{\theta_i} .$$

The sample target in Equation 6.9 is given by the pushforward transformation of a m-categorical distribution $(\eta(x'))$ followed by a categorical projection. As we demonstrated in Section 3.6, the projection of a such a transformed distribution can be expressed concisely from a set of coefficients $(\zeta_{i,j}(r) : i, j \in \{1, \dots, m\})$. In terms of the triangular and half-triangular kernels $(h_i)_{i=1}^m$ that define the categorical projection (Section 5.6), these coefficients are

$$\zeta_{i,j}(r) = h_i(\varsigma_m^{-1}(r + \gamma\theta_j - \theta_i)) . \tag{6.10}$$

With these coefficients, the update rule over the probability parameters is

$$p_i(x) \leftarrow (1-\alpha)p_i(x) + \alpha \sum_{j=1}^m \zeta_{i,j}(r) p_j(x') .$$

46. Although the action a is not needed to construct the sample target, we include it for consistency.

Our derivation illustrates how substituting random variables for their realizations directly leads to an incremental algorithm, provided we have the right operator to begin with. In many situations, this is simpler than the step-by-step process that we originally followed in Chapter 3. Because the random sample target is an unbiased estimate of the projected Bellman operator, it is also simpler to prove its convergence to the fixed point $\hat{\eta}^\pi_c$; in the second half of this chapter, we will in fact apply the same technique to analyze both temporal-difference learning and CTD.

Proof of Proposition 6.2. For a given $r \in \mathbb{R}$, $x' \in \mathcal{X}$, let us write

$$\tilde{\eta}(r, x') = (\mathrm{b}_{r,\gamma})_\# \eta(x') .$$

Fix $x \in \mathcal{X}$. For conciseness, let us define, for $z \in \mathbb{R}$,

$$\tilde{h}_i(z) = h_i(\varsigma_m^{-1}(z - \theta_i)) .$$

With this notation, we have

$$\begin{aligned}
\mathbb{E}_\pi \left[\Pi_c((\mathrm{b}_{R,\gamma})_\# \eta(X')) \mid X = x\right] &\overset{(a)}{=} \mathbb{E}_\pi \Big[\sum_{i=1}^m \delta_{\theta_i} \mathop{\mathbb{E}}_{Z \sim \tilde{\eta}(R,X')} [\tilde{h}_i(Z)] \mid X = x\Big] \\
&= \sum_{i=1}^m \delta_{\theta_i} \mathbb{E}_\pi \Big[\mathop{\mathbb{E}}_{Z \sim \tilde{\eta}(R,X')} [\tilde{h}_i(Z)] \mid X = x\Big] \\
&\overset{(b)}{=} \sum_{i=1}^m \delta_{\theta_i} \mathop{\mathbb{E}}_{Z' \sim (\mathcal{T}^\pi \eta)(x)} [\tilde{h}_i(Z')] \\
&= \Pi_c(\mathcal{T}^\pi \eta)(x) ,
\end{aligned}$$

where (a) follows by definition of the categorical projection in terms of the triangular and half-triangular kernels $(h_i)_{i=1}^m$ and (b) follows by noting that if the conditional distribution of $R + \gamma G(X')$ (where G is an instantiation of η independent of the sample transition (x, A, R, X')) given R, X' is $\tilde{\eta}(R, X') = (\mathrm{b}_{R,\gamma})_\# \eta(X')$, then the unconditional distribution of G when $X = x$ is $(\mathcal{T}^\pi \eta)(x)$. □

6.4 Quantile Temporal-Difference Learning

Quantile regression is a method for determining the quantiles of a probability distribution incrementally and from samples.[47] In this section, we develop an

47. More precisely, quantile regression is the problem of estimating a predetermined set of quantiles of a collection of probability distributions. By extension, in this book, we also use "quantile regression" to refer to the incremental method that solves this problem.

algorithm that aims to find the fixed point $\hat{\eta}^\pi_{\text{Q}}$ of the quantile-projected Bellman operator $\Pi_{\text{Q}}\mathcal{T}^\pi$ via quantile regression.

To begin, suppose that given $\tau \in (0, 1)$, we are interested in estimating the τth quantile of a distribution ν, corresponding to $F_\nu^{-1}(\tau)$. Quantile regression maintains an estimate θ of this quantile. Given a sample z drawn from ν, it adjusts θ according to

$$\theta \leftarrow \theta + \alpha(\tau - \mathbb{1}_{\{z<\theta\}}) . \tag{6.11}$$

One can show that quantile regression follows the negative gradient of the quantile loss[48]

$$\begin{aligned}\mathcal{L}_\tau(\theta) &= (\tau - \mathbb{1}_{\{z<\theta\}})(z-\theta) \\ &= |\mathbb{1}_{\{z<\theta\}} - \tau| \times |z-\theta| .\end{aligned} \tag{6.12}$$

In Equation 6.12, the term $|\mathbb{1}_{\{z<\theta\}} - \tau|$ is an *asymmetric step size* that is either τ or $1-\tau$, according to whether the sample z is greater or smaller than θ, respectively. When $\tau < 0.5$, samples greater than θ have a lesser effect on it than samples smaller than θ; the effect is reversed when $\tau > 0.5$. The update rule in Equation 6.11 will continue to adjust the estimate until the equilibrium point θ^* is reached (Exercise 6.4 asks you to visualize the behavior of quantile regression with different distributions). This equilibrium point is the location at which smaller and larger samples have an equal effect in expectation. At that point, letting $Z \sim \nu$, we have

$$\begin{aligned} 0 &= \mathbb{E}\left[\tau - \mathbb{1}_{\{Z<\theta^*\}}\right] \\ &= \tau - \mathbb{E}\left[\mathbb{1}_{\{Z<\theta^*\}}\right] \\ &= \tau - \mathbb{P}(Z<\theta^*) \\ \implies \mathbb{P}(Z<\theta^*) &= \tau \\ \implies \theta^* &= F_\nu^{-1}(\tau) . \end{aligned} \tag{6.13}$$

For ease of exposition, in the final line we assumed that there is a unique $z \in \mathbb{R}$ for which $F_\nu(z) = \tau$; Remark 6.1 discusses the general case.

Now, let us consider applying quantile regression to find a m-quantile approximation to the return-distribution function (ideally, the fixed point $\hat{\eta}^\pi_{\text{Q}}$). Recall that a m-quantile return-distribution function $\eta \in \mathscr{F}^{\mathcal{X}}_{\text{Q},m}$ is parameterized by the locations $((\theta_i(x))_{i=1}^m : x \in \mathcal{X})$:

$$\eta(x) = \frac{1}{m}\sum_{i=1}^{m} \delta_{\theta_i(x)} .$$

48. More precisely, Equation 6.11 updates θ in the direction of the negative gradient of $\mathcal{L}_\tau$ provided that $\mathbb{P}_{Z\sim\nu}(Z=\theta) = 0$. This holds trivially if ν is a continuous probability distribution.

Now, the quantile projection $\Pi_Q \nu$ of a probability distribution ν is given by

$$\Pi_Q \nu = \frac{1}{m} \sum_{i=1}^{m} \delta_{F_\nu^{-1}(\tau_i)}, \qquad \tau_i = \frac{2i-1}{2m} \text{ for } i = 1, \ldots, m .$$

Given a source state $x \in \mathcal{X}$, the general idea is to perform quantile regression for all location parameters $(\theta_i)_{i=1}^m$ simultaneously, using the quantile levels $(\tau_i)_{i=1}^m$ and samples drawn from $(\mathcal{T}^\pi \eta)(x)$. To this end, let us momentarily introduce a random variable J uniformly distributed on $\{1, \ldots, m\}$. By Proposition 4.11, we have

$$\mathcal{D}_\pi(R + \gamma \theta_J(X') \mid X = x) = (\mathcal{T}^\pi \eta)(x) . \tag{6.14}$$

Given a realized transition (x, a, r, x'), we may therefore construct m sample targets $(r + \gamma \theta_j(x'))_{j=1}^m$. Applying Equation 6.11 to these targets leads to the update rule

$$\theta_i(x) \leftarrow \theta_i(x) + \frac{\alpha}{m} \sum_{j=1}^{m} \Big(\tau_i - \mathbb{1}\{r + \gamma \theta_j(x') < \theta_i(x)\} \Big), \quad i = 1, \ldots m . \tag{6.15}$$

This is the *quantile temporal-difference learning* (QTD) algorithm. A concrete instantiation in the online case is summarized by Algorithm 6.1, by analogy with the presentation of categorical temporal-difference learning in Algorithm 3.4. Note that applying Equation 6.15 requires computing a total of m^2 terms per location; when m is large, an alternative is to instead use a single term from the sum in Equation 6.15, with j sampled uniformly at random. Interestingly enough, for m sufficiently small, the per-step cost of QTD is less than the cost of sorting the full distribution $(\mathcal{T}^\pi \eta)(x)$ (which has up to $N_\mathcal{X} N_\mathcal{R} m$ particles). This suggests that the quantile regression approach to the projection step may be useful even in the context of distributional dynamic programming.

The use of quantile regression to derive QTD can be seen as an instance of the principle introduced at the end of Section 6.2. Suppose that we consider an initial return function

$$\eta_0(x) = \frac{1}{m} \sum_{j=1}^{m} \delta_{\theta_j^0(x)} .$$

If we substitute the sample target in Equation 6.15 by a target constructed from this initial return function, we obtain the update rule

$$\theta_i(x) \leftarrow \theta_i(x) + \frac{\alpha}{m} \sum_{j=1}^{m} \Big(\tau_i - \mathbb{1}\{r + \gamma \theta_j^0(x') < \theta_i(x)\} \Big), \quad i = 1, \ldots m . \tag{6.16}$$

By inspection, we see that Equation 6.16 corresponds to quantile regression applied to the problem of determining, for each state $x \in \mathcal{X}$, the quantiles of

Algorithm 6.1: Online quantile temporal-difference learning

Algorithm parameters: step size $\alpha \in (0, 1]$,
policy $\pi : \mathcal{X} \to \mathscr{P}(\mathcal{A})$,
number of quantiles m,
initial locations $((\theta_i^0(x))_{i=1}^m : x \in \mathcal{X})$

$\theta_i(x) \leftarrow \theta_i^0(x)$ for $i = 1, \ldots, m$, $x \in \mathcal{X}$
$\tau_i \leftarrow \frac{2i-1}{2m}$ for $i = 1, \ldots, m$
Loop for each episode:
  Observe initial state x_0
  Loop for $t = 0, 1, \ldots$
    Draw a_t from $\pi(\cdot \mid x_t)$
    Take action a_t, observe r_t, x_{t+1}
    for $i = 1, \ldots, m$ **do**
      $\theta_i' \leftarrow \theta_i(x_t)$
      for $j = 1, \ldots, m$ **do**
        if x_{t+1} is terminal **then**
          $g \leftarrow r_t$
        else
          $g \leftarrow r_t + \gamma\theta_j(x_{t+1})$
        $\theta_i' \leftarrow \theta_i' + \frac{\alpha}{m}(\tau_i - \mathbb{1}\{g < \theta_i(x_t)\})$
      end for
    end for
    for $i = 1, \ldots, m$ **do**
      $\theta_i(x_t) \leftarrow \theta_i'$
    end for
  until x_{t+1} is terminal
end

the distribution $(\mathcal{T}^\pi\eta_0)(x)$. Consequently, one may think of quantile temporal-difference learning as performing an update that would converge to the quantiles of the target distribution, if that distribution were held fixed.

Based on this observation, we can verify that QTD is a reasonable distributional reinforcement learning algorithm by considering its behavior at the fixed point

$$\hat{\eta}_Q^\pi = \Pi_Q \mathcal{T}^\pi \hat{\eta}_Q^\pi ,$$

the solution found by quantile dynamic programming. Let us denote the parameters of this return function by $\hat{\theta}_i^\pi(x)$, for $i = 1, \ldots, m$ and $x \in \mathcal{X}$. For a given state x, consider the intermediate target

$$\tilde{\eta}(x) = (\mathcal{T}^\pi \hat{\eta}^\pi_{\text{Q}})(x) .$$

Now, by definition of the quantile projection operator, we have

$$\hat{\theta}^\pi_i(x) = F^{-1}_{\tilde{\eta}(x)}\left(\tfrac{2i-1}{2m}\right) .$$

However, by Equation 6.13, we also know that the quantile regression update rule applied at $\hat{\theta}^\pi_i(x)$ with $\tau_i = \frac{2i-1}{2m}$ leaves the parameter unchanged in expectation. In other words, the collection of locations $(\hat{\theta}^\pi_i(x))_{i=1}^m$ is a fixed point of the expected quantile regression update, and consequently the return function $\hat{\eta}^\pi_{\text{Q}}$ is a solution of the quantile temporal-difference learning algorithm. This gives some intuition that is it indeed a valid learning rule for distributional reinforcement learning with quantile representations.

Before concluding, it is useful to illustrate why the straightforward approach taken to derive categorical temporal-difference learning, based on unbiased operator estimation, cannot be applied to the quantile setting. Recall that the quantile-projected operator takes the form

$$(\Pi_{\text{Q}}\mathcal{T}^\pi\eta)(x) = \Pi_{\text{Q}}\, \mathbb{E}_\pi\left[(\mathrm{b}_{R,\gamma})_\# \eta^\pi(X') \mid X = x\right] . \tag{6.17}$$

As the following example shows, exchanging the expectation and projection results in a different operator, one whose fixed point is not $\hat{\eta}^\pi_{\text{Q}}$. Consequently, we cannot substitute random variables for their realizations, as was done in the categorical setting.

Example 6.3. Consider an MDP with a single state x, single action a, transition dynamics so that x transitions back to itself, and immediate reward distribution $\mathcal{N}(0,1)$. Given $\eta(x) = \delta_0$, we have $(\mathcal{T}^\pi\eta)(x) = \mathcal{N}(0,1)$, and hence the projection via Π_{Q} onto $\mathscr{F}_{\text{Q},m}$ with $m = 1$ returns a Dirac delta on the median of this distribution: δ_0.

In contrast, the sample target $(\mathrm{b}_{R,\gamma})_\#\eta(X')$ is δ_R, and so the projection of this target via Π_{Q} remains δ_R. We therefore have

$$\mathbb{E}_\pi[\Pi_{\text{Q}}(\mathrm{b}_{R,\gamma})_\#\eta)(X') \mid X = x] = \mathbb{E}_\pi[\delta_R \mid X = x] = \mathcal{N}(0,1) ,$$

which is distinct from the result of the projected operator, $(\Pi_{\text{Q}}\mathcal{T}^\pi\eta)(x) = \delta_0$. △

6.5 An Algorithmic Template for Theoretical Analysis

In the second half of this chapter, we present a theoretical analysis of a class of incremental algorithms that includes the incremental Monte Carlo algorithm (see Exercise 6.9), temporal-difference learning, and the CTD algorithm. This analysis builds on the contraction mapping theory developed in Chapter 4 but also accounts for the randomness introduced by the use of sample targets in the update rule, via stochastic approximation theory. Compared to the analysis of dynamic programming algorithms, the main technical challenge lies in characterizing the effect of this randomness on the learning process.

To begin, let us view the output of the temporal-difference learning algorithm after k updates as a value function estimate V_k. Extending the discussion from Section 6.1, this estimate is a random quantity because it depends on the particular sample transitions observed by the agent and possibly the randomness in the agent's choices.[49] We are particularly interested in the sequence of random estimates $(V_k)_{k\geq 0}$. From an initial estimate V_0, this sequence is formally defined as

$$V_{k+1}(X_k) = (1-\alpha_k)V_k(X_k) + \alpha_k(R_k + \gamma V_k(X'_k))$$
$$V_{k+1}(x) = V_k(x) \quad \text{if } x \neq X_k\,,$$

where $(X_k, A_k, R_k, X'_k)_{k\geq 0}$ is the sequence of random transitions used to calculate the TD updates. In our analysis, the object of interest is the limiting point of this sequence, and we seek to answer the question: does the algorithm's estimate converge to the value function V^π? We consider the limiting point because any single update may or may not improve the accuracy of the estimate V_k at the source state X_k. We will show that, under the right conditions, the sequence $(V_k)_{k\geq 0}$ converges to V^π. That is,

$$\lim_{k\to\infty} \left|V_k(x) - V^\pi(x)\right| = 0, \quad \text{for all } x \in \mathcal{X}\,.$$

More precisely, the above holds with probability 1: with overwhelming odds, the variables $X_0, R_0, X'_0, X_1, \ldots$ are drawn in such a way that $V_k \to V^\pi$.[50]

We will prove a more general result that holds for a family of incremental algorithms whose sequence of estimates can be expressed by the template

$$U_{k+1}(X_k) = (1-\alpha_k)U_k(X_k) + \alpha_k\hat{O}(U_k, X_k, Y_k)$$
$$U_{k+1}(x) = U_k(x) \quad \text{if } x \neq X_k\,. \tag{6.18}$$

Here, X_k is the (possibly random) source state at time k, $\hat{O}(U_k, X_k, Y_k)$ is the sample target, and α_k is an (also possibly random) step size. As in Section 6.2, the sample experience Y_k describes the collection of random variables used to construct the sample target: for example, a sample trajectory or a sample transition (X_k, A_k, R_k, and X'_k).

Under this template, the estimate U_k describes the collection of variables maintained by the algorithm and constitutes its "prediction." More specifically,

49. In this context, we even allow the step size α_k to be random.

50. Put negatively, there may be realizations of $X_0, R_0, X'_0, X_1, \ldots$ for which the sequence $(V_k)_{k\geq 0}$ does not converge, but the set of such realizations has zero probability.

it is a state-indexed collection of m-dimensional real-valued vectors, written $U_k \in \mathbb{R}^{\mathcal{X}\times m}$. In the case of the TD algorithm, $m = 1$ and $U_k = V_k$.

We assume that there is an operator $\mathcal{O} : \mathbb{R}^{\mathcal{X}\times m} \to \mathbb{R}^{\mathcal{X}\times m}$ whose unique fixed point is the quantity to be estimated by the incremental algorithm. If we denote this fixed point by U^*, this implies that

$$\mathcal{O}U^* = U^*.$$

We further assume the existence of a base norm $\|\cdot\|$ over $\mathbb{R}^m$, extended to the space of estimates according to

$$\|U\|_\infty = \sup_{x\in\mathcal{X}} \|U(x)\|,$$

such that $\mathcal{O}$ is a contraction mapping of modulus β with respect to the metric induced by $\|\cdot\|_\infty$. For TD learning, $\mathcal{O} = T^\pi$ and the base norm is simply the absolute value; the contractivity of T^π was established by Proposition 4.4.

Within this template, there is some freedom in how the source state X_k is selected. Formally, X_k is assumed to be drawn from a time-varying distribution ξ_k that may depend on all previously observed random variables up to but excluding time k, as well as the initial estimate U_0. That is,

$$X_k \sim \xi_k(X_{0:k-1}, Y_{0:k-1}, \alpha_{0:k-1}, U_0)\,.$$

This includes scenarios in which source states are drawn from a fixed distribution $\xi \in \mathscr{P}(\mathcal{X})$, enumerated in a round-robin manner, or selected in proportion to the magnitude of preceding updates (called *prioritized replay*; see Moore and Atkeson 1993; Schaul et al. 2016). It also accounts for the situation in which states are sequentially updated along a sampled trajectory, as is typical of online algorithms.

We further assume that the sample target is an unbiased estimate of the operator $\mathcal{O}$ applied to U_k and evaluated at X_k. That is, for all $x \in \mathcal{X}$ for which $\mathbb{P}(X_k = x) > 0$,

$$\mathbb{E}\,[\hat{\mathcal{O}}(U_k, X_k, Y_k) \mid X_{0:k-1}, Y_{0:k-1}, \alpha_{0:k-1}, U_0, X_k = x] = (\mathcal{O}U_k)(x)\,.$$

This implies that Equation 6.18 can be expressed in terms of a mean-zero noise w_k, similar to our derivation in Section 6.1:

$$U_{k+1}(X_k) = (1-\alpha_k)U_k(X_k) + \alpha_k\Big[(\mathcal{O}U_k)(X_k) + \underbrace{(\hat{\mathcal{O}}(U_k, X_k, Y_k) - (\mathcal{O}U_k)(X_k)}_{w_k}\Big]\,.$$

Because w_k is zero in expectation, this assumption guarantees that, on average, the incremental algorithm must make progress toward the fixed point U^*. That is, if we fix the source state $X_k = x$ and step size α_k, then

$$\mathbb{E}\,[U_{k+1}(x) \mid X_{0:k-1}, Y_{0:k-1}, \alpha_{0:k-1}, X_k = x, \alpha_k] \tag{6.19}$$

$$= (1 - \alpha_k)U_k(x) + \alpha_k \mathbb{E}\left[(OU_k)(x) + w_k \mid X_k = x\right]$$
$$= (1 - \alpha_k)U_k(x) + \alpha_k(OU_k)(x)\,.$$

By choosing an appropriate sequence of step sizes $(\alpha_k)_{k\geq 0}$ and under a few additional technical conditions, we can in fact provide the stronger guarantee that the sequence of iterates $(U_k)_{k\geq 0}$ converges to U^* w.p. 1, as the next section illustrates.

6.6 The Right Step Sizes

To understand the role of step sizes in the learning process, consider an abstract algorithm described by Equation 6.18 and for which

$$\hat{O}(U_k, X_k, Y_k) = (OU_k)(X_k)\,.$$

In this case, the noise term w_k is always zero and can be ignored: the abstract algorithm adjusts its estimate directly toward OU_k. Here we should take the step sizes $(\alpha_k)_{k\geq 0}$ to be large in order to make maximal progress toward U^*. For $\alpha_k = 1$, we obtain a kind of dynamic programming algorithm that updates its estimate one state at a time and whose convergence to U^* can be reasonably easily demonstrated; conversely, taking $\alpha_k < 1$ must in some sense slow down the learning process.

In general, however, the noise term is not zero and cannot be neglected. In this case, large step sizes amplify w_k and prevent the algorithm from converging to U^* (consider, in the extreme, what happens when $\alpha_k = 1$). A suitable choice of step size must therefore balance rapid learning progress and eventual convergence to the right solution.

To illustrate what "suitable choice" might mean in practice, let us distill the issue down to its essence and consider the process that estimates the mean of a distribution $\nu \in \mathscr{P}_1(\mathbb{R})$ according to the incremental update

$$V_{k+1} = (1 - \alpha_k)V_k + \alpha_k R_k\,, \tag{6.20}$$

where $(R_k)_{k\geq 0}$ are i.i.d. random variables distributed according to ν. For concreteness, let us assume that $\nu = \mathcal{N}(0, 1)$, so that we would like $(V_k)_{k\geq 0}$ to converge to 0.

Suppose that the initial estimate is $V_0 = 0$ (the desired solution) and consider three step size schedules: $\alpha_k = 0.1$, $\alpha_k = {}^{1}\!/_{k+1}$, and $\alpha_k = {}^{1}\!/_{(k+1)^2}$. Figure 6.1 illustrates the sequences of estimates obtained by applying the incremental update with each of these schedules and a single, shared sequence of realizations of the random variables $(R_k)_{k\geq 0}$.

The ${}^{1}\!/_{k+1}$ schedule corresponds to the right step size schedule for the incremental Monte Carlo algorithm (Section 3.2), and accordingly, we observe that it

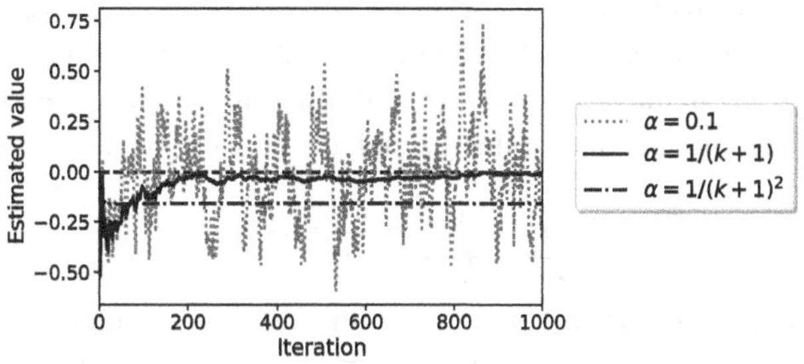

Figure 6.1
The behavior of a simple incremental update rule 6.20 for estimating the expected value of a normal distribution. Different curves represent the sequence of estimates obtained from different step size schedules. The ground truth ($V = 0$) is indicated by the dashed line.

is converging to the correct expected value.[51] By contrast, the constant schedule continues to exhibit variations over time, as the noise is not sufficiently averaged out. The quadratic schedule ($1/(k+1)^2$) decreases too quickly and the algorithm settles on an incorrect prediction.

To prove the convergence of algorithms that fit the template described in Section 6.5, we will require that the sequence of step sizes satisfies the *Robbins–Monro conditions* (Robbins and Monro 1951). These conditions formalize the range of step sizes that are neither too small nor too large and hence guarantee that the algorithm must eventually find the solution U^*. As with the source state X_k, the step size α_k at a given time k may be random, and its distribution may depend on X_k, $X_{0:k-1}$, $\alpha_{0:k-1}$, and $Y_{0:k-1}$ but not the sample experience Y_k. As in the previous section, these conditions should hold with probability 1.

Condition 1: not too small. In the example above, taking $\alpha_k = 1/(k+1)^2$ results in premature convergence of the estimate (to the wrong solution). This is because when the step sizes decay too quickly, the updates made by the algorithm may not be of large enough magnitude to reach the fixed point of interest. To avoid this situation, we require that $(\alpha_k)_{k\geq 0}$ satisfy

$$\sum_{k\geq 0} \alpha_k \mathbb{1}_{\{X_k = x\}} = \infty, \qquad \text{for all } x \in \mathcal{X}.$$

Implicit in this assumption is also the idea that every state should be updated infinitely often. This assumption is violated, for example, if there is a state x

51. In our example, V_k is the average of k i.i.d. normal random variables and is itself normally distributed. Its standard deviation can be computed analytically and is equal to $1/\sqrt{k}$ ($k \geq 1$). This implies that after $k = 1000$ iterations, we expect V_k to be in the range $\pm 3 \times 1/\sqrt{k} = \pm 0.111$, because 99.7 percent of a normal random variable's probability is within three standard deviations of its mean. Compare with Figure 6.1.

and time K after which $X_k \neq x$, for all $k \geq K$. For a reasonably well-behaved distribution of source states, this condition is satisfied for constant step sizes, including $\alpha_k = 1$: in the absence of noise, it is possible to make rapid progress toward the fixed point. On the other hand, it disallows $\alpha_k = 1/(k+1)^2$, since

$$\sum_{k=0}^{\infty} \frac{1}{(k+1)^2} = \frac{\pi^2}{6} < \infty .$$

Condition 2: not too large. Figure 6.1 illustrates how, with a constant step size and in the presence of noise, the estimate $U_k(x)$ continues to vary substantially over time. To avoid this issue, the step size should be decreased so that individual updates result in progressively smaller changes in the estimate. To achieve this, the second requirement on the step size sequence $(\alpha_k)_{k\geq 0}$ is

$$\sum_{k\geq 0} \alpha_k^2 \mathbb{1}_{\{X_k = x\}} < \infty , \qquad \text{for all } x \in \mathcal{X} .$$

In reinforcement learning, a simple step size schedule that satisfies both of these conditions is

$$\alpha_k = \frac{1}{N_k(X_k) + 1} , \tag{6.21}$$

where $N_k(x)$ is the number of updates to a state x up to but not including algorithm time k. We encountered this schedule in Section 3.2 when deriving the incremental Monte Carlo algorithm. As will be shown in the following sections, this schedule is also sufficient for the convergence of TD and CTD algorithms.[52] Exercise 6.7 asks you to verify that Equation 6.21 satisfies the Robbins–Monro conditions and investigates other step size sequences that also satisfy these conditions.

6.7 Overview of Convergence Analysis

Provided that an incremental algorithm satisfies the template laid out in Section 6.5, with a step size schedule that satisfies the Robbins–Monro conditions, we can prove that the sequence of estimates produced by this algorithm must converge to the fixed point U^* of the implied operator $\mathcal{O}$. Before presenting the proof in detail, we illustrate the main bounding-box argument underlying the proof.

Let us consider a two-dimensional state space $\mathcal{X} = \{x_1, x_2\}$ and an incremental algorithm for estimating a 1-dimensional quantity ($m = 1$). As per the template, we consider a contractive operator $\mathcal{O} : \mathbb{R}^{\mathcal{X}} \to \mathbb{R}^{\mathcal{X}}$ given by $\mathcal{O}U =$

52. Because the process of bootstrapping constructs sample targets that are not in general unbiased with regards to the value function V^π, the optimal step size schedule for TD learning decreases at a rate that is slower than $1/k$. See bibliographical remarks.

$(0.8U(x_2), 0.8U(x_2))$; note that the fixed point of $\mathcal{O}$ is $U^* = (0, 0)$. At each time step, a source state (x_1 or x_2) is chosen uniformly at random and the corresponding estimate is updated. The step sizes are $\alpha_k = (k+1)^{-0.7}$, satisfying the Robbins–Monro conditions.

Suppose first that the sample target is noiseless. That is,

$$\hat{\mathcal{O}}(U_k, X_k, Y_k) = 0.8U_k(x_2).$$

In this case, each iteration of the algorithm contracts along a particular coordinate. Figure 6.2a illustrates a sequence $(U_k)_{k\geq 0}$ defined by the update equations

$$U_{k+1}(X_k) = (1-\alpha_k)U_k(X_k) + \alpha_k \hat{\mathcal{O}}(X_k, U_k, Y_k)$$
$$U_{k+1}(x) = U_k(x), \quad x \neq X_k .$$

As shown in the figure, the algorithm makes steady (if not direct) progress toward the fixed point with each update. To prove that $(U_k)_{k\geq 0}$ converges to U^*, we first show that the error $\|U_k - U^*\|_\infty$ is bounded by a fixed quantity for all $k \geq 0$ (indicated by the outermost dashed-line square around the fixed point $U^* = 0$ in Figure 6.2a). The argument then proceeds inductively: if U_k lies within a given radius of the fixed point for all k greater than some K, then there is some $K' \geq K$ for which, for all $k \geq K'$, it must lie within the next smallest dashed-line square. We will see that this follows by contractivity of $\mathcal{O}$ and the first Robbins–Monro condition. Provided that the diameter of these squares shrinks to zero, then this establishes convergence of U_k to U^*.

Now consider adding noise to the sample target, such that

$$\hat{\mathcal{O}}(U_k, X_k, Y_k) = 0.8U_k(y) + w_k .$$

For concreteness, let us take w_k to be an independent random variable with distribution $\mathcal{U}([-1, 1])$. In this case, the behavior of the sequence $(U_k)_{k\geq 0}$ is more complicated (Figure 6.2b). The sequence $(U_k)_{k\geq 0}$ no longer follows a neat path to the fixed point but can behave somewhat more erratically. Nevertheless, the *long-term* behavior exhibited by the algorithm bears similarity to the noiseless case: overall, progress is made toward the fixed point U^*.

The proof of convergence follows the same pattern as for the noiseless case: prove inductively that if $\|U_k - U^*\|_\infty$ is eventually bounded by some fixed quantity $B_l \in \mathbb{R}$, then $\|U_k - U^*\|_\infty$ is eventually bounded by a smaller quantity B_{l+1}. As in the noiseless case, this argument is depicted by the concentric squares in Figure 6.2c. Again, if these diameters shrink to zero, this also establishes convergence of U_k to U^*.

Because noise can increase the error between the estimate U_k and the fixed point U^* at any given time step, to guarantee convergence we need to progressively decrease the step size α_k. The second Robbins–Monro condition is sufficient for this purpose, and with it the inductive step can be proven with a more delicate argument. One additional challenge is that the base case (that $\sup_{k\geq 0} \|U_k - U^*\|_\infty < \infty$) is no longer immediate; a separate argument is required to establish this fact. This property is called the *stability* of the sequence $(U_k)_{k\geq 0}$ and is often one of the harder aspects of the proof of convergence of incremental algorithms.

We conclude this section with a result that is crucial in understanding the influence of noise in the algorithm. In the analysis carried out in this chapter, it is the only result whose proof requires tools from advanced probability theory.[53]

Proposition 6.4. Let $(Z_k)_{k\geq 0}$ be a sequence of random variables taking values in $\mathbb{R}^m$ and $(\alpha_k)_{k\geq 0}$ be a collection of step sizes. Given $\bar{Z}_0 = 0$, consider the sequence defined by

$$\bar{Z}_{k+1} = (1 - \alpha_k)\bar{Z}_k + \alpha_k Z_k .$$

Suppose that the following conditions hold with probability 1:

$$\mathbb{E}[Z_k \mid Z_{0:k-1}, \alpha_{0:k}] = 0 , \quad \sup_{k\geq 0} \mathbb{E}[\|Z_k\|^2 \mid Z_{0:k-1}, \alpha_{0:k}] < \infty ,$$

$$\sum_{k=0}^{\infty} \alpha_k = \infty , \quad \sum_{k=0}^{\infty} \alpha_k^2 < \infty .$$

Then $\bar{Z}_k \to 0$ with probability 1. △

The proof is given in Remark 6.2; here, we provide some intuition that can be gleaned without consulting the proof.

First, parallels can be drawn with the strong law of large numbers. Expanding the definition of $\bar{Z}_k$ yields

$$\bar{Z}_k = (1 - \alpha_k) \cdots (1 - \alpha_1)\alpha_0 Z_0 + (1 - \alpha_k) \cdots (1 - \alpha_2)\alpha_1 Z_1 + \cdots + \alpha_k Z_k .$$

Thus, $\bar{Z}_k$ is a weighted average of the mean-zero terms $(Z_l)_{l=0}^k$. If $\alpha_k = 1/k+1$, then we obtain the usual uniformly weighted average that appears in the strong law of large numbers. We also note that unlike the standard strong law of large numbers, the noise terms $(Z_l)_{l=0}^k$ are not necessarily independent. Nevertheless, it seems reasonable that this sequence should exhibit similar behavior to the averages that appear in the strong law of large numbers. This also provides further intuition

53. Specifically, the supermartingale convergence theorem; the result is a special case of the Robbins–Siegmund theorem (Robbins and Siegmund 1971).

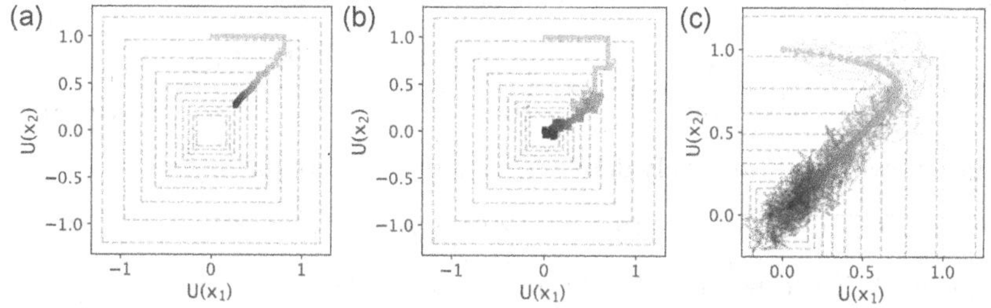

Figure 6.2
(a) Example behavior of the iterates $(U_k)_{k\geq 0}$ in the noiseless case. The shading indicates the iteration number from light ($k = 0$) through to dark ($k = 1000$). **(b)** Example behavior of the iterates $(U_k)_{k\geq 0}$ in the general case. **(c)** Behavior of iterates for ten random seeds, with the noiseless (expected) behavior overlaid.

for the conditions of Proposition 6.4. If the variance of individual noise terms $\alpha_k Z_k$ is too great, the weighted average may not "settle down" as the number of terms increases. Similarly, if $\sum_{k=0}^{\infty} \alpha_k$ is too small, the initial noise term Z_0 will have too large an influence over the weighted average, even as $k \to \infty$.

Second, for readers familiar with stochastic gradient descent, we can rewrite the update scheme as

$$\bar{Z}_{k+1} = \bar{Z}_k + \alpha_k(-\bar{Z}_k + Z_k)\,.$$

This is a stochastic gradient update for the loss function $\mathcal{L}(\bar{Z}_k) = 1/2\|\bar{Z}_k\|^2$ (the minimizer of which is the origin). The negative gradient of this loss is $-\bar{Z}_k$, Z_k is a mean-zero perturbation of this gradient, and α_k is the step size used in the update. Proposition 6.4 can therefore be interpreted as stating that stochastic gradient descent on this specific loss function converges to the origin, under the required conditions on the step sizes and noise. It is perhaps surprising that understanding the behavior of stochastic gradient descent in this specific setting is enough to understand the general class of algorithms expressed by Equation 6.18.

6.8 Convergence of Incremental Algorithms*

We now provide a formal run-through of the arguments of the previous section and explain how they apply to temporal-difference learning algorithms. We begin by introducing some notation to simplify the argument. We first define a per-state step size that incorporates the choice of source state X_k:

$$\alpha_k(x) = \begin{cases} \alpha_k & \text{if } x = X_k \\ 0 & \text{otherwise,} \end{cases} \qquad w_k(x) = \begin{cases} \hat{O}(U_k, X_k, Y_k) - (OU_k)(X_k) & \text{if } x = X_k \\ 0 & \text{otherwise.} \end{cases}$$

This allows us to recursively define $(U_k)_{k\geq 0}$ in a single equation:

$$U_{k+1}(x) = (1 - \alpha_k(x))U_k(x) + \alpha_k(x)((OU_k)(x) + w_k(x)) \,. \tag{6.22}$$

Equation 6.22 encapsulates all of the random variables – $(X_k)_{k\geq 0}, (Y_k)_{k\geq 0}, (\alpha_k)_{k\geq 0}$ – which together determine the sequence of estimates $(U_k)_{k\geq 0}$.

It is useful to separate the effects of the noise into two separate cases: one in which the noise has been "processed" by an application of the contractive mapping O and one in which the noise has not been passed through this mapping. To this end, we introduce the cumulative external noise vectors $(W_k(x) : k \geq 0, x \in \mathcal{X})$. These are random vectors, with each $W_k(x)$ taking values in $\mathbb{R}^m$, defined by

$$W_0(x) = 0 \,, \quad W_{k+1}(x) = (1 - \alpha_k(x))W_k(x) + \alpha_k(x)w_k(x) \,.$$

We also introduce the sigma-algebras $\mathcal{F}_k = \sigma(X_{0:k}, \alpha_{0:k}, Y_{0:k-1})$; these encode the information available to the learning agent just prior to sampling Y_k and applying the update rule to produce U_{k+1}.

We now list several assumptions we will require of the algorithm to establish the convergence result. Recall that $\|\cdot\|$ is the base norm identified in Section 6.5, which gives rise to the supremum extension $\|\cdot\|_\infty$. In particular, we assume that O is a β-contraction mapping in the metric induced by $\|\cdot\|_\infty$.

Assumption 6.5 (Robbins–Monro conditions)**.** For each $x \in \mathcal{X}$, the step sizes $(\alpha_k(x))_{k\geq 0}$ satisfy $\sum_{k\geq 0} \alpha_k(x) = \infty$ and $\sum_{k\geq 0} \alpha_k(x)^2 < \infty$ with probability 1. △

The second assumption encompasses the mean-zero condition described in Section 6.5 and introduces an additional condition that the variance of this noise, conditional on the state of the algorithm, does not grow too quickly.

Assumption 6.6. The noise terms $(w_k(x) : k \geq 0, x \in \mathcal{X})$ satisfy $\mathbb{E}[w_k(x) \mid \mathcal{F}_k] = 0$ with probability 1, and $\mathbb{E}[\|w_k(x)\|^2 \mid \mathcal{F}_k] \leq C_1 + C_2\|U_k\|_\infty^2$ w.p. 1, for some constants $C_1, C_2 \geq 0$, for all $x \in \mathcal{X}$ and $k \geq 0$. △

We would like to use Proposition 6.4 to show that the cumulative external noise $(W_k(x))_{k\geq 0}$ is well behaved, and then use this intermediate result to establish the convergence of the sequence $(U_k)_{k\geq 0}$ itself. The proposition is *almost* applicable to the sequence $(W_k(x))_{k\geq 0}$. The difficulty is that the proposition stipulates that the individual noise terms Z_k have *bounded* variance, whereas Assumption 6.6 only bounds the conditional expectation of $\|w_k(x)\|^2$ in terms of $\|U_k\|_\infty^2$, which a priori may be unbounded. Unfortunately, in temporal-difference learning algorithms, the update variance typically does scale with the magnitude of current estimates, so this is not an assumption that we can weaken. To get around this difficulty, we first establish the boundedness of the sequence $(U_k)_{k\geq 0}$, as described informally in the previous section, often referred to as *stability* in the stochastic approximation literature.

Proposition 6.7. Suppose Assumptions 6.5 and 6.6 hold. Then there is a finite random variable B such that $\sup_{k\geq 0} \|U_k\|_\infty < B$ with probability 1. △

Proof. The idea of the proof is to work with a "renormalized" version of the noises $(w_k(x))_{k\geq 0}$ to which Proposition 6.4 can be applied. First, we show that the contractivity of O means that when U is sufficiently far from 0, O contracts the iterate back toward 0. To make this precise, we first observe that

$$\|OU\|_\infty \leq \|OU - U^*\|_\infty + \|U^*\|_\infty \leq \beta\|U - U^*\|_\infty + \|U^*\|_\infty \leq \beta\|U\|_\infty + D,$$

where $D = (1+\beta)\|U^*\|_\infty$. Let $\bar{B} > D/1-\beta$ so that $\bar{B} > \beta\bar{B} + D$, and define $\psi \in (\beta, 1)$ by $\beta\bar{B} + D = \psi\bar{B}$. Now that that for any U with $\|U\|_\infty \geq \bar{B}$, we have

$$\|OU\|_\infty \leq \beta\|U\|_\infty + D = \beta\|U\|_\infty + (\psi - \beta)\bar{B} \leq \beta\|U\|_\infty + (\psi - \beta)\|U\|_\infty = \psi\|U\|_\infty .$$

Second, we construct a sequence of bounds $(\bar{B}_k)_{k\geq 0}$ related to the iterates $(U_k)_{k\geq 0}$ as follows. It will be convenient to introduce $1 + \varepsilon = \psi^{-1}$, the inverse of the contraction factor ψ above. Take $\bar{B}_0 = \max(\bar{B}, \|U_0\|_\infty)$, and iteratively define

$$\bar{B}_{k+1} = \begin{cases} \bar{B}_k & \text{if } \|U_{k+1}\|_\infty \leq (1+\varepsilon)\bar{B}_k \\ \min\{(1+\varepsilon)^l \bar{B}_0 : l \in \mathbb{N}^+, \|U_{k+1}\|_\infty \leq (1+\varepsilon)^l \bar{B}_0\} & \text{otherwise} . \end{cases}$$

Thus, the $(\bar{B}_k)_{k\geq 0}$ define a kind of soft "upper envelope" on $(\|U_k\|_\infty)_{k\geq 0}$, which are only updated when a norm exceeds the previous bound $\bar{B}_k$ by a factor at least $(1+\varepsilon)$. Note that $(\|U_k\|_\infty)_{k\geq 0}$ is unbounded if and only if $\bar{B}_k \to \infty$.

We now use the $(\bar{B}_k)_{k\geq 0}$ to define a "renormalized" noise sequence $(\tilde{w}_k)_{k\geq 0}$ to which Proposition 6.4 can be applied. We set $\tilde{w}_k = w_k / \bar{B}_k$, and define $\tilde{W}_k$ iteratively by $\tilde{W}_0 = w_0$, and

$$\tilde{W}_{k+1}(x) = (1 - \alpha_k(x))\tilde{W}_k(x) + \alpha_k(x)\tilde{w}_k(x) .$$

By Assumption 6.6, we still have $\mathbb{E}[\tilde{w}_k \mid \mathcal{F}_k] = 0$ and obtain that $\mathbb{E}[\|\tilde{w}_k\|_\infty^2 \mid \mathcal{F}_k]$ is uniformly bounded. Using Assumption 6.5, Proposition 6.4 now applies, and we deduce that $\tilde{W}_k \to 0$ with probability 1.

In particular, there is a (random) time K such that $\|\tilde{W}_k(x)\| < \varepsilon$ for all $k \geq K$ and $x \in \mathcal{X}$. Now supposing $\bar{B}_k \to \infty$, we may also take K so that $\|U_K\|_\infty \leq \bar{B}_K$. We will now prove by induction that for all $k \geq K$, we have both $\|U_k\|_\infty \leq (1+\varepsilon)\bar{B}_K$ and $\|U_k - W_k\|_\infty < \bar{B}_K$; the base case is clear from the above. For the inductive step, suppose for some $k \geq K$, we have both $||U_k||_\infty \leq (1+\varepsilon)\bar{B}_K$ and $||U_k - W_k||_\infty < \bar{B}_K$. Now observe that

$$
\begin{aligned}
&||U_{k+1}(x) - W_{k+1}(x)|| \\
=&||(1-\alpha_k(x))U_k(x) + \alpha_k(x)(OU_k)(x) + \alpha_k(x)w_k(x) - W_{k+1}(x)|| \\
\leq&(1-\alpha_k(x))||U_k(x) - W_k(x)|| + \alpha_k(x)||(OU_k)(x)|| \\
\leq&(1-\alpha_k(x))\bar{B}_K + \alpha_k(x)(\beta||U_k||_\infty + D) \\
\leq&(1-\alpha_k(x))\bar{B}_K + \alpha_k(x)\psi(1+\varepsilon)\bar{B}_K \\
\leq&\bar{B}_K\,.
\end{aligned}
$$

And additionally,

$$||U_{k+1}||_\infty \leq ||U_{k+1} - W_{k+1}||_\infty + ||W_{k+1}||_\infty \leq \bar{B}_K + \varepsilon\bar{B}_K = (1+\varepsilon)\bar{B}_K$$

as required. □

We can now establish the convergence of the cumulative external noise.

Proposition 6.8. Suppose Assumptions 6.5 and 6.6 hold. Then the external noise $W_k(x)$ converges to 0 with probability 1, for each $x \in \mathcal{X}$. △

Proof. By Proposition 6.7, there exists a finite random variable B such that $||U_k||_\infty \leq B$ for all $k \geq 0$. We therefore have $\mathbb{E}[||w_k(x)||^2 \mid \mathcal{F}_k] \leq C_1 + C_2B^2 =: B'$ w.p. 1 for all $x \in \mathcal{X}$ and $k \geq 0$, by Assumption 6.6. Proposition 6.4 therefore applies to give the conclusion. □

With this result in hand, we now prove the central result of this section, using the stability result as the base case for the inductive argument intuitively explained above.

Theorem 6.9. Suppose Assumptions 6.5 and 6.6 hold. Then $U_k \to U^*$ with probability 1. △

Proof. By Proposition 6.7, there is a finite random variable B_0 such that $||U_k - U^*||_\infty < B_0$ for all $k \geq 0$ w.p. 1. Let $\varepsilon > 0$ such that $\beta + 2\varepsilon < 1$; we will show by induction that if $B_l = (\beta + 2\varepsilon)B_{l-1}$ for all $l \geq 1$, then for each $l \geq 0$, there is a (possibly random) finite time K_l such that $||U_k - U^*||_\infty < B_l$ for all $k \geq K_l$, which proves the theorem.

To prove this claim by induction, let $l \geq 0$ and suppose there is a random finite time K_l such that $||U_k - U^*||_\infty < B_l$ for all $k \geq K_l$ w.p. 1. Now let $x \in \mathcal{X}$, and

$k \geq K_l$. We have

$$\begin{aligned}
U_{k+1}(x) - U^*(x) - W_{k+1}(x) \\
&= (1 - \alpha_k(x))U_k(x) + \alpha_k(x)((OU_k)(x) + w_k(x)) \\
&\qquad - U^*(x) - (1 - \alpha_k(x))W_k(x) - \alpha_k(x)w_k(x) \\
&= (1 - \alpha_k(x))(U_k(x) - U^*(x) - W_k(x)) + \alpha_k(x)((OU_k)(x) - U^*(x)).
\end{aligned}$$

Since O is a contraction mapping under $\|\cdot\|_\infty$ with fixed point U^* and contraction modulus β, we have $\|(OU_k)(x) - U^*(x)\| \leq \beta\|U_k - U^*\|_\infty < \beta B_l$, and so

$$\|U_{k+1}(x) - U^*(x) - W_{k+1}(x)\| \leq (1 - \alpha_k(x))\|U_k(x) - U^*(x) - W_k(x)\| + \alpha_k(x)\beta B_l .$$

Letting $\Delta_k(x) = \|U_k(x) - U^*(x) - W_k(x)\|$, we then have

$$\begin{aligned}
&\Delta_{k+1}(x) \leq (1 - \alpha_k(x))\Delta_k(x) + \alpha_k(x)\beta B_l \\
\implies &\Delta_{k+1}(x) - \beta B_l \leq (1 - \alpha_k(x))(\Delta_k(x) - \beta B_l).
\end{aligned}$$

Telescoping this inequality from K_l to k yields

$$\Delta_{k+1}(x) - \beta B_l \leq \left[\prod_{s=K_l}^{k}(1 - \alpha_s(x))\right](\Delta_{K_l}(x) - \beta B_l).$$

If $\Delta_{K_l}(x) - \beta B_l \leq 0$, then $\Delta_k(x) \leq \beta B_l$ for all $k \geq K_l$. If not, then we can use the inequality $1 - x \leq e^{-x}$ (applied to $x = \alpha_k \geq 0$) to deduce

$$\Delta_{k+1}(x) - \beta B_l \leq \exp\left(-\sum_{s=K_l}^{k} \alpha_k(x)\right)(\Delta_{K_l}(x) - \beta B_l),$$

and since $\sum_{s\geq 0} \alpha_s(x) = \infty$ by assumption, the right-hand side tends to 0. Therefore, there exists a random finite time after which $\Delta_k(x) \leq (\beta + \varepsilon)B_l$. Since $\mathcal{X}$ is finite, there is a random finite time after which this holds for all $x \in \mathcal{X}$. Finally, since $W_k(x) \to 0$ under $\|\cdot\|$ for all $x \in \mathcal{X}$ w.p. 1 by Proposition 6.8, there is a random finite time after which $\|W_k(x)\| \leq \varepsilon B_l$ for all $x \in \mathcal{X}$. Letting $K_{l+1} \geq K_l$ be the maximum of all these random times, we therefore have that for $k \geq K_{l+1}$, for all $x \in \mathcal{X}$,

$$\|U_k(x) - U^*(x)\| \leq \|U_k(x) - U^*(x) - W_k(x)\| + \|W_k(x)\| \leq (\beta + \varepsilon)B_l + \varepsilon B_l = B_{l+1},$$

as required. □

6.9 Convergence of Temporal-Difference Learning*

We can now apply Theorem 6.9 to demonstrate the convergence of the sequence of value function estimates produced by TD learning. Formally, we consider a stream of sample transitions $(X_k, A_k, R_k, X'_k)_{k\geq 0}$, along with associated step sizes

$(\alpha_k)_{k\geq 0}$, that satisfy the Robbins–Monro conditions (Assumption 6.5) and give rise to zero-mean noise terms (Assumption 6.6). More precisely, we assume there are sequences of functions $(\xi_k)_{k\geq 0}$ and $(\nu_k)_{k\geq 0}$ such that our sample model takes the following form (for $k \geq 0$):

$$X_k \,|\, (X_{0:k-1}, A_{0:k-1}, R_{0:k-1}, X'_{0:k-1}, \alpha_{0:k-1}) \sim \xi_k(X_{0:k-1}, A_{0:k-1}, R_{0:k-1}, X'_{0:k-1}, \alpha_{0:k-1});$$
$$\alpha_k \,|\, (X_{0:k}, A_{0:k-1}, R_{0:k-1}, X'_{0:k-1}, \alpha_{0:k-1}) \sim \nu_k(X_{0:k}, A_{0:k-1}, R_{0:k-1}, X'_{0:k-1}, \alpha_{0:k-1});$$
$$A_k \,|\, (X_{0:k}, A_{0:k-1}, R_{0:k-1}, X'_{0:k-1}, \alpha_{0:k}) \sim \pi(\cdot \mid X_k);$$
$$R_k \,|\, (X_{0:k}, A_{0:k}, R_{0:k-1}, X'_{0:k-1}, \alpha_{0:k}) \sim P_{\mathcal{R}}(\cdot \mid X_k, A_k);$$
$$X'_k \,|\, (X_{0:k}, A_{0:k}, R_{0:k}, X'_{0:k-1}, \alpha_{0:k}) \sim P_{\mathcal{X}}(\cdot \mid X_k, A_k)\,. \tag{6.23}$$

A generative, or algorithmic, perspective on this model is that at each update step k, a source state X_k and step size α_k are selected on the basis of all previously observed random variables (possibly using an additional source of randomness to make this selection), and the variables (A_k, R_k, X'_k) are sampled according to π and the environment dynamics, conditionally independent of all random variables already observed given X_k. Readers may compare this with the model equations in Section 2.3 describing the joint distribution of a trajectory generated by following the policy π. As discussed in Sections 6.5 and 6.6, this is fairly flexible model that allows us to analyze a variety of learning schemes.

Theorem 6.10. Consider the value function iterates $(V_k)_{k\geq 0}$ defined by some initial estimate V_0 and satisfying

$$V_{k+1}(X_k) = (1 - \alpha_k)V_k(X_k) + \alpha_k(R_k + V_k(X'_k))$$
$$V_{k+1}(x) = V_k(x) \quad \text{if } x \neq X_k\,,$$

where $(X_k, A_k, R_k, X'_k)_{k\geq 0}$ is a sequence of transitions. Suppose that:

(a) The source states $(X_k)_{k\geq 0}$ and step sizes $(\alpha_k)_{k\geq 0}$ satisfy the Robbins–Monro conditions: w.p. 1, for all $x \in \mathcal{X}$,

$$\sum_{k=0}^{\infty} \alpha_k \mathbb{1}_{\{X_k = x\}} = \infty\,, \qquad \sum_{k=0}^{\infty} \alpha_k^2 \mathbb{1}_{\{X_k = x\}} < \infty\,.$$

(b) The joint distribution of $(X_k, A_k, R_k, X'_k)_{k\geq 0}$ is an instance of the sampling model expressed in Equation 6.23.

(c) The reward distributions for all state-action pairs have finite variance.

Then $V_k \to V^\pi$ with probability 1. △

Theorem 6.10 gives formal meaning to our earlier assertion that the convergence of incremental reinforcement learning algorithms can be guaranteed for a variety of source state distributions. Interestingly enough, the condition on the source state distribution appears only implicitly, through the Robbins–Monro conditions: effectively, what matters is not so much when the states are updated but rather the "total amount of step size" by which the estimate may be moved.

Proof. We first observe that the temporal-difference algorithm described in the statement is an instance of the abstract algorithm described in Section 6.5, by taking $U_k = V_k$, $m = 1$, $O = T^\pi$, $Y_k = (A_k, R_k, X'_k)$, and $\hat{O}(U_k, X_k, Y_k) = R_k + \gamma V_k(X'_k)$. The base norm $\|\cdot\|$ is simply the absolute value on $\mathbb{R}$. In this case, O is a γ-contraction on $\mathbb{R}^\mathcal{X}$ by Proposition 4.4, with fixed point V^π, and the noise w_k is equal to $R_k + \gamma V_k(X'_k) - (T^\pi U_k)(X_k)$, by the decomposition in Equation 6.5. It therefore remains to check that Assumptions 6.5 and 6.6 hold; Theorem 6.9 then applies to give the result. Assumption 6.5 is immediate from the conditions of the theorem. To see that Assumption 6.6 holds, first note that

$$\begin{aligned}\mathbb{E}[w_k \mid \mathcal{F}_k] &= \mathbb{E}_\pi[R_k + \gamma V_k(X'_k) - (T^\pi V_k)(X_k) \mid \mathcal{F}_k] \\ &= \mathbb{E}_\pi[R_k + \gamma V_k(X'_k) - (T^\pi V_k)(X_k) \mid X_k, V_k] \\ &= 0\,,\end{aligned}$$

since conditional on (X_k, V_k), the expectation of $R + \gamma V_k(X'_k)$ is $(T^\pi V_k)(X_k)$. Additionally, we note that

$$\begin{aligned}\mathbb{E}[|w_k|^2 \mid \mathcal{F}_k] &= \mathbb{E}[|w_k|^2 \mid X_k, V_k] \\ &= \mathbb{E}[|R + \gamma V_k(X'_k) - (T^\pi V_k)(X_k)|^2 \mid X_k, V_k] \\ &\leq 2\Big(\mathbb{E}[|R + \gamma V_k(X'_k)|^2 \mid X_k, V_k] + (T^\pi V_k)(X_k)^2\Big) \\ &\leq C_1 + C_2\|V_k\|_\infty^2\,,\end{aligned}$$

for some $C_1, C_2 > 0$. □

6.10 Convergence of Categorical Temporal-Difference Learning*

Let us now consider proving the convergence of categorical TD learning by means of Theorem 6.9. Writing CTD in terms of a sequence of return distribution estimates, we have

$$\begin{aligned}\eta_{k+1}(X_k) &= (1 - \alpha_k)\eta_k(X_k) + \alpha_k(\Pi_c(\mathrm{b}_{R_k,\gamma})_\#\eta(X'_k)) \\ \eta_{k+1}(x) &= \eta_k(x) \quad \text{if } x \neq X_k\,.\end{aligned} \tag{6.24}$$

Following the principles of the previous section, we may decompose the update at X_k into an operator term and a noise term:

$$\begin{aligned}\eta_{k+1}(X_k) = (1 - \alpha_k)\eta_k(X_k) \\ + \alpha_k\Big(\underbrace{(\Pi_c\mathcal{T}^\pi\eta)(X_k)}_{(OU)(X_k)} + \underbrace{\Pi_c(\mathrm{b}_{R_k,\gamma})_\#\eta(X'_k) - (\Pi_c\mathcal{T}^\pi\eta)(X_k)}_{w_k}\Big)\,.\end{aligned} \tag{6.25}$$

Assuming that R_k and X'_k are drawn appropriately, this decomposition is sensible by virtue of Proposition 6.2, in the sense that the expectation of the sample target is the projected distributional Bellman operator:

$$\mathbb{E}_\pi \left[\Pi_c((\mathrm{b}_{R_k,\gamma})_\# \eta(X'_k)) \mid X_k = x\right] = (\Pi_c \mathcal{T}^\pi \eta)(x) \quad \text{for all } x\,. \tag{6.26}$$

With this decomposition, the noise term is not a probability distribution but rather a *signed distribution*; this is illustrated in Figure 6.3. Informally speaking, a signed distribution may assign negative "probabilities" and may not integrate to one (we will revisit signed distributions in Chapter 9). Based on Proposition 6.2, we may intuit that w_k is mean-zero noise, where "zero" here is to be understood as a special signed distribution.

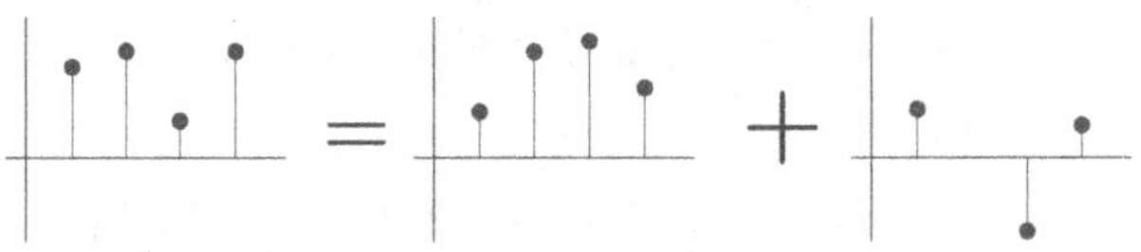

Figure 6.3
The sample target in a categorical TD update **(a)** can be decomposed into an expected update specified by the operator $\Pi_c \mathcal{T}^\pi$ **(b)** and a mean-zero signed distribution **(c)**.

However, expressing the CTD update rule in terms of signed distributions is not sufficient to apply Theorem 6.9. This is because the theorem requires that the iterates $(U_k(x))_{k \geq 0}$ be elements of $\mathbb{R}^m$, whereas $(\eta_k(x))_{k \geq 0}$ are probability distributions. To address this issue, we leverage the fact that categorical distributions are represented by a finite number of parameters and view their cumulative distribution functions as in vectors in $\mathbb{R}^m$.

Recall that $\mathscr{F}^{\mathcal{X}}_{\mathrm{C},m}$ is the space of m-categorical return distribution functions. To invoke Theorem 6.9, we construct an *isometry* $\mathcal{I}$ between $\mathscr{F}^{\mathcal{X}}_{\mathrm{C},m}$ and a certain subset of $\mathbb{R}^{\mathcal{X} \times m}$. For a categorical return function $\eta \in \mathscr{F}^{\mathcal{X}}_{\mathrm{C},m}$, write

$$\mathcal{I}(\eta) = \left(F_{\eta(x)}(\theta_i) : x \in \mathcal{X}, i \in \{1, \ldots, m\}\right) \in \mathbb{R}^{\mathcal{X} \times m}\,,$$

where as before $\theta_1, \ldots, \theta_m$ denotes the locations of the m particles whose probabilities are parameterized in $\mathscr{F}_{\mathrm{C},m}$. The isometry $\mathcal{I}$ maps return functions to elements of $\mathbb{R}^{\mathcal{X} \times m}$ describing the corresponding cumulative distribution functions (CDFs), evaluated at these particles. The image of $\mathscr{F}^{\mathcal{X}}_{\mathrm{C},m}$ under $\mathcal{I}$ is

$$\mathbb{R}^{\mathcal{X}}_{\mathcal{I}} = \{z \in \mathbb{R}^m : 0 \leq z_1 \leq \cdots \leq z_m = 1\}^{\mathcal{X}}.$$

The inverse map

$$\mathcal{I}^{-1} : \mathbb{R}_{\mathcal{I}}^{\mathcal{X}} \to \mathscr{F}_{\mathrm{C},m}^{\mathcal{X}}$$

maps vectors describing the cumulative distribution functions of categorical return functions back to their distributions.

With this construction, the metric induced by the L^2 norm $\|\cdot\|_2$ over $\mathbb{R}^m$ is proportional to the Cramér distance between probability distributions and is readily extended to $\mathbb{R}^{\mathcal{X}\times m}$. That is, for $\eta, \eta' \in \mathscr{F}_{\mathrm{C},m}^{\mathcal{X}}$, we have

$$\left\|\mathcal{I}(\eta) - \mathcal{I}(\eta')\right\|_{2,\infty} = \sup_{x\in\mathcal{X}} \left\|(\mathcal{I}(\eta))(x) - (\mathcal{I}(\eta'))(x)\right\|_2 = \frac{1}{\varsigma_m}\bar{\ell}_2(\eta, \eta').$$

We will prove the convergence of the sequence $(\eta_k)_{k\geq 0}$ defined by Equation 6.24 to the fixed point of the projected operator $\Pi_{\mathrm{c}}\mathcal{T}^\pi$ by applying Theorem 6.9 to the sequence $(\mathcal{I}(\eta_k))_{k\geq 0}$ and the L^2 metric and arguing (by isometry) that the original sequence must also converge. An important additional property of $\mathcal{I}$ is that it commutes with expectations, in the following sense.

Lemma 6.11. The isometry $\mathcal{I} : \mathscr{F}_{\mathrm{C},m}^{\mathcal{X}} \to \mathbb{R}_{\mathcal{I}}^{\mathcal{X}}$ is an affine map. That is, for any $\eta, \eta' \in \mathscr{F}_{\mathrm{C},m}^{\mathcal{X}}$ and $\alpha \in [0, 1]$,

$$\mathcal{I}(\alpha\eta + (1-\alpha)\eta') = \alpha\mathcal{I}(\eta) + (1-\alpha)\mathcal{I}(\eta').$$

As a result, if η is a random return-distribution function, then we have

$$\mathbb{E}[\mathcal{I}(\eta)] = \mathcal{I}(\mathbb{E}[\eta]).$$

△

Theorem 6.12. Let $m \geq 2$ and consider the return function iterates $(\eta_k)_{k\geq 0}$ generated by Equation 6.24 from some possibly random η_0. Suppose that:

(a) The source states $(X_k)_{k\geq 0}$ and step sizes $(\alpha_k)_{k\geq 0}$ satisfy the Robbins–Monro conditions: w.p. 1, for all $x \in \mathcal{X}$,

$$\sum_{k=0}^{\infty} \alpha_k \mathbb{1}_{\{X_k = x\}} = \infty, \qquad \sum_{k=0}^{\infty} \alpha_k^2 \mathbb{1}_{\{X_k = x\}} < \infty.$$

(b) The joint distribution of $(X_k, A_k, R_k, X_k')_{k\geq 0}$ is an instance of the sampling model expressed in Equation 6.23.

Then, with probability 1, $\eta_k \to \hat{\eta}_{\mathrm{c}}^\pi$ with respect to the supremum Cramér distance $\bar{\ell}_2$, where $\hat{\eta}_{\mathrm{c}}^\pi$ is the unique fixed point of the projected operator $\Pi_{\mathrm{c}}\mathcal{T}^\pi$:

$$\hat{\eta}_{\mathrm{c}}^\pi = \Pi_{\mathrm{c}}\mathcal{T}^\pi\hat{\eta}_{\mathrm{c}}^\pi.$$

△

Proof. We begin by constructing a sequence $(U_k)_{k\geq 0}$, $U_k \in \mathbb{R}^{\mathcal{X}}_{\mathcal{I}}$ that parallels the sequence of return functions $(\eta_k)_{k\geq 0}$. Write

$$\mathcal{O} = \mathcal{I} \circ \Pi_c \mathcal{T}^\pi \circ \mathcal{I}^{-1}$$

and define, for each $k \in \mathbb{N}$, $U_k = \mathcal{I}(\eta_k)$. By Lemma 5.24, $\mathcal{O}$ is a contraction with modulus $\gamma^{1/2}$ in $\|\cdot\|_{2,\infty}$ and we have

$$\begin{aligned} &U_{k+1}(X_k) \\ &\quad = (1-\alpha_k)U_k(X_k) + \alpha_k \mathcal{I}(\Pi_c(\mathrm{b}_{R_k,\gamma})_\# \eta(X'_k)) \\ &\quad = (1-\alpha_k)U_k(X_k) + \alpha_k\big((\mathcal{O}U_k)(X_k) + \underbrace{\mathcal{I}\Pi_c(\mathrm{b}_{R_k,\gamma})_\#(\mathcal{I}^{-1}U_k)(X'_k) - (\mathcal{O}U_k)(X_k)}_{w_k}\big). \end{aligned}$$

To see that Assumption 6.6 (bounded, mean-zero noise) holds, first note that by Proposition 6.2 and affineness of $\mathcal{I}$ and $\mathcal{I}^{-1}$ from Lemma 6.11, we have $\mathbb{E}[w_k \mid \mathcal{F}_k] = 0$. Furthermore, w_k is a bounded random variable, because each coordinate is a difference of two probabilities and hence in the interval $[-1, 1]$. Hence, we have $\mathbb{E}[\|w_k\|^2 \mid \mathcal{F}_k] < C$ for some $C > 0$, as required.

By Banach's theorem, the operator $\mathcal{O}$ has a unique fixed point U^*. We can thus apply Theorem 6.9 to conclude that the sequence $(U_k)_{k\geq 0}$ converges to U^*, satisfying

$$U^* = \mathcal{I}\Pi_c \mathcal{T}^\pi \mathcal{I}^{-1} U^*. \tag{6.27}$$

Because $\mathcal{I}$ is an isometry, this implies that $(\eta_k)_{k\geq 0}$ converges to $\eta^* = \mathcal{I}^{-1}U^*$. Applying $\mathcal{I}^{-1}$ to both sides of Equation 6.27, we obtain

$$\mathcal{I}^{-1}U^* = \Pi_c \mathcal{T}^\pi \mathcal{I}^{-1} U^*.$$

Since $\Pi_c \mathcal{T}^\pi$ has a unique fixed point, we conclude that $\eta^* = \hat{\eta}^\pi_c$. □

The proof of Theorem 6.12 illustrates how the parameters of categorical distributions are by design bounded, so that stability (i.e., Proposition 6.7) is immediate. In fact, stability is also immediate for TD learning when the reward distributions have bounded support.

6.11 Technical Remarks

Remark 6.1. Given a probability distribution $\nu \in \mathscr{P}(\mathbb{R})$ and a level $\tau \in (0, 1)$, quantile regression finds a value $\theta^* \in \mathbb{R}$ such that

$$F_\nu(\theta^*) = \tau. \tag{6.28}$$

In some situations, for example when ν is a discrete distribution, there are multiple values satisfying Equation 6.28. Let us write

$$S = \{\theta : F_\nu(\theta) = \tau\}.$$

Then one can show that S forms an interval. We can argue that quantile regression converges to this set by noting that, for $\tau \in (0,1)$ the expected quantile loss

$$\mathcal{L}_\tau(\theta) = \mathop{\mathbb{E}}_{Z\sim\nu}\left[|\mathbb{1}_{\{Z<\theta\}} - \tau| \times |Z - \theta|\right]$$

is convex in θ. In addition, for this loss, we have that for any $\theta, \theta' \in S$ and $\theta'' \notin S$,

$$\mathcal{L}_\tau(\theta) = \mathcal{L}_\tau(\theta') < \mathcal{L}_\tau(\theta'').$$

Convergence follows under appropriate conditions by appealing to standard arguments regarding the convergence of stochastic gradient descent; see, for example, Kushner and Yin (2003). △

Remark 6.2 (Proof of Proposition 6.4). Our goal is to show that $\|\bar{Z}_k\|_2^2$ behaves like a nonnegative supermartingale, from which convergence would follow from the supermartingale convergence theorem (see, e.g., Billingsley 2012). We begin by expressing the squared Euclidean norms of the sequence elements recursively, writing $\mathcal{F}_k = \sigma(Z_{0:k-1}, \alpha_{0:k})$:

$$\begin{aligned}
\mathbb{E}[\|\bar{Z}_{k+1}(x)\|_2^2 \mid \mathcal{F}_k] &= \mathbb{E}[\|(1-\alpha_k)\bar{Z}_k + \alpha_k Z_k\|_2^2 \mid \mathcal{F}_k] \\
&\overset{(a)}{=} (1-\alpha_k)^2\|\bar{Z}_k\|_2^2 + \alpha_k^2 \mathbb{E}[\|Z_k\|_2^2 \mid \mathcal{F}_k] \\
&\overset{(b)}{\leq} (1-\alpha_k)^2\|\bar{Z}_k\|_2^2 + \alpha_k^2 B \\
&\leq (1-\alpha_k)\|\bar{Z}_k\|_2^2 + \alpha_k^2 B\,.
\end{aligned} \tag{6.29}$$

Here, (a) follows by expanding the squared norm and using $\mathbb{E}[Z_k \mid \mathcal{F}_k] = 0$, and (b) follows from the boundedness of the conditional variance of the $(Z_k)_{k\geq 0}$, where B is a bound on such variances.

This inequality does not establish the supermartingale property, due to the presence of the additive term $\alpha_k^2 B$ on the right-hand side. However, the ideas behind the Robbins–Siegmund theorem (Robbins and Siegmund 1971) can be applied to deal with this term. The argument first constructs the sequence

$$\Lambda_k = \|\bar{Z}_k\|_2^2 + \sum_{s=0}^{k-1} \alpha_k \|\bar{Z}_s\|_2^2 - B\sum_{s=0}^{k-1} \alpha_s^2\,.$$

Inequality 6.29 above then shows that $(\Lambda_k)_{k\geq 0}$ is a supermartingale but may not be uniformly bounded below, meaning the supermartingale convergence theorem still cannot be applied. Defining the stopping times $t_q = \inf\{k \geq 0 : B' \sum_{s=0}^{k} \alpha_s^2 > q\}$ for each $q \in \mathbb{N}$ (with the convention that $\inf \emptyset = \infty$), each stopped process $(\Lambda_{k\wedge t_q})_{k\geq 0}$ is a supermartingale bounded below by $-q$, and hence each such process converges w.p. 1 by the supermartingale convergence theorem. However, $B' \sum_{s=0}^{\infty} \alpha_s^2 < \infty$ w.p. 1 by assumption, so w.p. 1 $t_q = \infty$ for sufficiently large q, and hence Λ_k converges w.p. 1. Since $\sum_{k=0}^{\infty} \alpha_k = \infty$ w.p. 1 by assumption, it must be the case that $\|\bar{Z}_k\|_2^2 \to 0$ as $k \to 0$, in order for Λ_k to have a finite limit, and hence we are done. Although somewhat involved, Exercise 6.13 demonstrates the necessity of this argument. △

6.12 Bibliographical Remarks

The focus of this chapter has been in developing and analyzing single-step temporal-difference algorithms. Further algorithmic developments include the use of multistep returns (Sutton 1988), off-policy corrections (Precup et al. 2000), and gradient-based algorithms (Sutton et al. 2009; Sutton et al. 2008a); the exercises in this chapter develop a few such approaches.

6.1–6.2. This chapter analyzes incremental algorithms through the lens of approximating the application of dynamic programming operators. Temporal-difference algorithms have a long history (Samuel 1959), and the idea of incremental approximations to dynamic programming formed motivation for several general-purpose temporal-difference learning algorithms (Sutton 1984, 1988; Watkins 1989).

Although early proofs of particular kinds of convergence for these algorithms did not directly exploit this connection with dynamic programming (Watkins 1989; Watkins and Dayan 1992; Dayan 1992), later a strong theoretical connection was established that viewed these algorithms through the lens of stochastic approximation theory, allowing for a unified approach to proving almost-sure convergence (Gurvits et al. 1994; Dayan and Sejnowski 1994; Tsitsiklis 1994; Jaakkola et al. 1994; Bertsekas and Tsitsiklis 1996; Littman and Szepesvári 1996). The unbiased estimation framework presented comes from these works, and the second principle is based on the ideas behind two-timescale algorithms (Borkar 1997, 2008). A broader framework based on asymptotically approximating the trajectories of differential equations is a central theme of algorithm design and stochastic approximation theory more generally (Ljung 1977; Kusher and Clark 1978; Borkar and Meyn 2000; Kushner and Yin 2003; Borkar 2008; Benveniste et al. 2012; Meyn 2022).

In addition to the CTD and QTD algorithms described in this chapter, several other approaches to incremental learning of return distributions have been proposed. Morimura et al. (2010b) propose to update parametric density models by taking gradients of the Kullback-Leibler divergence between the current estimates and the result of applying the Bellman operator to these estimates. Barth-Maron et al. (2018) also take this approach, using a representation based on mixtures of Gaussians. Nam et al. (2021) also use mixtures of Gaussians and minimize the Cramér distance from a multistep target, incorporating ideas from TD(λ) (Sutton 1984, 1988). Gruslys et al. (2018) combine CTD with

Retrace(λ), a multistep off-policy evaluation algorithm (Munos et al. 2016). Nguyen et al. (2021) combine the quantile representation with a loss based on the MMD metrics described in Chapter 4. Martin et al. (2020) propose a proximal update scheme for the quantile representation based on (regularized) Wasserstein flows (Jordan et al. 1998; Cuturi 2013; Peyré and Cuturi 2019).

Example 6.1 is from Bellemare et al. (2016).

6.3. The categorical temporal-difference algorithm as a mixture update was presented by Rowland et al. (2018). This is a variant of the C51 algorithm introduced by Bellemare et al. (2017a), which uses a projection in a mixture of Kullback–Leibler divergence and Cramér distance. Distributional versions of gradient temporal-difference learning (Sutton et al. 2008a; Sutton et al. 2009) based on the categorical representation have also been explored by Qu et al. (2019).

6.4. The QTD algorithm was introduced by Dabney et al. (2018b). Quantile regression itself is a long-established tool within statistics, introduced by Koenker and Bassett (1978); Koenker (2005) is a classic reference on the subject. The incremental rule for estimating quantiles of a fixed distribution was in fact proposed by Robbins and Monro (1951), in the same paper that launched the field of stochastic approximation.

6.5. The discussion of sequences of learning rates that result in convergence goes back to Robbins and Monro (1951), who introduced the field of stochastic approximation. Szepesvári (1998), for example, considers this framework in their study of the asymptotic convergence rate of Q-learning. A fine-grained analysis in the case of temporal-difference learning algorithms, taking finite-time concentration into account, was undertaken by Even-Dar and Mansour (2003); see also Azar et al. (2011).

6.6–6.10. Our proof of Theorem 6.9 (via Propositions 6.4, 6.7, and 6.8) closely follows the argument given by Bertsekas and Tsitsiklis (1996) and Tsitsiklis (1994). Specifically, we adapt this argument to deal with distributional information, rather than a single scalar value. Proposition 6.4 is a special case of the Robbins–Siegmund theorem (Robbins and Siegmund 1971), and a particularly clear exposition of this and related material is given by Walton (2021). We note also that this result can also be established via earlier results in the stochastic approximation literature (Dvoretzky 1956), as noted by Jaakkola et al. (1994). Theorem 6.10 is classical, and results of this kind can be found in Bertsekas and Tsitsiklis (1996). Theorem 6.12 was first proven by Rowland et al. (2018), albeit with a monotonicity argument based on that of Tsitsiklis (1994); the argument here is based on a contraction mapping argument to match the analysis of the temporal-difference algorithm. For further background on signed measures, see Doob (1994).

6.13 Exercises

Exercise 6.1. In this chapter, we argued for a correspondence between operators and incremental algorithms. This also holds true for the incremental Monte Carlo algorithm introduced in Section 3.2. What is peculiar about the corresponding operator? △

Exercise 6.2. Exercise 3.2 asked you to derive an incremental algorithm from the n-step Bellman equation

$$V^\pi(x) = \mathbb{E}_\pi\left[\sum_{t=0}^{n-1} \gamma^t R_t + \gamma^n V^\pi(X_n) \mid X_0 = x\right].$$

Describe this process in terms of the method where we substitute random variables with their realizations, then derive the corresponding incremental algorithm for state-action value functions. △

Exercise 6.3 (*). The n-step random-variable Bellman equation for a policy π is given by

$$G^\pi(x) \stackrel{\mathcal{D}}{=} \sum_{t=0}^{n-1} \gamma^t R_t + \gamma^n G^\pi(X_n), \quad X_0 = x,$$

where the trajectory $(X_0 = x, A_0, R_0, \ldots, X_n, A_n, R_n)$ is distributed according to $\mathbb{P}_\pi(\cdot \mid X_0 = x)$.

(i) Write down the distributional form of this equation and the corresponding n-step distributional Bellman operator.

(ii) Show that it is a contraction on a suitable subset of $\mathscr{P}(\mathbb{R})^{\mathcal{X}}$ with respect to an appropriate metric.

(iii) Further show that the composition of this operator with either the categorical projection or the quantile projection is also a contraction mapping in the appropriate metric.

(iv) Using the approach described in this chapter, derive n-step versions of categorical and quantile temporal-difference learning.

(v) In the case of n-step CTD, describe an appropriate set of conditions that allow for Theorem 6.9 to be used to obtain convergence to the projected operator fixed point with probability 1. What happens to the fixed points of the projected operators as $n \to \infty$?

△

Exercise 6.4. Implement the quantile regression update rule (Equation 6.11). Given an initial estimate $\theta_0 = 0$, visualize the sequence of estimates produced by quantile regression for $\tau \in \{0.01, 0.1, 0.5\}$ and a constant step size $\alpha = 0.01$, given samples from

(i) a normal distribution $\mathcal{N}(1, 2)$;
(ii) a Bernoulli distribution $\mathcal{U}(\{0, 1\})$;
(iii) the mixture distribution $\frac{1}{3}\delta_1 + \frac{2}{3}\mathcal{U}([2, 3])$. △

Exercise 6.5. Let η_0 be a m-quantile return-distribution function, and let (x, a, r, x') denote a sample transition. Find a Markov decision process for which the update rule

$$\eta(x) \leftarrow \Pi_Q\Big[(1 - \alpha)\eta(x) + \alpha\eta_0(x')\Big]$$

does not converge. △

Exercise 6.6. Implement the TD, CTD, and QTD algorithms, and use these algorithms to approximate the value (or return) function of the quick policy on the Cliffs domain (Example 2.9). Compare their accuracy to the ground-truth value function and return-distribution function estimated using many Monte Carlo rollouts, both in terms of an appropriate metric and by visually comparing the approximations to the ground-truth functions.

Investigate how this accuracy is affected by different choices of constant step sizes and sequences of step sizes that satisfy the requirements laid out in Section 6.6. What do you notice about the relative performance of these algorithms as the degree of action noise p is varied?

Investigate what happens when we modify the TD algorithm by restricting value function estimates on the interval $[-C, C]$, for a suitable $C \in \mathbb{R}^+$. Does this restriction affect the performance of the algorithm differently from the restriction to $[\theta_1, \theta_m]$ that is intrinsic to CTD? △

Exercise 6.7. Let $(X_k, A_k, R_k, X'_k)_{k\geq 0}$ be a random sequence of transitions such that for each $x \in \mathcal{X}$, we have $X_k = x$ for infinitely many $k \geq 0$. Show that taking

$$\alpha_k = \frac{1}{N_k(X_k) + 1}.$$

satisfies Assumption 6.5. △

Exercise 6.8. Theorems 6.10 and 6.12 establish convergence for state-indexed value functions and return-distribution functions under TD and CTD learning, respectively. Discuss how Theorem 6.9 can be used to establish convergence of the corresponding state-action-indexed algorithms. △

Exercise 6.9 (*). Theorem 6.10 establishes that temporal-difference learning converges for a reasonably wide parameterization of the distribution of source states and step size schedules. Given a source state X_k, consider the incremental Monte Carlo update

$$V_{k+1}(X_k) = (1 - \alpha_k)V_k(X_k) + \alpha_k G_k$$

$$V_{k+1}(x) = V_k(x) \quad \text{if } x \neq X_k\,,$$

where $G_k \sim \eta^\pi(X_k)$ is a random return. Explain how Theorem 6.10 and its proof should be adapted to prove that the sequence $(V_k)_{k\geq 0}$ converges to V^π. △

Exercise 6.10 (Necessity of conditions for convergence of TD learning). The purpose of this exercise is to explore the behavior of TD learning when the assumptions of Theorem 6.10 do not hold.

(i) Write down an MDP with a single state x, from which trajectories immediately terminate, and a sequence of positive step sizes $(\alpha_k(x))_{k\geq 0}$ satisfying $\sum_{k\geq 0} \alpha_k(x) < \infty$, with the property that the TD update rule applied with these step sizes produces a sequence of estimates $(V_k)_{k\geq 0}$ that does not converge to V^π.

(ii) For the same MDP, write down a sequence of positive step sizes $(\alpha_k(x))_{k\geq 0}$ such that $\sum_{k\geq 0} \alpha_k^2(x) = \infty$, and show that the sequence of estimates $(V_k)_{k\geq 0}$ generated by TD learning with these step sizes does not converge to V^π.

(iii) Based on your answers to the previous two parts, for which values of $\beta \geq 0$ do step size sequences of the form

$$\alpha_k = \frac{1}{(N_k(X_k) + 1)^\beta}$$

lead to guaranteed TD converge, assuming all states are visited infinitely often?

(iv) Consider an MDP with a single state x, from which trajectories immediately terminate. Suppose the reward distribution at x is a standard Cauchy distribution, with density

$$f(z) = \frac{1}{\pi(1 + z^2)}\,.$$

Show that if V_0 also has a Cauchy distribution with median 0, then for any positive step sizes $(\alpha_k(x))_{k\geq 0}$, V_k has a Cauchy distribution with median 0, and hence the sequence does not converge to a constant. *Hint.* The characteristic function of the Cauchy distribution is given by $s \mapsto \exp(-|s|)$.

(v) (*) In the proof of Theorem 6.9, the inequality $1 - u \leq \exp(-u)$ was used to deduce that the condition $\sum_{k\geq 0} \alpha_k(x) = \infty$ w.p. 1 is sufficient to guarantee that

$$\prod_{l=0}^{k} (1 - \alpha_l(x)) \to 0$$

w.p. 1. Show that if $\alpha_k(x) \in [0, 1]$, the condition $\sum_{k\geq 0} \alpha_k(x) = \infty$ w.p. 1 is necessary as well as sufficient for the sequence $\prod_{l=0}^{k}(1 - \alpha_l(x))$ to also converge to zero w.p. 1. △

Exercise 6.11. Using the tools from this chapter, prove the convergence of the undiscounted, finite-horizon categorical Monte Carlo algorithm (Algorithm 3.3). △

Exercise 6.12. Recall the no-loop operator introduced in Example 4.6:

$$(T^{\pi}_{\text{NL}} V)(x) = \mathbb{E}_{\pi}\left[R + \gamma V(X')\mathbb{1}_{\{X' \neq x\}} \mid X = x\right].$$

Denote its fixed point by V^{π}_{NL}. For a transition (x, a, r, x') and time-varying step size $\alpha_k \in [0, 1)$, consider the no-loop update rule:

$$V(x) \leftarrow \begin{cases} (1-\alpha_k)V(x) + \alpha_k(r + \gamma V(x')) & \text{if } x' \neq x, \\ (1-\alpha_k)V(x) + \alpha_k r & \text{if } x' = x \end{cases}.$$

(i) Demonstrate that this update rule can be derived by substitution applied to the no-loop operator.

(ii) In words, describe how you would modify the online, incremental first-visit Monte Carlo algorithm (Algorithm 3.1) to learn V^{π}_{NL}.

(iii) (*) Provide conditions under which the no-loop update converges to V^{π}_{NL}, and prove that it does converge under those conditions. △

Exercise 6.13. Assumption 6.5 requires that the sequence of step sizes $(\alpha_k(x) : k \geq 0, x \in \mathcal{X})$ satisfy

$$\sum_{k=0}^{\infty} \alpha_k(x)^2 < \infty \tag{6.30}$$

with probability 1. For $k \geq 0$, let $N_k(x)$ be the number of times that x has been updated, and let $u_k(x)$ be the most recent time at which $x_l = x$, $l < k$ with the convention that $u_k(x) = 1$ if $N_k(x) = 0$. Consider the step size schedule

$$\alpha_{k+1} = \begin{cases} \frac{1}{N_k(X_k)+1} & \text{if } u_k(X_k) \leq \frac{k}{2} \\ 1 & \text{otherwise.} \end{cases}$$

This schedule takes larger steps for states whose estimate has not been recently updated. Suppose that $X_k \sim \xi$ for some distribution ξ that puts positive probability mass on all states. Show that the sequence $(\alpha_k)_{k\geq 0}$ satisfies Equation 6.30 w.p. 1, yet there is no $B \in \mathbb{R}$ for which

$$\sum_{k=0}^{\infty} \alpha_k(x)^2 < B$$

with probability 1. This illustrates the need for the care taken in the proof of Proposition 6.4. △

7 Control

A superpressure balloon is a kind of aircraft whose altitude is determined by the relative pressure between its envelope and the ambient atmosphere and which can be flown high in the stratosphere. Like a submarine, a superpressure balloon ascends when it becomes lighter and descends when it becomes heavier. Once in flight, superpressure balloons are passively propelled by the winds around them, so that their direction of travel can be influenced simply by changing their altitude. This makes it possible to steer such a balloon in an energy-efficient manner and have it operate autonomously for months at a time. Determining the most efficient way to control the flight of a superpressure balloon by means of altitude changes is an example of a *control problem*, the topic of this chapter.

In reinforcement learning, control problems are concerned with finding policies that achieve or maximize specified objectives. This is in contrast with prediction problems, which are concerned with characterizing or quantifying the consequences of following a particular policy. The study of control problems involves not only the design of algorithms for learning optimal policies but also the study of the behavior of these algorithms under different conditions, such as when learning occurs one sample at a time (as per Chapter 6), when noise is injected into the process, or when only a finite amount of data is made available for learning. Under the distributional perspective, the dynamics of control algorithms exhibit a surprising complexity. This chapter gives a brief and necessarily incomplete overview of the control problem. In particular, our treatment of control differs from most textbooks in that we focus on the distributional component and for conciseness omit some traditional material such as policy iteration and λ-return algorithms.

7.1 Risk-Neutral Control

The problem of finding a policy that maximizes the agent's expected return is called the *risk-neutral control problem*, as it is insensitive to the deviations of returns from their mean. We have already encountered risk-neutral control when we introduced the Q-learning algorithm in Section 3.7. We begin this chapter by providing a theoretical justification for this algorithm.

Problem 7.1 (Risk-neutral control)**.** Given an MDP $(\mathcal{X}, \mathcal{A}, \xi_0, P_{\mathcal{X}}, P_{\mathcal{R}})$ and discount factor $\gamma \in [0, 1)$, find a policy π maximizing the objective function

$$J(\pi) = \mathbb{E}_\pi\Big[\sum_{t=0}^{\infty} \gamma^t R_t\Big]. \qquad \triangle$$

A solution π^* that maximizes J is called an *optimal policy.*

Implicit in the definition of risk-neutral control and our definition of a policy in Chapter 2 is the fact that the objective J is maximized by a policy that only depends on the current state: that is, one that takes the form

$$\pi : \mathcal{X} \to \mathscr{P}(\mathcal{A}).$$

As noted in Section 2.2, policies of this type are more properly called *stationary Markov policies* and are but a subset of possible decision rules. With stationary Markov policies, the action A_t is independent of the random variables $X_0, A_0, R_0, \ldots, X_{t-1}, A_{t-1}, R_{t-1}$ given X_t. In addition, the distribution of A_t, conditional on X_t, is the same for all time indices t.

By contrast, *history-dependent policies* select actions on the basis on the entire trajectory up to and including X_t (the history). Formally, a history-dependent policy is a time-indexed collection of mappings

$$\pi_t : (\mathcal{X} \times \mathcal{A} \times \mathbb{R})^{t-1} \times \mathcal{X} \to \mathscr{P}(\mathcal{A}).$$

In this case, we have that

$$A_t \mid (X_{0:t}, A_{0:t-1}, R_{0:t-1}) \sim \pi_t(\cdot \mid X_0, A_0, R_0, \ldots, A_{t-1}, R_{t-1}, X_t). \tag{7.1}$$

When clear from context, we omit the time subscript to π_t and write $\mathbb{P}_\pi$, $\mathbb{E}_\pi$, G^π, and η^π to denote the joint distribution, expectation, return-variable function, and return-distribution function implied by the generative equations but with Equation 7.1 substituting the earlier definition from Section 2.2. We write $\boldsymbol{\pi}_{\text{MS}}$ for the space of stationary Markov policies and $\boldsymbol{\pi}_{\text{H}}$ for the space of history-dependent policies.

It is clear that every stationary Markov policy is a history-dependent policy, though the converse is not true. In risk-neutral control, however, the added degree of freedom provided by history-dependent policies is not needed to achieve optimality; this is made formal by the following proposition (recall that a policy is deterministic if it always selects the same action for a given state or history).

Proposition 7.2. Let $J(\pi)$ be as in Problem 7.1. There exists a deterministic stationary Markov policy $\pi^* \in \boldsymbol{\pi}_{\text{MS}}$ such that

$$J(\pi^*) \geq J(\pi) \quad \forall \pi \in \boldsymbol{\pi}_{\text{H}}\,. \qquad \triangle$$

Proposition 7.2 is a central result in reinforcement learning – from a computational point of view, for example, it is easier to deal with deterministic policies (there are finitely many of them) than stochastic policies. Remark 7.1 discusses some other beneficial consequences of Proposition 7.2. Its proof involves a surprising amount of detail; we refer the interested reader to Puterman (2014, Section 6.2.4).

7.2 Value Iteration and Q-Learning

The main consequence of Proposition 7.2 is that when optimizing the risk-neutral objective, we can restrict our attention to deterministic stationary Markov policies. In turn, this makes it possible to find an optimal policy π^* by computing the *optimal state-action value function* Q^*, defined as

$$Q^*(x,a) = \sup_{\pi \in \boldsymbol{\pi}_{\text{MS}}} \mathbb{E}_\pi \Big[\sum_{t=0}^{\infty} \gamma^t R_t \mid X = x, A = a \Big]. \tag{7.2}$$

Just as the value function V^π for a given policy π satisfies the Bellman equation, Q^* satisfies the *Bellman optimality equation*:

$$Q^*(x,a) = \mathbb{E}\,[R + \gamma \max_{a' \in \mathcal{A}} Q^*(X', a') \mid X = x, A = a]\,. \tag{7.3}$$

The optimal state-action value function describes the expected return obtained by acting so as to maximize the risk-neutral objective when beginning from the state-action pair (x, a). Intuitively, we may understand Equation 7.3 as describing this maximizing behavior recursively. While there might be multiple optimal policies, they must (by definition) achieve the same objective value in Problem 7.1. This value is

$$\mathbb{E}_\pi[V^*(X_0)]\,,$$

where V^* is the optimal value function:

$$V^*(x) = \max_{a \in \mathcal{A}} Q^*(x,a)\,.$$

Given Q^*, an optimal policy is obtained by acting greedily with respect to Q^* – that is, choosing in state x any action a that maximizes $Q^*(x, a)$.

Value iteration is a procedure for finding the optimal state-action value function Q^* iteratively, from which π^* can then be recovered by choosing actions that have maximal state-action values. In Chapter 5, we discussed a procedure for computing V^π based on repeated applications of the Bellman operator T^π. Value iteration replaces the Bellman operator T^π in this procedure with the *Bellman optimality operator*

$$(TQ)(x,a) = \mathbb{E}_\pi \left[R + \gamma \max_{a' \in \mathcal{A}} Q(X',a') \mid X = x, A = a\right]. \tag{7.4}$$

Let us define the L^∞ norm of a state-action value function $Q \in \mathbb{R}^{\mathcal{X}\times\mathcal{A}}$ as

$$\|Q\|_\infty = \sup_{x\in\mathcal{X}, a\in\mathcal{A}} |Q(x,a)|.$$

As the following establishes, the Bellman optimality operator is a contraction mapping in this norm.

Lemma 7.3. The Bellman optimality operator T is a contraction in L^∞ norm with modulus γ. That is, for any $Q, Q' \in \mathbb{R}^{\mathcal{X}\times\mathcal{A}}$,

$$\|TQ - TQ'\|_\infty \le \gamma \|Q - Q'\|_\infty .$$ △

Corollary 7.4. The optimal state-action value function Q^* is the only value function that satisfies the Bellman optimality equation. Further, for any $Q_0 \in \mathbb{R}^{\mathcal{X}\times\mathcal{A}}$, the sequence $(Q_k)_{k\ge 0}$ defined by $Q_{k+1} = TQ_k$ (for $k \ge 0$) converges to Q^*. △

Corollary 7.4 is an immediate consequence of Lemma 7.3 and Proposition 4.7. Before we give the proof of Lemma 7.3, it is instructive to note that despite a visual similarity and the same contractive property in supremum norm, the optimality operator behaves somewhat differently from the fixed-policy operator T^π, defined for state-action value functions as

$$(T^\pi Q)(x,a) = \mathbb{E}_\pi \left[R + \gamma Q(X',A') \mid X = x, A = a\right], \tag{7.5}$$

where conditional on the random variables $(X = x, A = a, R, X')$, we have $A' \sim \pi(\cdot \mid X')$. In the context of this chapter, we call T^π a *policy evaluation operator*. Such an operator is said to be *affine*: for any two Q-functions Q, Q' and $\alpha \in [0,1]$, it satisfies

$$T^\pi(\alpha Q + (1-\alpha)Q') = \alpha T^\pi Q + (1-\alpha) T^\pi Q' . \tag{7.6}$$

Equivalently, the difference between $T^\pi Q$ and $T^\pi Q'$ can be expressed as

$$T^\pi Q - T^\pi Q' = \gamma P^\pi (Q - Q') .$$

The optimality operator, on the other hand, is not affine. While affine operators can be analyzed almost as if they were linear,[54] the optimality operator is generally a nonlinear operator. As such, its analysis requires a slightly different approach.

54. Consider the proof of Proposition 4.4.

Proof of Lemma 7.3. The proof relies on a special property of the maximum function. For $f_1, f_2 : \mathcal{A} \to \mathbb{R}$, it can be shown that

$$\left| \max_{a \in \mathcal{A}} f_1(a) - \max_{a \in \mathcal{A}} f_2(a) \right| \le \max_{a \in \mathcal{A}} \left| f_1(a) - f_2(a) \right| .$$

Now let $Q, Q' \in \mathbb{R}^{\mathcal{X} \times \mathcal{A}}$, and fix $x \in \mathcal{X}, a \in \mathcal{A}$. Let us write $\mathbb{E}_{xa,\pi}[\cdot] = \mathbb{E}_\pi[\cdot \mid X = x, A = a]$. By linearity of expectations, we have

$$\begin{aligned}
|(TQ)(x,a) - (TQ')(x,a)| &= \left| \mathbb{E}_{xa,\pi}[R + \gamma \max_{a' \in \mathcal{A}} Q(X', a') - R - \gamma \max_{a' \in \mathcal{A}} Q'(X', a')] \right| \\
&= \left| \mathbb{E}_{xa,\pi}[\gamma \max_{a' \in \mathcal{A}} Q(X', a') - \gamma \max_{a' \in \mathcal{A}} Q'(X', a')] \right| \\
&= \gamma \left| \mathbb{E}_{xa,\pi}[\max_{a' \in \mathcal{A}} Q(X', a') - \max_{a' \in \mathcal{A}} Q'(X', a')] \right| \\
&\le \gamma \, \mathbb{E}_{xa,\pi} \left[| \max_{a' \in \mathcal{A}} Q(X', a') - \max_{a' \in \mathcal{A}} Q'(X', a') | \right] \\
&\le \gamma \, \mathbb{E}_{xa,\pi} \left[\max_{a' \in \mathcal{A}} |Q(X', a') - Q'(X', a')| \right] \\
&\le \gamma \max_{(x',a') \in \mathcal{X} \times \mathcal{A}} |Q(x', a') - Q'(x', a')| \\
&= \gamma \|Q - Q'\|_\infty .
\end{aligned}$$

Since the bound holds for any (x, a) pair, it follows that

$$\|TQ - TQ'\|_\infty \le \gamma \|Q - Q'\|_\infty . \qquad \square$$

Corollary 7.5. For any initial state-action value function $Q_0 \in \mathbb{R}^{\mathcal{X} \times \mathcal{A}}$, the sequence of iterates $Q_{k+1} = TQ_k$ converges to Q^* in the L^∞ norm. △

We can use the unbiased estimation method of Section 6.2 to derive an incremental algorithm for learning Q^*, since the contractive operator T is expressible as an expectation over a sample transition. Given a realization (x, a, r, x'), we construct the sample target

$$r + \gamma \max_{a' \in \mathcal{A}} Q(x', a') .$$

We then incorporate this target to an update rule to obtain the Q-learning algorithm first encountered in Chapter 3:

$$Q(x, a) \leftarrow (1 - \alpha) Q(x, a) + \alpha (r + \gamma \max_{a' \in \mathcal{X}} Q(x', a')) .$$

Under appropriate conditions, the convergence of Q-learning to Q^* can be established with Lemma 7.3 and a suitable extension of the analysis of Chapter 6 to the space of action-value functions.

7.3 Distributional Value Iteration

Analogous to value iteration, we can devise a distributional dynamic programming procedure for the risk-neutral control problem; such a procedure determines an approximation to the return function of an optimal policy. As we will see, in some circumstances, this can be accomplished without complications, giving some theoretical justification for the distributional algorithm presented in Section 3.7.

As in Chapter 5, we perform distributional dynamic programming by implementing the combination of a projection step with a distributional optimality operator.[55] Because it is not possible to "directly maximize" a probability distribution, we instead define the operator via a *greedy selection rule* $\mathcal{G}$.

We can view the expected-value optimality operator T as substituting the expectation over the next-state action A' in Equation 7.5 by a maximization over a'. As such, it can be rewritten as a particular policy evaluation operator T^π whose policy π depends on the operand Q; the mapping from Q to π is what we call a greedy selection rule.

Definition 7.6. A greedy selection rule is a mapping $\mathcal{G} : \mathbb{R}^{\mathcal{X}\times\mathcal{A}} \to \pi_{\text{MS}}$ with the property that for any $Q \in \mathbb{R}^{\mathcal{X}\times\mathcal{A}}$, $\mathcal{G}(Q)$ is greedy with respect to Q. That is,

$$\mathcal{G}(Q)(a \mid x) > 0 \implies Q(x,a) = \max_{a' \in \mathcal{A}} Q(x,a') .$$

We extend $\mathcal{G}$ to return functions by defining, for $\eta \in \mathscr{P}_1(\mathbb{R})^{\mathcal{X}\times\mathcal{A}}$, the *induced state-action value function*

$$Q_\eta(x,a) = \mathop{\mathbb{E}}_{Z\sim\eta(x,a)}[Z] ,$$

and then letting

$$\mathcal{G}(\eta) = \mathcal{G}(Q_\eta) . \qquad \triangle$$

A greedy selection rule may produce stochastic policies, for example when assigning equal probability to two equally valued actions. However, it must select actions that are maximally valued according to Q. Given a greedy selection rule, we may rewrite the Bellman optimality operator as

$$TQ = T^{\mathcal{G}(Q)}Q . \tag{7.7}$$

In the distributional setting, we must make explicit the dependency of the operator on the greedy selection rule $\mathcal{G}$. Mirroring Equation 7.7, the distributional

55. As before, this implementation is at its simplest when there are finitely many possible states, actions, and rewards, and the projection step can be computed efficiently. Alternatives include a sample-based approach and, as we will see in Chapter 9, function approximation.

$$\begin{array}{ccc} \eta & \xrightarrow{\Pi_{\mathscr{F}}\mathcal{T}^{\mathcal{G}}} & \eta' \\ \downarrow & & \downarrow \\ Q & \xrightarrow{T} & Q' \end{array}$$

Figure 7.1
When the projection step $\Pi_{\mathscr{F}}$ is mean-preserving, distributional value iteration produces the same sequence of state-action value functions as standard value iteration.

Bellman optimality operator derived from $\mathcal{G}$ is

$$\mathcal{T}^{\mathcal{G}}\eta = \mathcal{T}^{\mathcal{G}(\eta)}\eta\,.$$

We will see that, unlike the expected-value setting, different choices of the greedy selection rule result in different operators with possibly different dynamics – we thus speak of the distributional Bellman optimality operators, in the plural.

Distributional value iteration algorithms combine a greedy selection rule $\mathcal{G}$ and a projection step (described by the operator $\Pi_{\mathscr{F}}$ and implying a particular choice of representation $\mathscr{F}$) to compute an approximate optimal return function. Given some initial return function $\eta_0 \in \mathscr{F}^{\mathcal{X}\times\mathcal{A}}$, distributional value iteration implements the iterative procedure

$$\eta_{k+1} = \Pi_{\mathscr{F}}\mathcal{T}^{\mathcal{G}(\eta_k)}\eta_k. \tag{7.8}$$

In words, distributional value iteration selects a policy that at each state x is greedy with respect to the expected values of $\eta_k(x, \cdot)$ and computes the return function resulting from a single step of distributional dynamic programming with that policy.

The induced value function Q_{η_k} plays an important role in distributional value iteration as it is used to derive the greedy policy $\pi_k = \mathcal{G}(\eta_k)$. When $\Pi_{\mathscr{F}}$ is mean-preserving (Section 5.11), Q_{η_k} behaves as if it had been computed from standard value iteration (Figure 7.1). That is,

$$Q_{\eta_{k+1}} = TQ_{\eta_k}\,.$$

By induction, distributional value iteration then produces the same sequence of state-action value functions as regular value iteration. That is, given the initial condition

$$Q_0(x, a) = \mathop{\mathbb{E}}_{Z\sim\eta_0(x,a)}[Z]\,, \quad (x, a) \in \mathcal{X}\times\mathcal{A}\,,$$

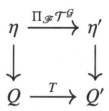

and the recursion

$$Q_{k+1} = TQ_k ,$$

we have

$$Q_{\eta_k} = Q_k , \text{ for all } k \geq 0 .$$

As a consequence, these two processes also produce the same greedy policies:

$$\pi_k = \mathcal{G}(Q_k) .$$

In the following section, we will use this equivalence to argue that distributional value iteration finds an approximation to an optimal return function. When $\Pi_{\mathscr{F}}$ is not mean-preserving, however, Q_{η_k} may deviate from Q_k. If that is the case, the greedy policy π_k is likely to be different from $\mathcal{G}(Q_k)$, and distributional value iteration may converge to the return function of a suboptimal policy. We discuss this point in Remark 7.2.

Before moving on, let us remark on an alternative procedure for approximating an optimal return function. This procedure first performs standard value iteration to obtain an approximation $\hat{Q}$ to Q^*. A greedy policy $\hat{\pi}$ is then extracted from $\hat{Q}$. Finally, distributional dynamic programming is used to approximate the return function $\eta^{\hat{\pi}}$. If $\hat{\pi}$ is an optimal policy, this directly achieves our stated aim and suggests doing away with the added complexity of distributional value iteration. In larger problems, however, it is difficult or undesirable to wait until Q^* has been determined before learning or computing the return function, or it may not be possible to decouple value and return predictions. In these situations, it is sensible to consider distributional value iteration.

7.4 Dynamics of Distributional Optimality Operators

In this section and the next, we analyze the behavior of distributional optimality operators; recall from Section 7.3 that there is one such operator for each choice of greedy selection rule $\mathcal{G}$, in contrast to non-distributional value iteration. Combined with a mean-preserving projection, our analysis also informs the behavior of distributional value iteration. As we will see, even in the absence of approximation due to a finite-parameter representation, distributional optimality operators exhibit complex behavior.

Thus far, we have analyzed distributional dynamic programming algorithms by appealing to contraction mapping theory. Demonstrating that an operator is a contraction provides a good deal of understanding about algorithms implementing its repeated application: they converge at a geometric rate to the operator's fixed point (when such a fixed point exists). Unfortunately, distributional optimality operators are not contraction mappings, as the following shows.

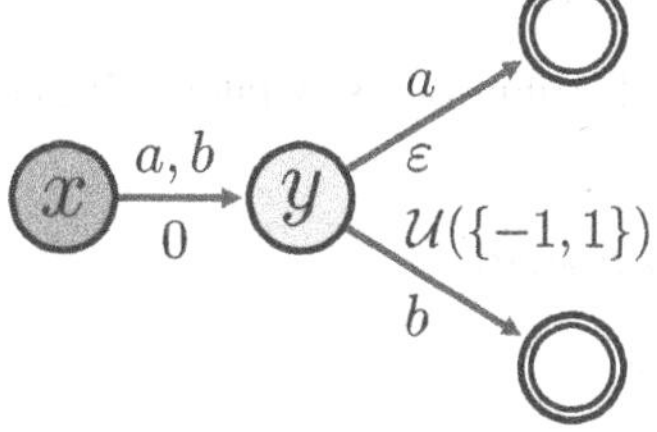

	η	$\mathcal{T}^{\mathcal{G}}\eta$	η^*
$x, \cdot$	$\gamma\varepsilon$	$\pm\gamma$	$\gamma\varepsilon$
y, a	$-\varepsilon$	ε	ε
y, b	± 1	± 1	± 1

Figure 7.2
Left: A Markov decision process for which no distributional Bellman optimality operator is a contraction mapping. **Right**: The optimal return-distribution function η^* and initial return-distribution function η used in the proof of Proposition 7.7 (expressed in terms of their support). Given a c-homogeneous probability metric d, the proof chooses ε so as to make $\overline{d}(\mathcal{T}^{\mathcal{G}}\eta, \eta^*) > \overline{d}(\eta, \eta^*)$.

Proposition 7.7. Consider a probability metric d and let $\overline{d}$ be its supremum extension. Suppose that for any $z, z' \in \mathbb{R}$,

$$d(\delta_z, \delta_{z'}) < \infty .$$

If d is c-homogeneous, there exist a Markov decision process and two return-distribution functions η, η' such that for any greedy selection rule $\mathcal{G}$ and any discount factor $\gamma \in (0, 1]$,

$$\overline{d}(\mathcal{T}^{\mathcal{G}}\eta, \mathcal{T}^{\mathcal{G}}\eta') > \overline{d}(\eta, \eta') . \tag{7.9}$$

△

Proof. Consider an MDP with two nonterminal states x, y and two actions a, b (Figure 7.2). From state x, all actions transition to state y and yield no reward. In state y, action a results in a reward of $\varepsilon > 0$, while action b results in a reward that is -1 or 1 with equal probability; both actions are terminal. We will argue that ε can be chosen to make $\mathcal{T}^{\mathcal{G}}$ satisfy Equation 7.9.

First note that any optimal policy must choose action a in state y, and all optimal policies share the same return-distribution function η^*. Consider another return-distribution function η that is equal to η^* at all state-action pairs except that $\eta(y, a) = \delta_{-\varepsilon}$ (Figure 7.2, right). This implies that the greedy selection rule must select b in y, and hence

$$(\mathcal{T}^{\mathcal{G}}\eta)(x, a) = (\mathcal{T}^{\mathcal{G}}\eta)(x, b) = (\mathrm{b}_{0,\gamma})_{\#}\eta(y, b) = (\mathrm{b}_{0,\gamma})_{\#}\eta^*(y, b) .$$

Because both actions are terminal from y, we also have that

$$(\mathcal{T}^{\mathcal{G}}\eta)(y, a) = \eta^*(y, a)$$

$$(\mathcal{T}^{\mathcal{G}}\eta)(y,b) = \eta^*(y,b).$$

Let us write $\nu = \frac{1}{2}\delta_{-1} + \frac{1}{2}\delta_1$ for the reward distribution associated with (y, b). We have

$$\begin{aligned}\overline{d}(\eta,\eta^*) &= d(\eta(y,a),\eta^*(y,a))\\ &= d(\delta_{-\varepsilon},\delta_\varepsilon)\,,\end{aligned} \tag{7.10}$$

and

$$\begin{aligned}\overline{d}(\mathcal{T}^{\mathcal{G}}\eta,\mathcal{T}^{\mathcal{G}}\eta^*) &= d((\mathcal{T}^{\mathcal{G}}\eta)(x,a),(\mathcal{T}^{\mathcal{G}}\eta^*)(x,a))\\ &= d((\mathrm{b}_{0,\gamma})_\#\eta^*(y,b),(\mathrm{b}_{0,\gamma})_\#\eta^*(y,a))\\ &= \gamma^c d(\nu,\delta_\varepsilon)\,,\end{aligned} \tag{7.11}$$

where the last line follows by c-homogeneity (Definition 4.22). We will show that for sufficiently small $\varepsilon > 0$, we have

$$d(\nu,\delta_\varepsilon) > \gamma^{-c} d(\delta_{-\varepsilon},\delta_\varepsilon)\,,$$

from which the result follows by Equations 7.10 and 7.11. To this end, note that

$$d(\delta_{-\varepsilon},\delta_\varepsilon) = \varepsilon^c d(\delta_{-1},\delta_1). \tag{7.12}$$

Since d is finite for any pair of Dirac deltas, we have that $d(\delta_{-1},\delta_1) < \infty$ and so

$$\lim_{\varepsilon\to 0} d(\delta_{-\varepsilon},\delta_\varepsilon) = 0\,. \tag{7.13}$$

On the other hand, by the triangle inequality we have

$$d(\nu,\delta_\varepsilon) + d(\delta_0,\delta_\varepsilon) \geq d(\nu,\delta_0) > 0\,, \tag{7.14}$$

where the second inequality follows because ν and δ_0 are different distributions. Again by c-homogeneity of d, we deduce that

$$\lim_{\varepsilon\to 0} d(\delta_0,\delta_\varepsilon) = 0\,,$$

and hence

$$\liminf_{\varepsilon\to 0} d(\nu,\delta_\varepsilon) \geq d(\nu,\delta_0) > 0\,.$$

Therefore, for $\varepsilon > 0$ sufficiently small, we have

$$d(\nu,\delta_\varepsilon) > \gamma^{-c} d(\delta_{-\varepsilon},\delta_\varepsilon)\,,$$

as required. □

Proposition 7.7 establishes that for any metric d that is sufficiently well behaved to guarantee that policy evaluation operators are contraction mappings (finite on bounded-support distributions, c-homogeneous), optimality operators are not generally contraction mappings with respect to $\overline{d}$. As a consequence, we cannot directly apply the tools of Chapter 4 to characterize distributional optimality operators. The issue, which is implied in the proof above, is that it is possible for distributions to have similar expectations yet differ substantially, for example, in their variance.

With a more careful line of reasoning, we can still identify situations in which the iterates $\eta_{k+1} = \mathcal{T}^{\mathcal{G}}\eta_k$ do converge. The most common scenario is when there is a unique optimal policy. In this case, the analysis is simplified by the existence of an action gap in the optimal value function.[56]

Definition 7.8. Let $Q \in \mathbb{R}^{\mathcal{X}\times\mathcal{A}}$. The *action gap* at a state x is the difference between the highest-valued and second highest-valued actions:

$$\textsc{gap}(Q, x) = \min\left\{Q(x, a^*) - Q(x, a) : a^*, a \in \mathcal{A}, a^* \neq a, Q(x, a^*) = \max_{a' \in \mathcal{A}} Q(x, a')\right\}.$$

The action gap of Q is the smallest gap over all states:

$$\textsc{gap}(Q) = \min_{x \in \mathcal{X}} \textsc{gap}(Q, x).$$

By extension, the action gap for a return function η is

$$\textsc{gap}(\eta) = \min_{x \in \mathcal{X}} \textsc{gap}(Q_\eta, x), \quad Q_\eta(x, a) = \mathop{\mathbb{E}}_{Z \sim \eta(x,a)}[Z]. \qquad \triangle$$

Definition 7.8 is such that if two actions are optimal at a given state, then $\textsc{gap}(Q^*) = 0$. When $\textsc{gap}(Q^*) > 0$, however, there is a unique action that is optimal at each state (and vice versa). In this case, we can identify an iteration K from which $\mathcal{G}(\eta_k) = \pi^*$ for all $k \geq K$. From that point on, any distributional optimality operator reduces to the π^* evaluation operator, which enjoys now-familiar convergence guarantees.

Theorem 7.9. Let $\mathcal{T}^{\mathcal{G}}$ be the distributional Bellman optimality operator instantiated with a greedy selection rule $\mathcal{G}$. Suppose that there is a unique optimal policy π^* and let $p \in [1, \infty]$. Under Assumption 4.29(p) (well-behaved reward distributions), for any initial return-distribution function $\eta_0 \in \mathscr{P}(\mathbb{R})$, the sequence of iterates defined by

$$\eta_{k+1} = \mathcal{T}^{\mathcal{G}}\eta_k$$

converges to η^{π^*} with respect to the metric $\overline{w}_p$. $\triangle$

Theorem 7.9 is stated in terms of supremum p-Wasserstein distances for conciseness. Following the conditions of Theorem 4.25 and under a different set of assumptions, we can of course also establish the convergence in, say, the supremum Cramér distance.

56. Example 6.1 introduced an update rule that increases this action gap.

Corollary 7.10. Suppose that $\Pi_{\mathscr{F}}$ is a mean-preserving projection for some representation $\mathscr{F}$ and that there is a unique optimal policy π^*. If Assumption 5.22(w_p) holds and $\Pi_{\mathscr{F}}$ is a nonexpansion in $\overline{w}_p$, then under the conditions of Theorem 7.9, the sequence

$$\eta_{k+1} = \Pi_{\mathscr{F}} \mathcal{T}^{\mathcal{G}} \eta_k$$

produced by distributional value iteration converges to the fixed point

$$\hat{\eta}^{\pi^*} = \Pi_{\mathscr{F}} \mathcal{T}^{\pi^*} \hat{\eta}^{\pi^*} . \qquad \triangle$$

Proof of Theorem 7.9. As there is a unique optimal policy, it must be that the action gap of Q^* is strictly greater than zero. Fix $\varepsilon = \frac{1}{2}\text{GAP}(Q^*)$. Following the discussion of the previous section, we have that

$$Q_{\eta_{k+1}} = T Q_{\eta_k} .$$

Because T is a contraction in L^∞ norm, we know that there exists a $K_\varepsilon \in \mathbb{N}$ after which

$$\|Q_{\eta_k} - Q^*\|_\infty < \varepsilon \qquad \forall\, k \geq K_\varepsilon . \tag{7.15}$$

For a fixed x, let a^* be the optimal action in that state. From Equation 7.15, we deduce that for any $a \neq a^*$,

$$\begin{aligned} Q_{\eta_k}(x, a^*) &\geq Q^*(x, a^*) - \varepsilon \\ &\geq Q^*(x, a) + \text{GAP}(Q^*) - \varepsilon \\ &> Q_{\eta_k}(x, a) + \text{GAP}(Q^*) - 2\varepsilon \\ &= Q_{\eta_k}(x, a) . \end{aligned}$$

Thus, the greedy action in state x after time K_ε is the optimal action for that state. Thus, $\mathcal{G}(\eta_k) = \pi^*$ for $k \geq K_\varepsilon$ and

$$\eta_{k+1} = \mathcal{T}^{\pi^*} \eta_k \qquad k \geq K_\varepsilon .$$

We can treat η_{K_ε} as a new initial condition η'_0, and apply Proposition 4.30 to conclude that $\eta_k \to \eta^{\pi^*}$. $\square$

In Section 3.7, we introduced the categorical Q-learning algorithm in terms of a deterministic greedy policy. Generalized to an arbitrary greedy selection rule $\mathcal{G}$, categorical Q-learning is defined by the update

$$\eta(x, a) \leftarrow (1 - \alpha)\eta(x, a) + \alpha \sum_{a' \in \mathcal{A}} \mathcal{G}(\eta)(a' \mid x')\Big(\Pi_c (\mathrm{b}_{r,\gamma})_\# \eta(x', a')\Big).$$

Because the categorical projection is mean-preserving, its induced state-value function follows the update

$$Q_\eta(x, a) \leftarrow (1-\alpha)\eta(x, a) + \alpha \underbrace{\sum_{a' \in \mathcal{A}} \mathcal{G}(\eta)(a' \mid x')(r + \gamma Q_\eta(x', a'))}_{r+\gamma \max_{a' \in \mathcal{A}} Q_\eta(x', a')}.$$

Using the tools of Chapter 6, one can establish the convergence of categorical Q-learning under certain conditions, including the assumption that there is a unique optimal policy π^*. The proof essentially combines the insight that, under certain conditions, the sequence of greedy policies tracked by the algorithm matches that of Q-learning and hence eventually converges to π^*, at which point the algorithm is essentially performing categorical policy evaluation of π^*. The actual proof is somewhat technical; we omit it here and refer the interested reader to Rowland et al. (2018).

7.5 Dynamics in the Presence of Multiple Optimal Policies*

In the value-based setting, it does not matter which greedy selection rule is used to represent the optimality operator: By definition, any greedy selection rule must be equivalent to directly maximizing over $Q(x, \cdot)$. In the distributional setting, however, different rules usually result in different operators. As a concrete example, compare the rule "among all actions whose expected value is maximal, pick the one with smallest variance" to "assign equal probability to actions whose expected value is maximal."

Theorem 7.9 relies on the fact that, when there is a unique optimal policy π^*, we can identify a time after which the distributional optimality operator behaves like a policy evaluation operator. When there are multiple optimal policies, however, the action gap of the optimal value function Q^* is zero and the argument cannot be used. To understand why this is problematic, it is useful to write the iterates $(\eta_k)_{k\geq 0}$ more explicitly in terms of the policies $\pi_k = \mathcal{G}(\eta_k)$:

$$\eta_{k+1} = \mathcal{T}^{\pi_k}\eta_k = \mathcal{T}^{\pi_k}\mathcal{T}^{\pi_{k-1}}\eta_{k-1} = \mathcal{T}^{\pi_k}\cdots\mathcal{T}^{\pi_0}\eta_0. \tag{7.16}$$

When the action gap is zero, the sequence of policies $\pi_k, \pi_{k+1}, \ldots$ may continue to vary over time, depending on the greedy selection rule. Although all optimal actions have the same optimal value (guaranteeing the convergence of the expected values to Q^*), they may correspond to different distributions. Thus, distributional value iteration – even with a mean-preserving projection – may mix together the distribution of different optimal policies. Even if $(\pi_k)_{k\geq 0}$ converges, the policy it converges to may depend on initial conditions (Exercise 7.5). In the worst case, the iterates $(\eta_k)_{k\geq 0}$ might not even converge, as the following example shows.

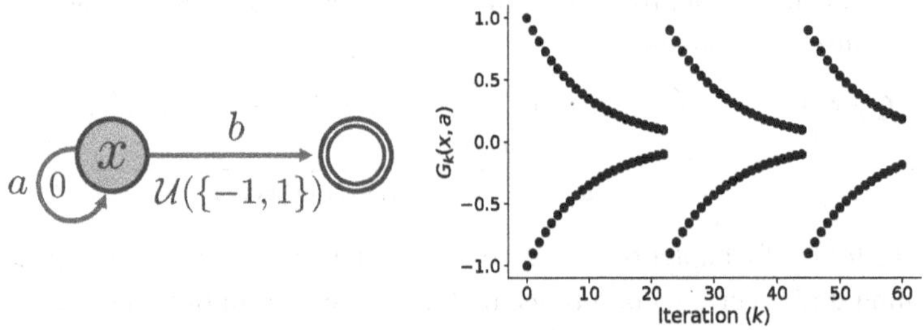

Figure 7.3
Left: A Markov decision process in which the sequence of return function estimates $(\eta_k)_{k\geq 0}$ does not converge (Example 7.11). **Right**: The return-distribution function estimate at (x, a) as a function of k and beginning with $c_0 = 1$. At each k, the pair of dots indicates the support of the distribution.

Example 7.11 (Failure to converge). Consider a Markov decision process with a single nonterminal state x and two actions, a and b (Figure 7.3, left). Action a gives a reward of 0 and leaves the state unchanged, while action b gives a reward of -1 or 1 with equal probability and leads to a terminal state. Note that the expected return from taking either action is 0.

Let $\gamma \in (0, 1)$. We will exhibit a sequence of return function estimates $(\eta_k)_{k\geq 0}$ that is produced by distributional value iteration and does not have a limit. We do so by constructing a greedy selection rule that achieves the desired behavior. Suppose that

$$\eta_0(x, b) = \tfrac{1}{2}(\delta_{-1} + \delta_1) .$$

For any initial parameter $c_0 \in \mathbb{R}$, if

$$\eta_0(x, a) = \tfrac{1}{2}(\delta_{-c_0} + \delta_{c_0}) ,$$

then by induction, there is a sequence of scalars $(c_k)_{k\geq 0}$ such that

$$\eta_k(x, a) = \tfrac{1}{2}(\delta_{-c_k} + \delta_{c_k}) \quad \forall k \in \mathbb{N} . \tag{7.17}$$

This is because

$$\eta_{k+1}(x, a) = (\mathrm{b}_{0,\gamma})_\# \eta_k(x, a) \quad \text{or} \quad \eta_{k+1}(x, a) = (\mathrm{b}_{0,\gamma})_\# \eta_k(x, b),\ k \geq 0 .$$

Define the following greedy selection rule: at iteration $k + 1$, choose a if $c_k \geq \frac{1}{10}$; otherwise, choose b. With some algebra, this leads to the recursion ($k \geq 0$)

$$c_{k+1} = \begin{cases} \gamma & \text{if } c_k < \frac{1}{10}, \\ \gamma c_k & \text{otherwise.} \end{cases}$$

Over a period of multiple iterations, the estimate $\eta_k(x, a)$ exhibits cyclical behavior (Figure 7.3, right). △

The example illustrates how, without additional constraints on the greedy selection rule, it is not possible to guarantee that the iterates converge. However, one can prove a weaker result based on the fact that only optimal actions must eventually be chosen.

Definition 7.12. For a given Markov decision process $\mathcal{M}$, the set of nonstationary Markov optimal return-distribution functions is

$$\boldsymbol{\eta}_{\text{NMO}} = \{\eta^{\bar{\pi}} : \bar{\pi} = (\pi_k)_{k\geq 0}, \pi_k \in \boldsymbol{\pi}_{\text{MS}} \text{ is optimal for } \mathcal{M}, \forall k \in \mathbb{N}\}. \qquad \triangle$$

In particular, any history-dependent policy $\bar{\pi}$ satisfying the definition above is also optimal for $\mathcal{M}$.

Theorem 7.13. Let $\mathcal{T}^{\mathcal{G}}$ be a distributional Bellman optimality operator instantiated with some greedy selection rule, and let $p \in [1, \infty]$. Under Assumption 4.29(p), for any initial condition $\eta_0 \in \mathscr{P}_p(\mathbb{R})^{\mathcal{X}}$ the sequence of iterates $\eta_{k+1} = \mathcal{T}^{\mathcal{G}}\eta_k$ converges to the set $\boldsymbol{\eta}_{\text{NMO}}$, in the sense that

$$\lim_{k\to\infty} \inf_{\eta\in\boldsymbol{\eta}_{\text{NMO}}} \overline{w}_p(\eta, \eta_k) = 0. \qquad \triangle$$

Proof. Along the lines of the proof of Theorem 7.9, there must be a time $K \in \mathbb{N}$ after which the greedy policies π_k, $k \geq K$ are optimal. For $l \in \mathbb{N}^+$, let us construct the history-dependent policy

$$\bar{\pi} = \pi_{K+l-1}, \pi_{K+l-2}, \ldots, \pi_{K+1}, \pi_K, \pi^*, \pi^*, \ldots,$$

where π^* is some stationary Markov optimal policy. Denote the return of this policy from the state-action pair (x, a) by $G^{\bar{\pi}}(x, a)$ and its return-distribution function by $\eta^{\bar{\pi}}$. Because $\bar{\pi}$ is an optimal policy, we have that $\eta^{\bar{\pi}} \in \boldsymbol{\eta}_{\text{NMO}}$. Let G_k be an instantiation of η_k, for each $k \in \mathbb{N}$. Now for a fixed $x \in \mathcal{X}$, $a \in \mathcal{A}$, let $(X_t, A_t, R_t)_{t=0}^{l}$ be the initial segment of a random trajectory generated by following $\bar{\pi}$ beginning with $X_0 = x, A_0 = a$. More precisely, for $t = 1, \ldots, l$, we have

$$A_t \mid (X_{0:t}, A_{0:t-1}, R_{0:t-1}) \sim \pi_{K+l-t}(\cdot \mid X_t).$$

From this, we write

$$G^{\bar{\pi}}(x, a) \overset{\mathcal{D}}{=} \sum_{t=0}^{l-1} \gamma^t R_t + \gamma^l G^{\pi^*}(X_l, A_l).$$

Because $\eta_{K+l} = \mathcal{T}^{\pi_{K+l-1}} \cdots \mathcal{T}^{\pi_K}\eta_K$, by inductively applying Proposition 4.11, we also have

$$G_{K+l}(x,a) \overset{\mathcal{D}}{=} \sum_{t=0}^{l-1} \gamma^t R_t + \gamma^l G_K(X_l, A_l).$$

Hence,

$$\begin{aligned} w_p(\eta_{K+l}(x,a), \eta^{\bar\pi}(x,a)) &= w_p(G_{K+l}(x,a), G^{\bar\pi}(x,a)) \\ &= w_p\Big(\sum_{t=0}^{l-1} \gamma^t R_t + \gamma^l G_K(X_l, A_l), \sum_{t=0}^{l-1} \gamma^t R_t + \gamma^l G^{\pi^*}(X_l, A_l)\Big). \end{aligned}$$

Consequently,

$$\begin{aligned} w_p(\eta_{K+l}(x,a), \eta^{\bar\pi}(x,a)) &\le w_p(\gamma^l G_K(X_l, A_l), \gamma^l G^{\pi^*}(X_l, A_l)) \\ &= \gamma^l w_p(G_K(X_l, A_l), G^{\pi^*}(X_l, A_l)) \\ &\le \gamma^l \overline{w}_p(G_K, G^{\pi^*}), \end{aligned}$$

following the arguments in the proof of Proposition 4.15. The result now follows by noting that $\overline{w}_p(G_K, G^{\pi^*}) < \infty$ under our assumptions, taking the supremum over (x, a) on the left-hand side, and taking the limit with respect to l. □

Theorem 7.13 shows that, even before the effect of the projection step is taken into account, the behavior of distributional value iteration is in general quite complex. When the iterates η_k are approximated (for example, because they are estimated from samples), nonstationary Markov return-distribution functions may also be produced by distributional value iteration – even when there is a unique optimal policy.

It may appear that the convergence issues highlighted by Example 7.11 and Theorem 7.13 are consequences of using the wrong greedy selection rule. To address these issues, one may be tempted to impose an ordering on policies (for example, always prefer the action with the lowest variance, at equal expected values). However, it is not clear how to do this in a satisfying way. One hurdle is that, to avoid the cyclical behavior demonstrated in Example 7.11, we would like a greedy selection rule that is continuous with respect to its input. This seems problematic, however, since we also need this rule to return the correct answer when there is a unique optimal (and thus deterministic) policy (Exercise 7.7). This suggests that when learning to control with a distributional approach, the learned return distributions may simultaneously reflect the random returns from multiple policies.

7.6 Risk and Risk-Sensitive Control

Imagine being invited to interview for a desirable position at a prestigious research institute abroad. Your plane tickets and the hotel have been booked weeks in advance. Now the night before an early morning flight, you make arrangements for a cab to pick you up from your house and take you to the airport. How long in advance of your plane's actual departure do you request the cab for? If someone tells you that, on average, a cab to your local airport takes an hour – is that sufficient information to make the booking? How does your answer change when the flight is scheduled around rush hour, rather than early morning?

Fundamentally, it is often desirable that our choices be informed by the variability in the process that produces outcomes from these choices. In this context, we call this variability *risk*. Risk may be inherent to the process or incomplete knowledge about the state of the world (including any potential traffic jams and the mechanical condition of the hired cab).

In contrast to risk-neutral behavior, decisions that take risk into account are called *risk-sensitive*. The language of distributional reinforcement learning is particularly well suited for this purpose, since it lets us reason about the full spectrum of outcomes, along with their associated probabilities. The rest of the chapter gives an overview of how one may account for risk in the decision-making process and of the computational challenges that arise when doing so. Rather than be exhaustive, here we take the much more modest aim of exposing the reader to some of the major themes in risk-sensitive control and their relation to distributional reinforcement learning; references to more extensive surveys are provided in the bibliographical remarks.

Recall that the risk-neutral objective is to maximize the expected return from the (possibly random) initial state X_0:

$$J(\pi) = \mathbb{E}_\pi\Big[\sum_{t=0}^{\infty}\gamma^t R_t\Big] = \mathbb{E}_\pi\big[G^\pi(X_0)\big].$$

Here, we may think of the expectation as mapping the random variable $G^\pi(X_0)$ to a scalar. Risk-sensitive control is the problem that arises when we replace this expectation by a *risk measure*.

Definition 7.14. A risk measure[57] is a mapping

$$\rho : \mathscr{P}_\rho(\mathbb{R}) \to [-\infty, \infty),$$

57. More precisely, this is a *static* risk measure, in that it is only concerned with the return from time $t = 0$. See bibliographical remarks.

defined on a subset $\mathscr{P}_\rho(\mathbb{R}) \subseteq \mathscr{P}(\mathbb{R})$ of probability distributions. By extension, for a random variable Z instantiating the distribution ν, we write $\rho(Z) = \rho(\nu)$. △

Problem 7.15 (Risk-sensitive control)**.** Given an MDP $(\mathcal{X}, \mathcal{A}, \xi_0, P_{\mathcal{X}}, P_{\mathcal{R}})$, a discount factor $\gamma \in [0, 1)$, and a risk measure ρ, find a policy $\pi \in \pi_{\text{H}}$ maximizing[58]

$$J_\rho(\pi) = \rho\Big(\sum_{t=0}^{\infty} \gamma^t R_t\Big). \tag{7.18}$$

△

In Problem 7.15, we assume that the distribution of the random return lies in $\mathscr{P}_\rho(\mathbb{R})$, similar to our treatment of probability metrics in Chapter 4. From a technical perspective, subsequent examples and results should be interpreted with this assumption in mind.

The risk measure ρ may take into account higher-order moments of the return distribution, be sensitive to rare events, and even disregard the expected value altogether. Note that according to this definition, $\rho = \mathbb{E}$ also corresponds to a risk-sensitive control problem. However, we reserve the term for risk measures that are sensitive to more than only expected values.

Example 7.16 (Mean-variance criterion)**.** Let $\lambda > 0$. The variance-penalized risk measure penalizes high-variance outcomes:

$$\rho^\lambda_{\text{MV}}(Z) = \mathbb{E}[Z] - \lambda \text{Var}(Z).$$

△

Example 7.17 (Entropic risk)**.** Let $\lambda > 0$. Entropic risk puts more weight on smaller-valued outcomes:

$$\rho^\lambda_{\text{ER}}(Z) = -\frac{1}{\lambda} \log \mathbb{E}[e^{-\lambda Z}].$$

△

Example 7.18 (Value-at-risk)**.** Let $\tau \in (0, 1)$. The value-at-risk measure (Figure 7.4) corresponds to the τth quantile of the return distribution:

$$\rho^\tau_{\text{VAR}}(Z) = F^{-1}_Z(\tau).$$

△

7.7 Challenges in Risk-Sensitive Control

Many convenient properties of the risk-neutral objective do not carry over to risk-sensitive control. As a consequence, finding an optimal policy is usually significantly more involved. This remains true even when the risk-sensitive

58. Typically, Problem 7.15 is formulated in terms of a risk to be *minimized*, which linguistically is a more natural objective. Here, however, we consider the maximization of $J_\rho(\pi)$ so as to keep the presentation unified with the rest of the book.

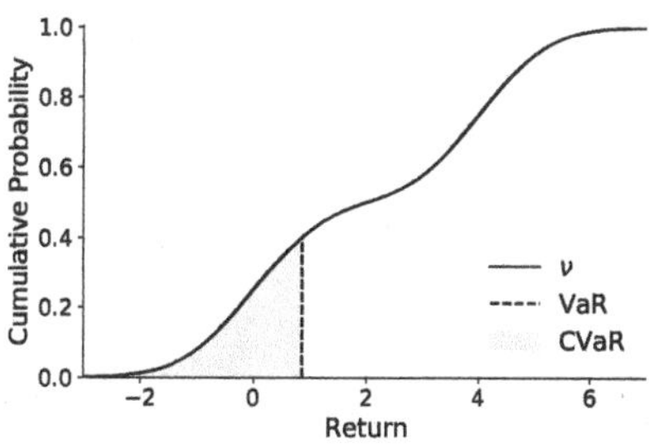

Figure 7.4
Illustration of value-at-risk (VaR) and conditional value-at-risk (CVaR). Depicted is the cumulative distribution function of the mixture of normal distributions $\nu = \frac{1}{2}\mathcal{N}(0, 1) + \frac{1}{2}\mathcal{N}(4, 1)$. The dashed line corresponds to VaR; CVaR ($\tau = 0.4$) can be determined from a suitable transformation of the shaded area (see Section 7.8 and Exercise 7.10).

objective (Equation 7.18) can be evaluated efficiently: for example, by using distributional dynamic programming to approximate the return-distribution function η^π. In this section, we illustrate some of these challenges by characterizing optimal policies for the *variance-constrained* control problem.

The variance-constrained problem introduces risk sensitivity by forbidding policies whose return variance is too high. Given a parameter $C \geq 0$, the objective is to

$$\begin{aligned} &\text{maximize} \quad \mathbb{E}_\pi[G^\pi(X_0)] \\ &\text{subject to} \quad \mathrm{Var}_\pi(G^\pi(X_0)) \leq C\,. \end{aligned} \tag{7.19}$$

Equation 7.19 can be shown to satisfy our definition of a risk-sensitive control problem if we express it in terms of a Lagrange multiplier:

$$J_{\text{VC}}(\pi) = \min_{\lambda \geq 0} \Big(\mathbb{E}_\pi \left[G^\pi(X_0)\right] - \lambda(\mathrm{Var}_\pi(G^\pi(X_0)) - C) \Big).$$

The variance-penalized and variance-constrained problems are related in that they share the *Pareto set* $\boldsymbol{\pi}_{\text{PAR}} \subseteq \boldsymbol{\pi}_{\text{H}}$ of possibly optimal solutions. A policy π is in the set $\boldsymbol{\pi}_{\text{PAR}}$ if we have that for all $\pi' \in \boldsymbol{\pi}_{\text{H}}$,

(a) $\mathrm{Var}(G^\pi(X_0)) > \mathrm{Var}(G^{\pi'}(X_0)) \implies \mathbb{E}[G^\pi(X_0)] > \mathbb{E}[G^{\pi'}(X_0)]$, and
(b) $\mathrm{Var}(G^\pi(X_0)) = \mathrm{Var}(G^{\pi'}(X_0)) \implies \mathbb{E}[G^\pi(X_0)] \geq \mathbb{E}[G^{\pi'}(X_0)]$.

In words, between two policies with equal variances, the one with lower expectation is never a solution to either problem. However, these problems are generally not equivalent (Exercise 7.8).

Proposition 7.1 establishes the existence of a solution of the risk-neutral control problem that is (a) deterministic, (b) stationary, and (c) Markov. By contrast, the solution to the variance-constrained problem may lack any or all of these properties.

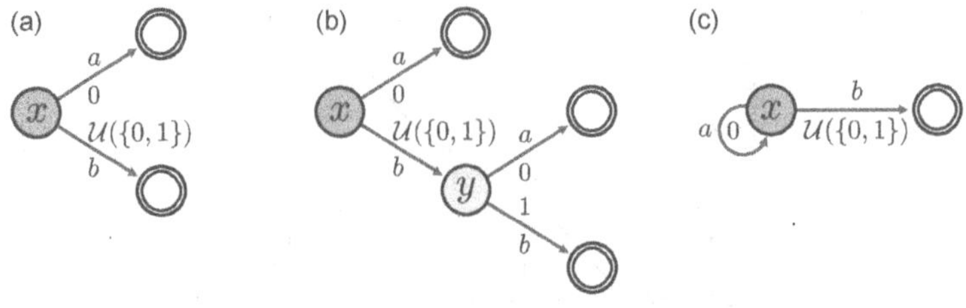

Figure 7.5
Examples demonstrating how the optimal policy for the variance-constrained control problem might not be **(a)** deterministic, **(b)** Markov, or **(c)** stationary.

Example 7.19 (The optimal policy may not be deterministic)**.** Consider the problem of choosing between two actions, a and b. Action a always yields a reward of 0, while action b yields a reward of 0 or 1 with equal probability (Figure 7.5a). If we seek the policy that maximizes the expected return subject to the variance constraint $C = 3/16$, the best deterministic policy respecting the variance constraint must choose a, for a reward of 0. On the other hand, the policy that selects a and b with equal probability achieves an expected reward of $1/4$ and a variance of $3/16$. △

Example 7.20 (The optimal policy may not be Markov)**.** Consider the Markov decision process in Figure 7.5b. Suppose that we seek a policy that maximizes the expected return from state x, now subject to the variance constraint $C = 0$. Let us assume $\gamma = 1$ for simplicity. Action a has no variance and is therefore a possible solution, with zero return. Action b gives a greater expected return, at the cost of some variance. Any policy π that depends on the state alone and chooses b in state x must incur this variance. On the other hand, the following history-dependent policy achieves a positive expected return without violating the variance constraint: in state x, choose b; if the first reward R_0 is 0, select action b in x; otherwise, select action a. In all cases, the return is 1, an improvement over the best Markov policy. △

Example 7.21 (The optimal policy may not be stationary)**.** In general, the optimal policy may require keeping track of time. Consider the problem of maximizing the expected return from the unique state x in Figure 7.5c, subject to $\mathrm{Var}_\pi(G^\pi(X_0)) \leq C$, for $C \leq 1/4$. Exercise 7.9 asks you to show that a simple time-dependent policy that chooses a for T_C steps and then selects b achieves an expected return of up to $\sqrt{C}$. This is possible because the variance of the return decays at a rate of γ^2, while its expected value decays at the slower rate of γ. By contrast, the best randomized stationary policy performs substantially worse

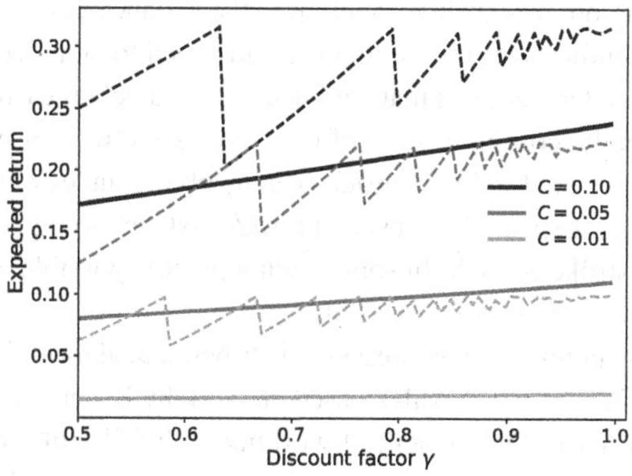

Figure 7.6
Expected return as a function of the discount factor γ and variance constraint C in Example 7.21. Solid and dashed lines indicate the expected return of the best stationary and time-dependent policies, respectively. The peculiar zigzag shape of the curve for the time-varying policy arises because the time T_C at which action b is taken must be an integer (see Exercise 7.9).

for a discount factor γ close to 1 and small values of C (Figure 7.6). Intuitively, a randomized policy must choose a with a sufficiently large probability to avoid receiving the random reward early, which prevents it from selecting b quickly beyond the threshold of T_C time steps. △

The last two examples establish that the variance-constrained risk measure is *time-inconsistent*: informally, the agent's preference for one outcome over another at time t may be reversed at a later time. Compared to the risk-neutral problem, the variance-constrained problem is more challenging because the space of policies that one needs to consider is much larger. Among other things, the lack of an optimal policy that is Markov with respect to the state alone also implies that the dependency on the initial distribution in Equation 7.19 is necessary to keep things well defined. The variance-constrained objective must be optimized for globally, considering the policy at all states at once; this is in contrast with the risk-neutral setting, where value iteration can make overall improvements to the policy by acting greedily with respect to the value function at individual states (see Remark 7.1). In fact, finding an optimal policy for the variance-constrained control problem is NP-hard (Mannor and Tsitsiklis 2011).

7.8 Conditional Value-At-Risk*

In the previous section, we saw that solutions to the variance-constrained control problem can take unintuitive forms, including the need to penalize better-than-expected outcomes. One issue is that variance only coarsely measures what we mean by "risk" in the common sense of the word. To refine our meaning, we may identify two types of risk: *downside risk*, involving undesirable outcomes such as greater-than-expected losses, and *upside risk*, involving what we may informally call a stroke of luck. In some situations, it is possible and useful to separately account for these two types of risk.

To illustrate this point, we now present a distributional algorithm for optimizing *conditional value-at-risk* (CVaR), based on work by Bäuerle and Ott (2011) and Chow et al. (2015). One benefit of working with full return distributions is that the algorithmic template we present here can be reasonably adjusted to deal with other risk measures, including the entropic risk measure described in Example 7.17. For conciseness, in what follows, we will state without proof a few technical facts about conditional value-at-risk that can be found in those sources and the work of Rockafellar and Uryasev (2002).

Conditional value-at-risk measures downside risk by focusing on the lower tail behavior of the return distribution, specifically the expected value of this tail. This expected value quantifies the magnitude of losses in extreme scenarios. Let Z be a random variable with cumulative and inverse cumulative distribution functions F_Z and F_Z^{-1}, respectively. For a parameter $\tau \in (0, 1)$, the CVaR of Z is

$$\mathrm{CVaR}_\tau(Z) = \frac{1}{\tau}\int_0^\tau F_Z^{-1}(u)\mathrm{d}u\,. \tag{7.20}$$

When the inverse cumulative distribution F_Z^{-1} is strictly increasing, the right-hand side of Equation 7.20 is equivalent to

$$\mathbb{E}[Z \mid Z \leq F_Z^{-1}(\tau)]\,. \tag{7.21}$$

In this case, CVaR quantifies the expected return, conditioned on the event that this return is no greater than the return's τth quantile – that is, is within the τth fraction of lowest returns.[59] In a reinforcement learning context, this leads to the risk-sensitive objective

$$J_{\mathrm{CVaR}}(\pi) = \mathrm{CVaR}_\tau\Big(\sum_{t=0}^{\infty}\gamma^t R_t\Big). \tag{7.22}$$

59. In other fields, CVaR is applied to losses rather than returns, in which case it measures the expected loss subject that this loss is above the τth percentile. For example, Equation 7.21 becomes $\mathbb{E}[Z \mid Z \geq F_Z^{-1}(\tau)]$, and the subsequent derivations need to be adjusted accordingly.

In general, there may not be an optimal policy that depends on the state x alone; however, one can show that optimality can be achieved with a deterministic, stationary Markov policy on an augmented state that incorporates information about the return thus far. In the context of this section, we assume that rewards are bounded in $[R_{\text{MIN}}, R_{\text{MAX}}]$. At a high level, we optimize the CVaR objective by

(a) defining the requisite *augmented state*;
(b) performing a form of *risk-sensitive value iteration* on this augmented state, using a suitable selection rule; and
(c) extracting the resulting policy.

We now explain each of these steps.

Augmented state. Central to the algorithm and to the state augmentation procedure is a reformulation of CVaR in terms of a desired minimum return or *target* $b \in \mathbb{R}$. Let $[x]^+$ denote the function that is 0 if $x < 0$ and x otherwise. For a random variable Z and $\tau \in (0, 1)$, Rockafellar and Uryasev (2002) establish that

$$\text{CVaR}_\tau(Z) = \max_{b \in \mathbb{R}} \left(b - \tau^{-1}\, \mathbb{E}\,[[b - Z]^+]\right). \tag{7.23}$$

When F_Z^{-1} is strictly increasing, the maximum-achieving b for Equation 7.23 is the quantile $F_Z^{-1}(\tau)$. In fact, taking the derivative of the expression inside the brackets with respect to α yields the quantile update rule (Equation 6.11; see Exercise 7.11). The advantage of this formulation is that it is more easily optimized in the context of a policy-dependent return. To see this, let us write

$$G^\pi = \sum_{t=0}^{\infty} \gamma^t R_t$$

to denote the random return from the initial state X_0, following some history-dependent policy $\pi \in \pi_{\text{H}}$. We then have that

$$\begin{aligned}\max_{\pi \in \pi_{\text{H}}} J_{\text{CVaR}}(\pi) &= \max_{\pi \in \pi_{\text{H}}} \max_{b \in \mathbb{R}} \left(b - \tau^{-1}\, \mathbb{E}\,[[b - G^\pi]^+]\right) \\ &= \max_{b \in \mathbb{R}} \left(b - \tau^{-1} \min_{\pi \in \pi_{\text{H}}} \mathbb{E}\,[[b - G^\pi]^+]\right). \end{aligned} \tag{7.24}$$

In words, the CVaR objective can be optimized by jointly finding an optimal target b and a policy that minimizes the expected shortfall $\mathbb{E}\,[[b - G^\pi]^+]$. For a fixed target b, we will see that it is possible to minimize the expected shortfall by means of dynamic programming. By adjusting b appropriately, one then obtains an optimal policy.

Based on Equation 7.24, let us now consider an augmented state (X_t, B_t), where X_t is as usual the current state and B_t takes on values in $\mathcal{B} = [V_{\text{MIN}}, V_{\text{MAX}}]$; we will describe its dynamics in a moment. With this augmented state, we may consider a class of stationary Markov policies, π_{CVaR}, which take the form $\pi : \mathcal{X} \times \mathcal{B} \to \mathscr{P}(\mathcal{A})$.[60]

We use the variable B_t to keep track of the amount of discounted reward that should be obtained from X_t onward in order to achieve a desired minimum return of $b_0 \in \mathbb{R}$ over the entire trajectory. The transition structure of the Markov decision process over the augmented state is defined by modifying the generative equations (Section 2.3):

$$B_0 = b_0$$
$$A_t \mid (X_{0:t}, B_{0:t}, A_{0:t-1}, R_{0:t-1}) \sim \pi(\cdot \mid X_t, B_t)$$
$$B_{t+1} \mid (X_{0:t+1}, B_{0:t}, A_{0:t}, R_{0:t}) = \frac{B_t - R_t}{\gamma}\,;$$

we similarly extend the sample transition model with the variables B and B'. This definition of the variables $(B_t)_{t\geq 0}$ can be understood by noting, for example, that a minimum return of b_0 is achieved over the whole trajectory if

$$\sum_{t=1}^{\infty} \gamma^{t-1} R_t \geq \frac{b_0 - R_0}{\gamma}\,.$$

If the reward R_t is small or negative, the new target B_{t+1} may of course be larger than B_t. Note that the value of b_0 is a parameter of the algorithm, rather than given by the environment.

Risk-sensitive value iteration. We next construct a method for optimizing the expected shortfall given a target $b_0 \in \mathbb{R}$. Let us write $\eta \in \mathscr{P}(\mathbb{R})^{\mathcal{X}\times\mathcal{B}\times\mathcal{A}}$ for a return-distribution function on the augmented state-action space, instantiated as a return-variable function G. For ease of exposition, we will mostly work with this latter form of the return function. As usual, we write G^π for the return-variable function associated with a policy $\pi \in \pi_{\text{CVaR}}$. With this notation in mind, we write

$$J_{\text{CVaR}}(\pi, b_0) = \max_{b\in\mathbb{R}} \left(b - \tau^{-1}\, \mathbb{E}\left[[b - G^\pi(X_0, b_0, A_0)]^+\right]\right) \tag{7.25}$$

to denote the conditional value-at-risk obtained by following policy π from the initial state (X_0, b_0), with $A_0 \sim \pi(\cdot \mid X_0, b_0)$.

Similar to distributional value iteration, the algorithm constructs a series of approximations to the return-distribution function by repeatedly applying the distributional Bellman operator with a policy derived from a greedy selection rule $\tilde{\mathcal{G}}$. Specifically, we write

$$\mathrm{a}_G(x, b) = \arg\min_{a\in\mathcal{A}} \mathbb{E}\left[[b - G(x, b, a)]^+\right] \tag{7.26}$$

60. Since B_t is a function of the trajectory up to time t, π_{CVaR} is a strict subset of π_{H}.

for the greedy action at the augmented state (x, b), breaking ties arbitrarily. The selection rule $\tilde{\mathcal{G}}$ is itself given by

$$\tilde{\mathcal{G}}(\eta)(a \mid x, b) = \tilde{\mathcal{G}}(G)(a \mid x, b) = \mathbb{1}\{a = \mathrm{a}_G(x, b)\}.$$

The algorithm begins by initializing $\eta_0(x, b, a) = \delta_0$ for all $x \in \mathcal{X}$, $b \in \mathcal{B}$, and $a \in \mathcal{A}$, and iterating

$$\eta_{k+1} = \mathcal{T}^{\tilde{\mathcal{G}}} \eta_k, \tag{7.27}$$

as in distributional value iteration. Expressed in terms of return-variable functions, this is

$$G_{k+1}(x, b, a) \overset{\mathcal{D}}{=} R + \gamma G_k(X', B', \mathrm{a}_{G_k}(X', B')), \quad X = x, B = b, A = a.$$

After k iterations, the policy $\pi_k = \tilde{\mathcal{G}}(\eta_k)$ can be extracted according to Equation 7.26. As suggested by Equation 7.25, a suitable choice of starting state is

$$b_0 = \arg\max_{b \in \mathcal{B}} \Big(b - \tau^{-1}\, \mathbb{E}\,[[b - G_k(X_0, b, \mathrm{a}_{G_k}(X_0, b))]^+] \Big). \tag{7.28}$$

As given, there are two hurdles to producing a tractable implementation of Equation 7.27: in addition to the usual concern that return distributions may need to be projected onto a finite-parameter representation, we also have to contend with a real-valued state variable B_t. Before discussing how this can be addressed, we first establish that Equation 7.27 is a sound approach to finding an optimal policy for the CVaR objective.

Theorem 7.22. Consider the sequence of return-distribution functions $(\eta_k)_{k \geq 0}$ defined by Equation 7.27. Then the greedy policy $\pi_k = \tilde{\mathcal{G}}(\eta_k)$ is such that for all $x \in \mathcal{X}$, $b \in \mathcal{B}$, and $a \in \mathcal{A}$,

$$\mathbb{E}\,[[b - G^{\pi_k}(x, b, a)]^+] \leq \min_{\pi \in \pi_{\mathrm{CVaR}}} \mathbb{E}\,[[b - G^{\pi}(X_0, b, a)]^+] + \frac{\gamma^k (V_{\mathrm{MAX}} - V_{\mathrm{MIN}})}{1 - \gamma}.$$

Consequently, we also have that the conditional value-at-risk of this policy satisfies (with b_0 given by Equation 7.28)

$$J_{\mathrm{CVaR}}(\pi_k, b_0) \geq \max_{b \in \mathcal{B}} \max_{\pi \in \pi_{\mathrm{CVaR}}} J_{\mathrm{CVaR}}(\pi, b) - \frac{\gamma^k (V_{\mathrm{MAX}} - V_{\mathrm{MIN}})}{\tau(1 - \gamma)}. \qquad \triangle$$

The proof is somewhat technical and is provided in Remark 7.3.

Theorem 7.22 establishes that with sufficiently many iterations, the policy π_k is close to optimal. Of course, when distribution approximation is introduced, the resulting policy will in general only approximately optimize the CVaR objective, with an error term that depends on the expressivity of the probability distribution representation (i.e., the parameter m in Chapter 5). To perform dynamic programming with the state variable B, one may use function approximation, the subject of the next chapter. Another solution is to consider a discrete number of values for B and to extend the operator $\mathcal{T}^{\tilde{\mathcal{G}}}$ to operate on this discrete set (Exercise 7.12).

7.9 Technical Remarks

Remark 7.1. In Section 7.7, we presented some of the challenges involved with finding an optimal policy for the variance-constrained objective. In some sense, these challenges should not be too surprising given that that we are looking to maximize a function J of an infinite-dimensional object (a history-dependent policy). Rather, what should be surprising is the relative ease with which one can obtain an optimal policy in the risk-neutral setting.

From a technical perspective, this ease is a consequence of Lemma 7.3, which guarantees that Q^* (and hence π^*) can be efficiently approximated. However, another important property of the risk-neutral setting is that the policy can be improved *locally*: that is, at each state simultaneously. To see this, consider a state-action value function Q^π for a given policy π and denote by π' a greedy policy with respect to Q^π. Then,

$$TQ^\pi = T^{\pi'}Q^\pi \geq T^\pi Q^\pi = Q^\pi . \tag{7.29}$$

That is, a single step of value iteration applied to the value function of a policy π results in a new value function that is at least as good as Q^π at all states – the Bellman operator is said to be *monotone*. Because this single step also corresponds to the value of a nonstationary policy that acts according to π' for one step and then switches to π, we can equivalently interpret it as constructing, one step at a time, a deterministic history-dependent policy for solving the risk-neutral problem.

By contrast, it is not possible to use a direct dynamic programming approach over the objective J to find the optimal policy for an arbitrary risk-sensitive control problem. A practical alternative is to perform the optimization instead with an ascent procedure (e.g., a policy gradient-type algorithm). Ascent algorithms can often be computed in closed form, and tend to be simpler to implement. On the other hand, convergence is typically only guaranteed to local optima, seemingly unavoidable when the optimization problem is known to be computationally hard. △

Remark 7.2. When the projection $\Pi_{\mathscr{F}}$ is not mean-preserving, distributional value iteration induces a state-action value function Q_{η_k} that is different from the value function Q_k determined by standard value iteration under equivalent initial conditions. Under certain conditions on the distributions of rewards, it is possible to bound this difference as $k \to \infty$. To do so, we use a standard error bound on approximate value iteration (see, e.g., Bertsekas 2012).

Lemma 7.23. Let $(Q_k)_{k\geq 0}$ be a sequence of iterates in $\mathbb{R}^{\mathcal{X}\times\mathcal{A}}$ satisfying

$$\|Q_{k+1} - TQ_k\|_\infty \leq \varepsilon$$

for some $\varepsilon > 0$ and where T is the Bellman optimality operator. Then,

$$\limsup_{k\to\infty} \|Q_k - Q^*\|_\infty \leq \frac{\varepsilon}{1-\gamma}. \qquad \triangle$$

In the context of distributional value iteration, we need to bound the difference

$$\|Q_{\eta_{k+1}} - TQ_{\eta_k}\|_\infty .$$

When the rewards are bounded on the interval $[R_{\text{MIN}}, R_{\text{MAX}}]$ and the projection step is Π_Q, the w_1-projection onto the m-quantile representation, a simple bound follows from an intermediate result used in proving Lemma 5.30 (see Exercise 5.20). In this case, for any ν bounded on $[V_{\text{MIN}}, V_{\text{MAX}}]$,

$$w_1(\Pi_Q\nu, \nu) \leq \frac{V_{\text{MAX}} - V_{\text{MIN}}}{2m};$$

Conveniently, the 1-Wasserstein distance bounds the difference of means between any distributions $\nu, \nu' \in \mathscr{P}_1(\mathbb{R})$:

$$\Big|\mathop{\mathbb{E}}_{Z\sim\nu}[Z] - \mathop{\mathbb{E}}_{Z\sim\nu'}[Z]\Big| \leq w_1(\nu, \nu').$$

This follows from the dual representation of the Wasserstein distance (Villani 2008). Consequently, for any (x, a),

$$\begin{aligned}\big|Q_{\eta_{k+1}}(x,a) - (T^{\pi_k}Q_{\eta_k})(x,a)\big| &\leq w_1(\eta_{k+1}(x,a), (\mathcal{T}^{\pi_k}\eta_k)(x,a))\\ &= w_1((\Pi_Q\mathcal{T}^{\pi_k}\eta_k)(x,a), (\mathcal{T}^{\pi_k}\eta_k)(x,a))\\ &\leq \frac{V_{\text{MAX}} - V_{\text{MIN}}}{2m}.\end{aligned}$$

By taking the maximum over (x, a) on the left-hand side of the above and combining with Lemma 7.23, we obtain

$$\limsup_{k\to\infty} \|Q_{\eta_k} - Q^*\|_\infty \leq \frac{V_{\text{MAX}} - V_{\text{MIN}}}{2m(1-\gamma)}. \qquad \triangle$$

Remark 7.3. Theorem 7.22 is proven from a few facts regarding partial returns, which we now give. Let us write $\mathrm{a}_k(x, b) = \mathrm{a}_{G_k}(x, b)$. We define the mapping $U_k : \mathcal{X} \times \mathcal{B} \times \mathcal{A} \to \mathbb{R}$ as

$$U_k(x, b, a) = \mathbb{E}\,[[b - G_k(x, b, a)]^+],$$

and for a policy $\pi \in \pi_{\text{CVaR}}$ similarly write

$$U^\pi(x, b, a) = \mathbb{E}\,[[b - G^\pi(x, b, a)]^+].$$

Lemma 7.24. For any $x \in \mathcal{X}$, $a \in \mathcal{A}$, $b \in \mathcal{B}$, and $k \in \mathbb{N}$, we have

$$U_{k+1}(x,b,a) = \gamma \mathbb{E}\left[U_k(X',B',\mathrm{a}_k(X',B')) \mid X=x, B=b, A=a\right].$$

In addition, if $\pi \in \pi_{\text{CVaR}}$ is a stationary Markov policy on $\mathcal{X} \times \mathcal{B}$, we have

$$U^\pi(x,b,a) = \gamma \mathbb{E}_\pi\left[U^\pi(X',B',A') \mid X=x, B=b, A=a\right]. \quad \triangle$$

Proof. The result follows by time-homogeneity and the Markov property. Consider the sample transition model (X,B,A,R,X',B',A'), with $A' = \mathrm{a}_k(X',B')$. Simultaneously, consider the partial trajectory $(X_t,B_t,A_t,R_t)_{t=0}^k$ for which $A_0 = A'$ and $A_t \sim \pi_{k-t}(\cdot \mid X_t, B_t)$ for $t > 0$. As $\gamma B' = B - R$, we have

$$\begin{aligned}
&\gamma \mathbb{E}\left[U_k(X',B',A') \mid X=x, B=b, A=a\right] \\
&= \gamma \mathbb{E}\left[\mathbb{E}\left[[B' - G_k(X',B',A')]^+\right] \mid X=x, B=b, A=a\right] \\
&= \gamma \mathbb{E}\left[\mathbb{E}\left[[B' - \sum_{t=0}^{k} \gamma^t R_t]^+ \mid X_0 = X', B_0 = B', A_0 = A'\right] \mid X=x, A=a\right] \\
&= \mathbb{E}\left[\mathbb{E}\left[[b - R - \gamma \sum_{t=0}^{k} \gamma^t R_t]^+ \mid X_0 = X', B_0 = B', A_0 = A'\right] \mid X=x, B=b, A=a\right] \\
&= \mathbb{E}\left[\mathbb{E}\left[[b - R - \sum_{t=1}^{k+1} \gamma^t R_t]^+ \mid X_1 = X', B_1 = B', A_1 = A'\right] \mid X=x, B=b, A=a\right] \\
&= \mathbb{E}\left[[b - \sum_{t=0}^{k+1} \gamma^t R_t]^+ \mid X_0 = x, B_0 = b, A_0 = a\right].
\end{aligned}$$

The second statement follows similarly. $\triangle$

Lemma 7.25. Suppose that $V_{\text{MIN}} \le 0$ and $V_{\text{MAX}} \ge 0$. Let $(R_t)_{t \ge 0}$ be a sequence of rewards in $[R_{\text{MIN}}, R_{\text{MAX}}]$. For any $b \in \mathbb{R}$ and $k \in \mathbb{N}$,

$$\mathbb{E}\,[[b - \sum_{t=0}^{\infty} \gamma^t R_t]^+] + \gamma^{k+1} V_{\text{MIN}} \le$$

$$\mathbb{E}\,[[b - \sum_{t=0}^{k} \gamma^t R_t]^+] \le \mathbb{E}\,[[b - \sum_{t=0}^{\infty} \gamma^t R_t]^+] + \gamma^{k+1} V_{\text{MAX}}. \quad \triangle$$

Proof. First note that, for any $b, z, z' \in \mathbb{R}$,

$$[b - z]^+ \le [b - z']^+ + [z' - z]^+. \tag{7.30}$$

To obtain the first inequality in the statement, we set

$$[b-\sum_{t=0}^{\infty}\gamma^t R_t]^+ \le [b-\sum_{t=0}^{k}\gamma^t R_t]^+ + [-\sum_{t=k+1}^{\infty}\gamma^t R_t]^+ .$$

Since rewards are bounded in $[R_{\text{MIN}}, R_{\text{MAX}}]$, we have that

$$-\sum_{t=k+1}^{\infty}\gamma^t R_t \le -\gamma^{k+1}\frac{R_{\text{MIN}}}{1-\gamma} = -\gamma^{k+1}V_{\text{MIN}}.$$

As we have assumed that $V_{\text{MIN}} \le 0$, it follows that

$$[b-\sum_{t=0}^{\infty}\gamma^t R_t]^+ \le [b-\sum_{t=0}^{k}\gamma^t R_t]^+ - \gamma^{k+1}V_{\text{MIN}}.$$

The second inequality in the statement is obtained analogously. △

Lemma 7.26. The sequence $(\eta_k)_{k\ge 0}$ defined by Equation 7.27 satisfies, for any $x \in \mathcal{X}$, $b \in \mathcal{B}$, and $a \in \mathcal{A}$,

$$U_k(x,b,a) = \min_{\pi\in\pi_{\text{CVaR}}} \mathbb{E}_\pi\,[[b-\sum_{t=0}^{k}\gamma^t R_t]^+ \mid X=x, B=b, A=a]. \tag{7.31}$$

△

Proof (sketch). Our choice of $G_0(x,b,a)=0$ guarantees that the statement is true for $k=0$. The result then follows by Lemma 7.24, the fact that the policy $\tilde{\mathcal{G}}(\eta_k)$ chooses the action minimizing the left-hand side of Equation 7.31, and by induction on k. △

Proof of Theorem 7.22. Let us assume that $V_{\text{MIN}} \le 0$ and $V_{\text{MAX}} \ge 0$ so that Lemma 7.25 can be applied. This is without loss of generality, as otherwise we may first construct a new sequence of rewards shifted by an appropriate constant C, such that $R_{\text{MIN}} = 0, R_{\text{MAX}} \ge 0$; by inspection, this transformation does not affect the statement of the Theorem 7.22.

Let $\pi^* \in \pi_{\text{CVaR}}$ be an optimal deterministic policy, in the sense that

$$U^{\pi^*}(x,b,a) = \min_{\pi\in\pi_{\text{CVaR}}} U^\pi(x,b,a).$$

Combining Lemmas 7.25 and 7.26, we have

$$U^{\pi^*}(x,b,a) + \gamma^{k+1}V_{\text{MIN}} \le U_k(x,b,a) \le U^{\pi^*}(x,b,a) + \gamma^{k+1}V_{\text{MAX}}. \tag{7.32}$$

Write $\mathbb{E}_{xba}[\cdot] = \mathbb{E}_\pi[\cdot \mid X=x, B=b, A=a]$. By Lemma 7.24,

$$U^{\pi_k}(x,b,a) - U_k(x,b,a) \tag{7.33}$$

$$= \gamma\,\mathbb{E}_{xba}\Big[U^{\pi_k}(X',B',\mathrm{a}_k(X',B')) - U_{k-1}(X',B',\mathrm{a}_{k-1}(X',B'))\Big]$$

$$= \gamma\, \mathbb{E}_{xba}\Big[U^{\pi_k}(X', B', \mathrm{a}_k(X', B')) - U_k(X', B', \mathrm{a}_k(X', B'))$$
$$+ U_k(X', B', \mathrm{a}_k(X', B')) - U_{k-1}(X', B', \mathrm{a}_{k-1}(X', B'))\Big]$$
$$\leq \gamma\, \mathbb{E}_{xba}\Big[U^{\pi_k}(X', B', \mathrm{a}_k(X', B')) - U_k(X', B', \mathrm{a}_k(X', B'))\Big] + \gamma^{k+1} V_{\text{MAX}} - \gamma^k V_{\text{MIN}}.$$

Now, the quantity

$$\varepsilon_k(x, b, a) = U^{\pi_k}(x, b, a) - U_k(x, b, a)$$

is bounded above and hence

$$\varepsilon_k = \sup_{x \in \mathcal{X}, b \in \mathcal{B}, a \in \mathcal{A}} \varepsilon_k(x, b, a)$$

exists. Taking the supremum over x, b and a on both sides of Equation 7.33, we have

$$\varepsilon_k \leq \gamma \varepsilon_k + \gamma^{k+1} V_{\text{MAX}} - \gamma^k V_{\text{MIN}}$$

and consequently for all x, b, a,

$$U^{\pi_k}(x, b, a) - U_k(x, b, a) \leq \frac{\gamma^k(\gamma V_{\text{MAX}} - V_{\text{MIN}})}{1 - \gamma}. \tag{7.34}$$

Because $U_k(x, b, a) \leq U^{\pi^*}(x, b, a) + \gamma^{k+1} V_{\text{MAX}}$, then

$$U^{\pi_k}(x, b, a) \leq U^{\pi^*}(x, b, a) + \gamma^{k+1} V_{\text{MAX}} + \frac{\gamma^k(\gamma V_{\text{MAX}} - V_{\text{MIN}})}{1 - \gamma}.$$

Now,

$$\gamma^{k+1} + \frac{\gamma^{k+1}}{1 - \gamma} = \gamma^k \frac{\gamma + \gamma(1 - \gamma)}{1 - \gamma} \leq \frac{\gamma^k}{1 - \gamma}.$$

Hence, we conclude that also

$$U^{\pi_k}(x, b, a) \leq \min_{\pi \in \pi_{\text{CVaR}}} U^{\pi}(x, b, a) + \frac{\gamma^k(V_{\text{MAX}} - V_{\text{MIN}})}{1 - \gamma},$$

as desired.

For the second statement, following Equation 7.28 we have

$$b_0 = \arg\max_{b \in \mathcal{B}} \Big(b - \tau^{-1}\, \mathbb{E}\big[[b - G_k(X_0, b, \mathrm{a}_k(X_0, b))]^+\big]\Big).$$

The algorithm begins in state (X_0, b_0), selects action $A_0 = \mathrm{a}_k(X_0, b_0)$, and then executes policy π_k from there on; its return is $G^{\pi_k}(X_0, b_0, A_0)$. In particular,

$$J_{\text{CVaR}}(\pi_k, b_0) = \max_{b \in \mathcal{B}} \Big(b - \tau^{-1}\, \mathbb{E}\big[[b - G^{\pi_k}(X_0, b_0, A_0)]^+\big]\Big)$$
$$\geq b_0 - \tau^{-1}\, \mathbb{E}\big[[b_0 - G^{\pi_k}(X_0, b_0, A_0)]^+\big]. \tag{7.35}$$

Write $\mathrm{a}_{\pi^*}(x, b)$ for the action selected by π^* in (x, b). Because Equation 7.32 holds for all x, b, and a, we have

$$\min_{a \in \mathcal{A}} U_k(x, b, a) \le \min_{a \in \mathcal{A}} U^{\pi^*}(x, b, a) + \gamma^{k+1} V_{\text{MAX}}$$
$$\implies U_k(x, b, \mathrm{a}_k(x, b)) \le U^{\pi^*}(x, b, \mathrm{a}_{\pi^*}(x, b)) + \gamma^{k+1} V_{\text{MAX}},$$

since $\mathrm{a}_k(x, b)$ is the action a that minimizes $U_k(x, b, a)$. Hence, for any state x, we have

$$\max_{b \in \mathcal{B}} (b - \tau^{-1} \min_{a \in \mathcal{A}} U_k(x, b, a)) \ge \max_{b \in \mathcal{B}} (b - \tau^{-1} \min_{a \in \mathcal{A}} U^{\pi^*}(x, b, a)) - \frac{\gamma^{k+1} V_{\text{MAX}}}{\tau},$$

and so

$$b_0 - \tau^{-1} U_k(X_0, b_0, \mathrm{a}_k(X_0, b_0)) \ge \max_{b \in \mathcal{B}} (b - \tau^{-1} U^{\pi^*}(X_0, b, \mathrm{a}_{\pi^*}(x, b))) - \frac{\gamma^{k+1} V_{\text{MAX}}}{\tau}$$
$$= \max_{b \in \mathcal{B}} J_{\text{CVaR}}(\pi^*, b) - \frac{\gamma^{k+1} V_{\text{MAX}}}{\tau}.$$

Combined with Equations 7.34 and 7.35, this yields

$$J_{\text{CVaR}}(\pi_k, b_0) \ge \max_{b \in \mathcal{B}} \max_{\pi \in \pi_{\text{CVaR}}} J_{\text{CVaR}}(\pi, b) - \frac{\gamma^k (V_{\text{MAX}} - V_{\text{MIN}})}{\tau(1 - \gamma)}. \qquad \triangle$$

7.10 Bibliographical Remarks

7.0. The balloon navigation example at the beginning of the chapter is from Bellemare et al. (2020). Sutton and Barto (2018) separate "control problem" from "prediction problem"; the latter figures more predominantly in this book. In earlier literature, the control problem comes first (see, e.g., Bellman 1957b) and prediction is typically used as a subroutine for control (Howard 1960).

7.1. Time-dependent policies are common in finite-horizon scenarios and are studied at length by Puterman (2014). The technical core of Proposition 7.2 involves demonstrating that any feasible value function can be attained by a stationary Markov policy; see the results by Puterman (2014, Theorem 5.5.1), Altman (1999) and the discussion by Szepesvári (2020).

In reinforcement learning, history-dependent policies are also used to deal with partially observable environments, in which the agent receives an observation o at each time step rather than the identity of its state. For example, McCallum (1995) uses a variable-length history to represent state-action values, while Veness et al. (2011) use a history-based probabilistic model to learn a model of the environment. History-dependent policies also play a central role in the study of optimality in the fairly large class of computable environments (Hutter 2005).

7.2. The canonical reference for value iteration is the book by Bellman (1957b); see also Bellman (1957a) for an asymptotic analysis in the undiscounted setting. Lemma 7.3 is standard and can be found in most reinforcement learning textbooks (Bertsekas and Tsitsiklis 1996; Szepesvári 2010; Puterman 2014). State-action value functions were introduced along with the Q-learning algorithm (Watkins 1989) and subsequently used in the development of SARSA (Rummery and Niranjan 1994). Watkins and Dayan (1992) give a restricted result regarding the convergence of Q-learning, which is more thoroughly established by Jaakkola et al. (1994), Tsitsiklis (1994), and Bertsekas and Tsitsiklis (1996).

7.3–7.5. The expression of the optimality operator as a fixed-policy operator whose policy varies with the input is common in the analysis of control algorithms (see, e.g., Munos 2003; Scherrer 2014). The view of value iteration as constructing a history-dependent policy is taken by Scherrer and Lesner (2012) to derive more accurate value learning algorithms in the approximate setting.

The extension to distributional value iteration and Theorem 7.13 are from Bellemare et al. (2017a). The correspondence between standard value iteration and distributional value iteration with a mean-preserving projection is given by Lyle et al. (2019).

The notion of action gap plays an important role in understanding the relationship between value function estimates and policies, in particular when estimates are approximate. Farahmand (2011) gives a gap-dependent bound on the expected return obtained by a policy derived from an approximate value function. Bellemare et al. (2016) derive an algorithm for increasing the action gap so as to improve performance in the approximate setting.

An example of a selection rule that explicitly incorporates distributional information is the lexicographical rule of Jaquette (1973), which orders policies according to the magnitude of their moments.

7.6. The notion of risk and risk-sensitive decisions can be traced back to Markowitz (1952), who introduced the concept of trading off expected gains and variations in those gains in the context of constructing an investment portfolio; see also Steinbach (2001) for a retrospective. Artzner et al. (1999) propose a collection of desirable characteristics that make a risk measure *coherent* in the sense that it satisfies certain preference axioms. Of the risk measures mentioned here, CVaR is coherent but the variance-constrained objective is not. Artzner et al. (2007) discuss coherent risk measures in the context of sequential decisions. Ruszczyński (2010) introduces the notion of dynamic risk measures for Markov decision processes, which are amenable to optimization via Bellman-style recursions; see also Chow (2017) for a discussion of static and dynamic risk measures as well as time consistency. Jiang and Powell (2018) develop sample-based optimization methods for dynamic risk measures based on quantiles.

Howard and Matheson (1972) considered the optimization of an exponential utility function applied to the random return by means of policy iteration. The same objective is given a distributional treatment by Chung and Sobel (1987). Heger (1994) considers optimizing for worst-case returns. Haskell and Jain (2015) study the use of occupancy measures over augmented state spaces as an approach for finding optimal policies for risk-sensitive control; similarly, an occupancy measure-based approach to CVaR optimization is studied by Carpin et al. (2016). Mihatsch and Neuneier (2002) and Shen et al. (2013) extend Q-learning to the optimization of recursive risk measures, where a base risk measure is applied at each time step. Recursive risk measures are more easily optimized than risk measures directly applied to the random return but are not as easily interpreted. Martin et al. (2020) consider combining distributional reinforcement learning with the notion of second-order stochastic dominance as a means of action selection. Quantile criteria are considered by Filar et al. (1995) in the case of average-reward MDPs and, more recently, by Gilbert et al. (2017) and Li et al. (2022). Delage and Mannor (2010) solve a risk-constrained optimization problem to handle uncertainty in a learned model's parameters. See Prashanth and Fu (2021) for a survey on risk-sensitive reinforcement learning.

7.7. Sobel (1982) establishes that an operator constructed directly from the variance-penalized objective does not have the monotone improvement property, making its optimization more challenging. The examples demonstrating the need for randomization and a history-dependent policy are adapted from Mannor and Tsitsiklis (2011), who also prove the NP-hardness of the problem of optimizing the variance-constrained objective. Tamar et al. (2012) propose a policy gradient algorithm for optimizing a mean-variance objective and for the CVaR objective (Tamar et al. 2015); see also Prashanth and Ghavamzadeh (2013) and Chow and Ghavamzadeh (2014) for actor-critic algorithms for these criteria. Chow et al. (2018) augment the state with the return-so-far in order to extend gradient-based algorithms to a broader class of risk measures.

7.8. The reformulation of the conditional value-at-risk (CVaR) of a random variable in terms of the (convex) optimization of a function of a variable $b \in \mathbb{R}$ is due to Rockafellar and Uryasev (2000); see also Rockafellar and Uryasev (2002) and Shapiro et al. (2009). Bäuerle and Ott (2011) provide an algorithm for optimizing the CVaR of the random return in Markov decision processes. Their work forms the basis for the algorithm presented in this section, although the treatment in terms of return-distribution functions is new here. Another closely related algorithm is due to Chow et al. (2015), who additionally provide an approximation error bound on the computed CVaR. Brown et al. (2020) apply

Rockafellar and Uryasev's approach to design an agent that is risk-sensitive with respect to a prior distribution over possible reward functions. Keramati et al. (2020) combine categorical temporal-difference learning with an exploration bonus derived from the Dvoretzky–Kiefer–Wolfowitz inequality to develop an algorithm to optimize for conditional value-at-risk.

7.11 Exercises

Exercise 7.1. Find a counterexample that shows that the Bellman optimality operator is not an affine operator. △

Exercise 7.2. Consider the Markov decision process depicted in Figure 2.4a. For which values of the discount factor $\gamma \in [0, 1)$ is there more than one optimal action from state x? Use this result to argue that the optimal policy depends on the discount factor. △

Exercise 7.3. Proposition 7.7 establishes that distributional Bellman optimality operators are not contraction mappings.

(i) Instantiate the result with the 1-Wasserstein distance. Provide a visual explanation for the result by drawing the relevant cumulative distribution functions before and after the application of the operator.
(ii) Discuss why it was necessary, in the proof of Proposition 7.7, to assume that the probability metric d is c-homogeneous. △

Exercise 7.4. Suppose that there is a unique optimal policy π^*, as per Section 7.4. Consider the use of a projection $\Pi_{\mathscr{F}}$ for a probability representation $\mathscr{F}$ and the iterates

$$\eta_{k+1} = \Pi_{\mathscr{F}} \mathcal{T} \eta_k \,. \tag{7.36}$$

Discuss under what conditions the sequence of greedy policies $(\mathcal{G}(\eta_k))_{k \geq 0}$ converges to π^* when $\Pi_{\mathscr{F}}$ is

(i) the m-categorical projection Π_{c};
(ii) the m-quantile projection Π_{Q}.

Where necessary, provide proofs of your statements. Does your answer depend on m or on $\theta_1, \ldots, \theta_m$ for the case of the categorical representation? △

Exercise 7.5. Give a Markov decision process for which the limit of the sequence of iterates defined by $\eta_{k+1} = \mathcal{T}^{\mathcal{G}} \eta_k$ depends on the initial condition η_0, irrespective of the greedy selection rule $\mathcal{G}$. *Hint.* Construct a scenario where the implied policy π_k is the same for all k but depends on η_0. △

Exercise 7.6. Consider the greedy selection rule that selects an action with minimal variance among those with maximal expected value, breaking ties uniformly at random. Provide an example Markov decision process in which this rule results in a sequence of return-distribution functions that does not converge, as per Example 7.11. *Hint.* Consider reward distributions of the form $\frac{1}{3}\sum_{i=1}^{3}\delta_{\theta_i}$. △

Exercise 7.7 (*). Consider the 1-Wasserstein distance w_1 and its supremum extension $\overline{w}_1$. In addition, let d be a metric on $\mathscr{P}(\mathcal{A})$. Suppose that we are given a mapping

$$\tilde{\mathcal{G}} : \mathscr{P}(\mathbb{R})^{\mathcal{X}\times\mathcal{A}} \to \boldsymbol{\pi}_{\text{MS}}$$

which is continuous at every state, in the sense that for any $\varepsilon > 0$, there exists a $\delta > 0$ such that for any return functions η, η',

$$\overline{w}_1(\eta, \eta') < \delta \implies \max_{x\in\mathcal{X}} d(\tilde{\mathcal{G}}(\eta)(\cdot \mid x), \tilde{\mathcal{G}}(\eta')(\cdot \mid x)) < \varepsilon.$$

Show that this mapping cannot be a greedy selection rule in the sense of Definition 7.6. △

Exercise 7.8. Consider the Markov decision process depicted in Figure 7.5a. Show that there is no $\lambda \geq 0$ such that the policy maximizing

$$J_{\text{MV}}(\pi) = \mathbb{E}_\pi\left[G^\pi(X_0)\right] - \lambda \text{Var}_\pi(G^\pi(X_0))$$

is stochastic. This illustrates how the variance-constrained and variance-penalized control problems are not equivalent. △

Exercise 7.9. Consider the Markov decision process depicted in Figure 7.5c.

(i) Solve for the optimal stopping time T_C maximizing the return of a time-dependent policy that selects action a for T_C time steps, then selects action b (under the constraint that the variance should be no greater than C).

(ii) Prove that this policy can achieve an expected return of up to $\sqrt{C}$.

(iii) Based on your conclusions, design a policy that improves on this strategy.

(iv) Show that the expectation and variance of the return of a randomized stationary policy that selects action b with probability p are given by

$$\mathbb{E}_\pi[G^\pi(x)] = \frac{p}{2(1-\gamma(1-p))}$$

$$\text{Var}_\pi(G^\pi(x)) = \frac{p}{4}\left(\frac{2}{1-\gamma^2(1-p)} - \frac{p}{(1-\gamma(1-p))^2}\right).$$

(v) Using your favorite visualization program, chart the returns achieved by the optimal randomized policy and the optimal time-dependent policy, for values of C and γ different from those shown in Figure 7.5c. What do you observe? *Hint.* Use a root-finding algorithm to determine the maximum expected return of a randomized policy under the constraint $\text{Var}_\pi(G^\pi(x)) \leq C$. △

Exercise 7.10. Explain the relationship between the shaded area in Figure 7.4 and conditional value-at-risk for the depicted distribution. △

Exercise 7.11. Following Equation 7.23, for a distribution $\nu \in \mathscr{P}(\mathbb{R})$, consider the function

$$f(\theta) = \theta - \tau^{-1} \mathop{\mathbb{E}}_{Z \sim \nu} [[\theta - Z]^+] .$$

Show that for $\tau \in (0, 1)$,

$$\theta \leftarrow \theta - \alpha\tau \frac{d}{d\theta} f(\theta)$$

is equivalent to the quantile update rule (Equation 6.11), in expectation. △

Exercise 7.12 (*). Consider a uniform discretization $\mathcal{B}_\varepsilon$ of the interval $[V_{\text{MIN}}, V_{\text{MAX}}]$ into intervals of width ε (endpoints included). For a return-distribution function η on the discrete space $\mathcal{X} \times \mathcal{B}_\varepsilon \times \mathcal{A}$, define its extension to $\mathcal{X} \times [V_{\text{MIN}}, V_{\text{MAX}}] \times \mathcal{A}$

$$\tilde{\eta}(x, b, a) = \eta(x, \varepsilon \left\lfloor \frac{b}{\varepsilon} \right\rfloor, a).$$

Suppose that probability distributions can be represented exactly (i.e., without needing to resort to a finite-parameter representation). For the CVaR objective (Equation 7.22), derive an upper bound for the suboptimality

$$\max_{\pi \in \pi_{\text{CVaR}}} J(\pi) - J(\tilde{\pi}),$$

where $\tilde{\pi}$ is found by the procedure of Section 7.8 applied to the discrete space $\mathcal{X} \times \mathcal{B}_\varepsilon \times \mathcal{A}$ and using the extension $\eta \mapsto \tilde{\eta}$ to implement the operator $\mathcal{T}^{\tilde{\mathcal{G}}}$. △

8 Statistical Functionals

The development of distributional reinforcement learning in previous chapters has focused on approximating the full return function with parameterized families of distributions. In our analysis, we quantified the accuracy of an algorithm's estimate according to its distance from the true return-distribution function, measured using a suitable probability metric.

Rather than try to approximate the full distribution of the return, we may instead select specific properties of this distribution and directly estimate these properties. Implicitly, this is the approach taken when estimating the expected return. Other common properties of interest include quantiles of the distributions, high-probability tail bounds, and the risk-sensitive objectives described in Chapter 7. In this chapter, we introduce the language of *statistical functionals* to describe such properties.

In some cases, the statistical functional approach allows us to obtain accurate estimates of quantities of interest, in a more straightforward manner. As a concrete example, there is a low-cost dynamic programming procedure to determine the variance of the return distribution.[61] By contrast, categorical and quantile dynamic programming usually under- or overestimate this variance.

This chapter develops the framework of *statistical functional dynamic programming* as a general method for approximately determining the values of statistical functionals. As we demonstrate in Section 8.4, it is in fact possible to interpret both categorical and quantile dynamic programming as operating over statistical functionals. We will see that while some characteristics of the return (including its variance) can be accurately estimated by an iterative procedure, in general, some care must be taken when estimating arbitrary statistical functionals.

61. In fact, the return variance can be determined to machine precision by solving a linear system of equations, similar to what was done in Section 5.1 for the value function.

8.1 Statistical Functionals

A *functional* maps functions to real values. By extension, a *statistical functional* maps probability distributions to the reals. In this book, we view statistical functionals as measuring a particular property or characteristic of a probability distribution. For example, the mapping

$$\nu \mapsto \mathbb{P}_{Z\sim\nu}(Z \geq 0), \quad \nu \in \mathscr{P}(\mathbb{R})$$

is a statistical functional that measures how much probability mass its argument ν puts on the nonnegative reals. Statistical functionals express quantifiable properties of probability distributions such as their mean and variance. The following formalizes this point.

Definition 8.1. A *statistical functional* ψ is a mapping from a subset of probability distributions $\mathscr{P}_\psi(\mathbb{R}) \subseteq \mathscr{P}(\mathbb{R})$ to the reals, written

$$\psi : \mathscr{P}_\psi(\mathbb{R}) \to \mathbb{R}\,.$$

We call the particular scalar $\psi(\nu)$ associated with a probability distribution ν a *functional value* and the set $\mathscr{P}_\psi(\mathbb{R})$ the *domain* of the functional. △

Example 8.2. The *mean functional* maps probability distributions to their expected values. As before, let

$$\mathscr{P}_1(\mathbb{R}) = \{\nu \in \mathscr{P}(\mathbb{R}) : \mathop{\mathbb{E}}_{Z\sim\nu}[|Z|] < \infty\}$$

be the set of distributions with finite first moment. For $\nu \in \mathscr{P}_1(\mathbb{R})$, the mean functional is

$$\mu_1(\nu) = \mathop{\mathbb{E}}_{Z\sim\nu}[Z]\,.$$

The restriction to $\mathscr{P}_1(\mathbb{R})$ is necessary to exclude from the definition distributions without a well-defined mean. △

The purpose of this chapter is to study how functional values of the return distribution can be approximated using dynamic programming procedures and incremental algorithms. In general, we will be interested in a collection of such functionals that exhibit desirable properties: for example, because they can be jointly determined by dynamic programming or because they provide complementary information about the return function. We call such a collection a *distribution sketch*.

Definition 8.3. A distribution sketch (or simply sketch) $\psi : \mathscr{P}_\psi(\mathbb{R}) \to \mathbb{R}^m$ is a vector-valued function specified by a tuple $(\psi_1, \ldots, \psi_m)$ of statistical functionals. Its domain is

$$\mathscr{P}_\psi(\mathbb{R}) = \bigcap_{i=1}^{m} \mathscr{P}_{\psi_i}(\mathbb{R})\,,$$

and it is defined as

$$\psi(\nu) = (\psi_1(\nu), \dots, \psi_m(\nu)), \quad \nu \in \mathscr{P}_\psi(\mathbb{R}).$$

Its *image* is

$$I_\psi = \{\psi(\nu) : \nu \in \mathscr{P}_\psi(\mathbb{R})\} \subseteq \mathbb{R}^m.$$

We also extend this notation to return-distribution functions:

$$\psi(\eta) = (\psi(\eta(x)) : x \in \mathcal{X}), \quad \eta \in \mathscr{P}_\psi(\mathbb{R})^{\mathcal{X}}.$$

△

Example 8.4. The *quantile functionals* are a family of statistical functionals indexed by $\tau \in (0, 1)$ and defined over $\mathscr{P}(\mathbb{R})$. The τ-quantile functional is defined in terms of the inverse cumulative distribution function of its argument (Definition 4.12):

$$\psi_\tau^{\mathrm{Q}}(\nu) = F_\nu^{-1}(\tau).$$

A finite collection of quantile functionals (say, for $\tau_1, \dots, \tau_m \in (0, 1)$) constitutes a sketch. △

Example 8.5. To prove the convergence of categorical temporal-difference learning (Section 6.10), we introduced the isometry $\mathcal{I} : \mathscr{F}_{\mathrm{C},m} \to \mathbb{R}_\mathcal{I}$ defined as

$$\mathcal{I}(\nu) = \big(F_\nu(\theta_i) : i \in \{1, \dots, m\}\big), \tag{8.1}$$

where $(\theta_i)_{i=1}^m$ is the set of locations for the categorical representation. This isometry is also a sketch in the sense of Definition 8.3. If we extend its domain to be $\mathscr{P}(\mathbb{R})$, Equation 8.1 still defines a valid sketch but it is no longer an isometry: it is not possible to recover the distribution ν from its functional values $\mathcal{I}(\nu)$. △

8.2 Moments

Moments are an especially important class of statistical functionals. For an integer $p \in \mathbb{N}^+$, the pth moment of a distribution $\nu \in \mathscr{P}_p(\mathbb{R})$ is given by

$$\mu_p(\nu) = \mathop{\mathbb{E}}_{Z \sim \nu}[Z^p].$$

In particular, the first moment of ν is its mean, while the variance of ν is the difference between its second moment and squared mean:

$$\mu_2(\nu) - (\mu_1(\nu))^2. \tag{8.2}$$

Moments are ubiquitous in mathematics. They form a natural way of capturing important aspects of a probability distribution, and the infinite sequence of moments $(\mu_p(\nu))_{p=1}^\infty$ uniquely characterizes many probability distributions of interest; see Remark 8.3.

Our goal in this section is to describe a dynamic programming approach to determining the moments of the return distribution. Fix a policy π, and consider a state $x \in \mathcal{X}$ and action $a \in \mathcal{A}$. The pth moment of the return distribution $\eta^\pi(x, a)$ is given by

$$\mathbb{E}_\pi\left[(G^\pi(x,a))^p\right],$$

where as before, $G^\pi(x, a)$ is an instantiation of $\eta^\pi(x, a)$. Although we can also study dynamic programming approaches to learning the pth moment of state-indexed return distributions,

$$\mathbb{E}_\pi\left[(G^\pi(x))^p\right],$$

this is complicated by a potential conditional dependency between the reward R and next state X' due to the action A. One solution is to assume independence of R and X', as we did in Section 5.4. Here, however, to avoid making this assumption, we work with functions indexed by state-action pairs.

To begin, let us fix $m \in \mathbb{N}^+$. The m-moment function M^π is

$$M^\pi(x,a,i) = \mathbb{E}_\pi[(G^\pi(x,a))^i] = \mu_i(\eta^\pi(x,a)), \quad \text{for } i = 1,\ldots,m. \tag{8.3}$$

As with value functions, we view M^π as the function (or vector) in $\mathbb{R}^{\mathcal{X}\times\mathcal{A}\times m}$ describing the collection of the first m moments of the random return. In particular, $M^\pi(\cdot,\cdot,1)$ is the usual state-action value function. As elsewhere in the book, to ensure that the expectation in Equation 8.3 is well defined, we assume that all reward distributions have finite pth moments, for $p = 1, \ldots, m$. In fact, it is sufficient to assume that this holds for $p = m$ (Assumption 4.29(m)).

As with the standard Bellman equation, from the *state-action random-variable Bellman equation*

$$G^\pi(x,a) = R + \gamma G^\pi(X',A'), \quad X = x, A = a$$

we can derive Bellman equations for the moments of the return distribution. To do so, we raise both sides to the ith power and take expectations with respect to both the random return variables G^π and the random transition $(X = x, A = a, R, X', A')$:

$$\mathbb{E}_\pi[(G^\pi(x,a))^i] = \mathbb{E}_\pi[(R + \gamma G^\pi(X',A'))^i \mid X = x, A = a].$$

From the binomial expansion of the term inside the expectation, we obtain

$$\mathbb{E}_\pi[(G^\pi(x,a))^i] = \mathbb{E}_\pi\left[\sum_{j=0}^{i} \gamma^{i-j}\binom{i}{j} R^j G^\pi(X',A')^{i-j} \,\middle|\, X = x, A = a\right].$$

Since R and $G^\pi(X', A')$ are independent given X and A, we can rewrite the above as

$$M^\pi(x, a, i) = \sum_{j=0}^{i} \gamma^{i-j} \binom{i}{j} \mathbb{E}_\pi[R^j \mid X = x, A = a] \, \mathbb{E}_\pi[M^\pi(X', A', i - j) \mid X = x, A = a],$$

where by convention we take $M^\pi(x', a', 0) = 1$ for all $x' \in \mathcal{X}$ and $a' \in \mathcal{A}$. This is a recursive characterization of the ith moment of a return distribution, analogous to the familiar Bellman equation for the mean. The recursion is cast into the familiar framework of operators with the following definition.

Definition 8.6. Let $m \in \mathbb{N}^+$. The m-moment Bellman operator $T^\pi_{(m)} : \mathbb{R}^{\mathcal{X}\times\mathcal{A}\times m} \to \mathbb{R}^{\mathcal{X}\times\mathcal{A}\times m}$ is given by

$$(T^\pi_{(m)} M)(x, a, i) = \sum_{j=0}^{i} \gamma^{i-j} \binom{i}{j} \mathbb{E}_\pi[R^j \mid X = x, A = a] \, \mathbb{E}_\pi \left[M(X', A', i - j) \mid X = x, A = a\right]. \quad \triangle \tag{8.4}$$

The collection of moments $(M^\pi(x, a, i) : (x, a) \in \mathcal{X} \times \mathcal{A}, i = 1, \ldots, m)$ is a fixed point of the operator $T^\pi_{(m)}$. In general, the m-moment Bellman operator is not a contraction mapping with respect to the L^∞ metric (except, of course, for $m = 1$; see Exercise 8.1). However, with a more nuanced analysis, we can still show that $T^\pi_{(m)}$ has a unique fixed point to which the iterates

$$M_{k+1} = T^\pi_{(m)} M_k \tag{8.5}$$

converge.

Proposition 8.7. Let $m \in \mathbb{N}^+$. Under Assumption 4.29(m), M^π is the unique fixed point of $T^\pi_{(m)}$. In addition, for any initial condition $M_0 \in \mathbb{R}^{\mathcal{X}\times\mathcal{A}\times m}$, the iterates of Equation 8.5 converge to M^π. $\triangle$

Proof. We begin by constructing a suitable notion of distance between m-moment functions $\mathbb{R}^{\mathcal{X}\times\mathcal{A}\times m}$. For $M \in \mathbb{R}^{\mathcal{X}\times\mathcal{A}\times m}$, let

$$\|M\|_{\infty,i} = \sup_{(x,a)\in\mathcal{X}\times\mathcal{A}} |M(x, a, i)|, \qquad \text{for } i = 1, \ldots, m$$

$$\|M\|_{\infty,<i} = \sup_{j=1,\ldots,i-1} \|M\|_{\infty,j}, \qquad \text{for } i = 2, \ldots, m.$$

Each of $\|\cdot\|_{\infty,i}$ (for $i = 1, \ldots, m$) and $\|\cdot\|_{\infty,<i}$ (for $i = 2, \ldots, m$) is a *semi-norm*; they fulfill the requirements of a norm, except that neither $\|M\|_{\infty,i} = 0$ nor $\|M\|_{\infty,<i} = 0$ implies that $M = 0$. From these semi-norms, we construct the pseudo-metrics

$$(M, M') \mapsto \|M - M'\|_{\infty,i},$$

noting that it is possible for the distance between M and M' to be zero even when M is different from M'.

The structure of the proof is to argue that $T^\pi_{(m)}$ is a contraction with modulus γ with respect to $\|\cdot\|_{\infty,1}$ and then to show inductively that it satisfies an inequality of the form

$$\|T^\pi_{(m)}M - T^\pi_{(m)}M'\|_{\infty,i} \leq C_i\|M - M'\|_{\infty,<i} + \gamma^i\|M - M'\|_{\infty,i}, \qquad (8.6)$$

for each $i = 2, \ldots, m$, and some constant C_i that depends on $P_{\mathcal{R}}$. Chaining these results together then leads to the convergence statement, and uniqueness follows as an immediate corollary.

To see that $T^\pi_{(m)}$ is a contraction with respect to $\|\cdot\|_{\infty,1}$, let $M \in \mathbb{R}^{\mathcal{X}\times\mathcal{A}\times m}$, and write $M_{(i)} = (M(x, a, i) : (x, a) \in \mathcal{X} \times \mathcal{A})$ for the function in $\mathbb{R}^{\mathcal{X}\times\mathcal{A}}$ corresponding to the ith moment function estimates given by M. By inspecting Equation 8.4 with $i = 1$, it follows that

$$(T^\pi_{(m)}M)_{(1)} = T^\pi M_{(1)},$$

where T^π is the usual Bellman operator. Furthermore, $\|M\|_{\infty,1} = \|M_{(1)}\|_\infty$, and so the statement that $T^\pi_{(m)}$ is a contraction with respect to the pseudo-metric implied by $\|\cdot\|_{\infty,1}$ is equivalent to the contractivity of T^π on $\mathbb{R}^{\mathcal{X}\times\mathcal{A}}$ with the respect to the L^∞ norm, which was shown in Proposition 4.4.

To see that $T^\pi_{(m)}$ satisfies the bound of Equation 8.6 for $i > 1$, let $L \in \mathbb{R}$ be such that

$$\big|\mathbb{E}[R^i \mid X = x, A = a]\big| \leq L, \quad \text{for all } x, a \in \mathcal{X} \times \mathcal{A} \text{ and } i = 1, \ldots, m.$$

Observe that

$$\begin{aligned}
&\big|(T^\pi_{(m)}M)(x, a, i) - (T^\pi_{(m)}M')(x, a, i)\big| \\
&= \Bigg|\sum_{j=0}^{i-1} \gamma^{i-j}\binom{i}{j}\mathbb{E}_\pi[R^j \mid X = x, A = a] \times \\
&\qquad \sum_{\substack{x' \in \mathcal{X} \\ a' \in \mathcal{A}}} P_{\mathcal{X}}(x' \mid x, a)\pi(a' \mid x')(M - M')(x', a', i - j)\Bigg| \\
&\leq \sum_{j=1}^{i-1} \gamma^{i-j}\binom{i}{j}\big|\mathbb{E}_\pi[R^j \mid X = x, A = a]\big| \times \|M - M'\|_{\infty,<i} + \gamma^i\|M - M'\|_{\infty,i} \\
&\leq L\sum_{j=1}^{i-1} \gamma^{i-j}\binom{i}{j}\|M - M'\|_{\infty,<i} + \gamma^i\|M - M'\|_{\infty,i} \\
&\leq (2^i - 2)L\|M - M'\|_{\infty,<i} + \gamma^i\|M - M'\|_{\infty,i}.
\end{aligned}$$

Taking $C_i = (2^i - 2)L$, we have

$$\|T^\pi_{(m)}M - T^\pi_{(m)}M'\|_{\infty,i} \leq C_i\|M - M'\|_{\infty,<i} + \gamma^i\|M - M'\|_{\infty,i}, \text{ for } i = 2, \dots, m.$$

To chain these results together, first observe that

$$\|M_k - M^\pi\|_{\infty,1} \to 0.$$

We next argue inductively that if, for a given $i < m$, $(M_k)_{k\geq 0}$ converges to M^π in the pseudo-metric induced by $\|\cdot\|_{\infty,<i}$, then also

$$\|M_k - M^\pi\|_{\infty,i} \to 0, \quad \text{and hence}$$
$$\|M_k - M^\pi\|_{\infty,<(i+1)} \to 0.$$

Let $y_k = \|M_k - M^\pi\|_{\infty,<i}$ and $z_k = \|M_k - M^\pi\|_{\infty,i}$. Then the generalized contraction result states that $z_{k+1} \leq C_i y_k + \gamma^i z_k$. Taking the limit superior on both sides yields

$$\limsup_{k\to\infty} z_k \leq \limsup_{k\to\infty} [C_i y_k + \gamma^i z_k] = \gamma^i \limsup_{k\to\infty} z_k\,,$$

where we have used the result $y_k \to 0$. From this, we deduce $\limsup_{k\to\infty} z_k \leq 0$, but since $(z_k)_{k\geq 0}$ is a nonnegative sequence, we therefore have $z_k \to 0$. This completes the inductive step, and we therefore obtain $\|M_k - M^\pi\|_{\infty,i} \to 0$, as required. □

In essence, Proposition 8.7 establishes that the m-moment Bellman operator behaves in a similar fashion to the usual Bellman operator, in the sense that its iterates converge to the fixed point M^π. From here, we may follow the derivations of Chapter 5 to construct a dynamic programming algorithm for learning these moments[62] or those of Chapter 6 to construct the corresponding incremental algorithm (Section 8.8). Although the proof above does not demonstrate the contractive nature of the moment Bellman operator, for $m = 2$, this can be achieved using a different norm and analysis technique (Exercise 8.4).

8.3 Bellman Closedness

In preceding chapters, our approach to distributional reinforcement learning considered approximations of the return distributions that could be tractably manipulated by algorithms. The m-moment Bellman operator, on the other hand, is not directly applied to probability distributions – compared to say, a m-categorical distribution, there is no immediate procedure for drawing a sample from a collection of m moments. Compared to the categorical and

62. When the reward distributions take on a finite number of values, in particular, the expectations of Definition 8.6 can be implemented as sums.

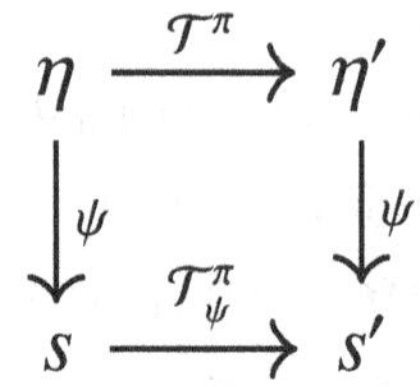

Figure 8.1
A sketch is Bellman closed if there is an operator $\mathcal{T}^\pi_\psi$ such that in the diagram above, the composite functions $\psi \circ \mathcal{T}^\pi$ and $\mathcal{T}^\pi_\psi \circ \psi$ coincide.

quantile projected operators, however, the m-moment operator yields an error-free dynamic programming procedure – with sufficiently many iterations and under some finiteness assumptions, we can determine the moments of the return function to any degree of accuracy. The concept of *Bellman closedness* formalizes this idea.

Definition 8.8. A sketch $\psi = (\psi_1, \ldots, \psi_m)$ is *Bellman closed* if, whenever its domain $\mathscr{P}_\psi(\mathbb{R})^{\mathcal{X}}$ is closed under the distributional Bellman operator:

$$\eta \in \mathscr{P}_\psi(\mathbb{R})^{\mathcal{X}} \implies \mathcal{T}^\pi \eta \in \mathscr{P}_\psi(\mathbb{R})^{\mathcal{X}},$$

there is an operator $\mathcal{T}^\pi_\psi : I_\psi^{\mathcal{X}} \to I_\psi^{\mathcal{X}}$ such that

$$\psi(\mathcal{T}^\pi \eta) = \mathcal{T}^\pi_\psi \psi(\eta) \quad \text{for all } \eta \in \mathscr{P}_\psi(\mathbb{R})^{\mathcal{X}}.$$

The operator $\mathcal{T}^\pi_\psi$ is said to be the Bellman operator for the sketch ψ. △

As was demonstrated in the preceding section, the collection of the m first moments $(\mu_1, \ldots, \mu_m)$ is a Bellman-closed sketch. Its associated operator is the m-moment operator $T^\pi_{(m)}$.

When a sketch ψ is Bellman closed, the operator T^π_ψ mirrors the application of the distributional Bellman operator to the return-distribution function η; see Figure 8.1. The concept of Bellman closedness is related to that of a diffusion-free projection (Chapter 5), and we will in fact establish an equivalence between the two in Section 8.4. In addition, Bellman-closed sketches are particularly interesting from a computational perspective because they support an exact dynamic programming procedure, as the following establishes.

Proposition 8.9. Let $\psi = (\psi_1, \ldots, \psi_m)$ be a Bellman-closed sketch and suppose that $\mathscr{P}_\psi(\mathbb{R})^{\mathcal{X}}$ is closed under $\mathcal{T}^\pi$. Then for any initial condition $\eta_0 \in \mathscr{P}_\psi(\mathbb{R})^{\mathcal{X}}$, and sequences $(\eta_k)_{k\geq 0}$, $(s_k)_{k\geq 0}$ defined by

$$\eta_{k+1} = \mathcal{T}^\pi \eta_k\,, \quad s_0 = \psi(\eta_0)\,, \quad s_{k+1} = \mathcal{T}^\pi_\psi s_k\,,$$

we have, for $k \geq 0$,

$$s_k = \psi(\eta_k)\,.$$

In addition, the functional values $s^\pi = \psi(\eta^\pi)$ of the return-distribution function are a fixed point of the operator $\mathcal{T}^\pi_\psi$. △

Proof. Both parts of the result follow immediately from the definition of the operator $\mathcal{T}^\pi_\psi$. First suppose that $s_k = \psi(\eta_k)$, for some $k \geq 0$. Then note that

$$s_{k+1} = \mathcal{T}^\pi_\psi s_k = \mathcal{T}^\pi_\psi \psi(\eta_k) = \psi(\mathcal{T}^\pi \eta_k) = \psi(\eta_{k+1})\,.$$

Thus, by induction, the first statement is proven. For the second statement, we have

$$s^\pi = \psi(\eta^\pi) = \psi(\mathcal{T}^\pi \eta^\pi) = \mathcal{T}^\pi_\psi \psi(\eta^\pi) = \mathcal{T}^\pi_\psi s^\pi\,. \qquad \square$$

Of course, dynamic programming is only feasible if the operator $\mathcal{T}^\pi_\psi$ can itself be implemented in a computationally tractable manner. In the case of the m-moment operator, we know this is possible under similar assumptions as were made in Chapter 5.

Proposition 8.9 illustrates how, when the sketch ψ is Bellman closed, we can do away with probability distributions and work exclusively with functional values. However, many sketches of interest fail to be Bellman closed, as the following examples demonstrate.

Example 8.10 (The median functional). A median of a distribution ν is its 0.5-quantile $F_\nu^{-1}(0.5)$.[63] Perhaps surprisingly, there is in general no way to determine the median of a return distribution based solely on the medians at the successor states. To see this, consider a state x that leads to state y_1 with probability $1/3$ and to state y_2 with probability $2/3$, with zero reward. The following are two scenarios in which the median returns at y_1 and y_2 are the same, but the median at x is different (see Figure 8.2):

63. As usual, there might be multiple values of z for which $\mathbb{P}_{Z\sim\nu}(Z \leq z) = 0.5$; recall that F^{-1} takes the smallest such value.

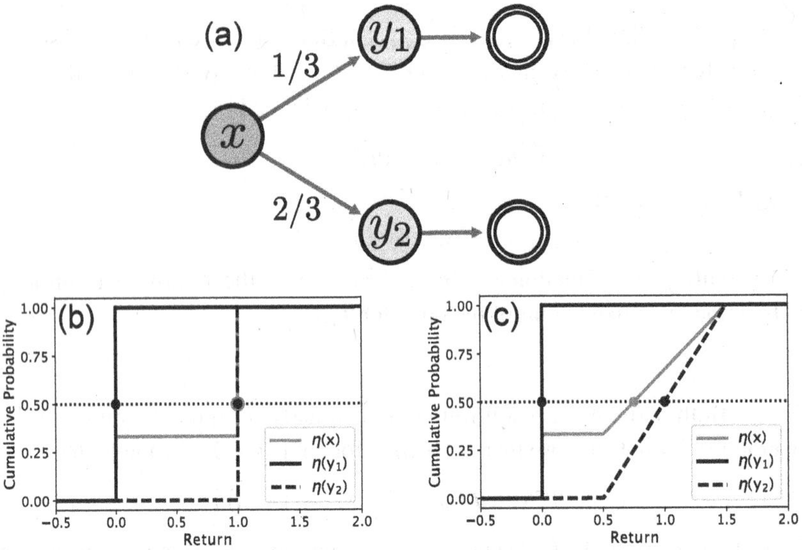

Figure 8.2
Illustration of Example 8.10. **(a)** A Markov decision process in which state x leads to states y_1 and y_2 with probability $1/3$ and $2/3$, respectively. **(b)** Case 1, in which the median of $\eta(x)$ matches the median of $\eta(y_2)$. **(c)** Case 2, in which the median of $\eta(x)$ differs from the median of $\eta(y_2)$.

Case 1. The return distributions at y_1 and y_2 are Dirac deltas at 0 and 1, respectively, and these are also the medians of these distributions. The median at x is also 1.

Case 2. The return distributions at y_1 and y_2 are a Dirac delta at 0 and the uniform distribution on [0.5, 1.5], respectively, and have the same medians as in Case 1. However, the median at x is now 0.75. △

Example 8.11 (At-least functionals). For $\nu \in \mathscr{P}(\mathbb{R})$ and $z \in \mathbb{R}$, let us define the *at-least functional*

$$\psi_{\geq z}(\nu) = \mathbb{1}\{\mathbb{P}_{Z\sim\nu}(Z \geq z) > 0\},$$

measuring whether ν assigns positive probability to values in $[z, \infty)$. Now consider a state x that deterministically leads to y, with no reward, and suppose that there is a single action a available. The statement "it is possible to obtain a return of at least 10 at state y" corresponds to

$$\psi_{\geq 10}(\eta^\pi(y)) = 1. \tag{8.7}$$

If Equation 8.7 holds, can we deduce whether or not a return of at least 10 is possible at state x? The answer is no. Suppose that $\gamma = 0.9$, and consider the following two situations:

Case 1. $\eta^\pi(y) = \delta_{10}$. Then $\psi_{\geq 10}(\eta^\pi(y)) = 1$, $\eta^\pi(x) = \delta_9$ and $\psi_{\geq 10}(\eta^\pi(x)) = 0$.

Case 2. $\eta^\pi(y) = \delta_{20}$. Then $\psi_{\geq 10}(\eta^\pi(y)) = 1$ still. However, $\eta^\pi(x) = \delta_{18}$ and $\psi_{\geq 10}(\eta^\pi(x)) = 1$. △

What goes wrong in the examples above is that we do not have sufficient information about the return distribution at the successor states to compute the functional values for the return distribution of state x. Consequently, we cannot use an iterative procedure to determine the functional values of η^π, at least not without error.

As it turns out, m-moment sketches are somewhat special in being Bellman closed. As the following theorem establishes, any sketch whose functionals are expectations of functions must encode the same information as a moment sketch.

Theorem 8.12. Let $\psi = (\psi_1, \ldots, \psi_m)$ be a sketch. Suppose that ψ is Bellman closed and that for each $i = 1, \ldots, m$, there is a function $f_i : \mathbb{R} \to \mathbb{R}$ for which

$$\psi_i(\nu) = \mathop{\mathbb{E}}_{Z \sim \nu} [f_i(Z)] .$$

Then, ψ is equivalent to the first n-moment functionals for some $n \leq m$, in the sense that there are real-valued coefficients (b_{ij}) and (c_{ij}) such that for any $\nu \in \mathscr{P}_\psi(\mathbb{R}) \cap \mathscr{P}_m(\mathbb{R})$,

$$\psi_i(\nu) = \sum_{j=1}^{n} b_{ij} \mu_j(\nu) + b_{i0} , \quad i = 1, \ldots, m ;$$

$$\mu_j(\nu) = \sum_{i=1}^{m} c_{ij} \psi_i(\nu) + c_{0j} , \quad j = 1, \ldots, n .$$ △

The proof is somewhat lengthy and is given in Remark 8.2 at the end of the chapter.

As a corollary, we may deduce that any sketch that can be expressed as an invertible function of the first m moments is also Bellman closed. More precisely, if ψ' is a sketch that is an invertible transformation of the sketch ψ corresponding to the first m moments, say $\psi' = h \circ \psi$, then ψ' is Bellman closed with corresponding Bellman operator $h \circ \mathcal{T}^\pi_\psi \circ h^{-1}$. Thus, for example, we may deduce that the sketch corresponding to the mean and variance functionals is Bellman closed, since the mean and variance are expressible as an invertible function of the mean and uncentered second moment. On the other hand, many other statistical functionals (including quantile functionals) are not covered by Theorem 8.12. In the latter case, this is because there is no function $f:\mathbb{R}\to\mathbb{R}$ whose expectation for an arbitrary distribution ν recovers the τth quantile of ν (Exercise 8.5). Still, as established in Example 8.10, quantile sketches are not Bellman closed.

8.4 Statistical Functional Dynamic Programming

When a sketch ψ is not Bellman closed, we lack an operator $\mathcal{T}^\pi_\psi$ that emulates the combination of the distributional Bellman operator and this sketch. This precludes a dynamic programming approach that bootstraps its functional value estimates directly from the previous estimates. However, approximate dynamic programming with arbitrary statistical functionals is still possible if we introduce an additional *imputation step* ι that reconstructs plausible probability distributions from functional values. As we will now see, this allows us to apply the distributional Bellman operator to the reconstructed distributions and then extract the functional values of the resulting return function estimate.

Definition 8.13. An *imputation strategy* for the sketch $\psi:\mathscr{P}_\psi(\mathbb{R})\to\mathbb{R}^m$ is a function $\iota: I_\psi\to\mathscr{P}_\psi(\mathbb{R})$. We say that it is *exact* if for any valid functional values $(s_1,\ldots,s_m)\in I_\psi$, we have

$$\psi_i(\iota(s_1,\ldots,s_m))=s_i,\quad i=1,\ldots,m.$$

Otherwise, we say that it is *approximate*.

By extension, we write $\iota(s)\in\mathscr{P}_\psi(\mathbb{R})^{\mathcal{X}}$ for the return-distribution function corresponding to the collection of functional values $s\in I_\psi^{\mathcal{X}}$. △

In other words, if ι is an exact imputation strategy for the sketch $\psi=(\psi_1,\ldots,\psi_m)$, then for any valid values $s_1,\ldots,s_m$ of the functionals $\psi_1,\ldots,\psi_m$, we have that $\iota(s_1,\ldots,s_m)$ is a probability distribution with the required values under each functional. In a certain sense, ι is a pseudo-inverse to the vector-valued map $\psi:\nu\mapsto(\psi_1(\nu),\ldots,\psi_m(\nu))$. Note that a true inverse to ψ does not exist, as ψ generally does not capture all aspects of the distribution ν.

Once an imputation strategy has been selected, it is possible to write down an approximate dynamic programming algorithm for the functional values under consideration. An abstract framework is given in Algorithm 8.1. In effect, such an algorithm recursively computes the iterates

$$s_{k+1}=\psi(\mathcal{T}^\pi\iota(s_k)) \tag{8.8}$$

from an initial $s_0\in I_\psi^{\mathcal{X}}$. Procedures that implement the iterative process described by Equation 8.8 are referred to as *statistical functional dynamic programming* (SFDP) algorithms. When the sketch ψ is Bellman closed and its imputation strategy ι is exact, the sequence of iterates $(s_k)_{k\geq 0}$ converges to $\psi(\eta^\pi)$, so long as ψ is continuous (with respect to a Wasserstein metric).

Algorithm 8.1: Statistical functional dynamic programming

Algorithm parameters: statistical functionals $\psi_1, \ldots, \psi_m$, imputation strategy ι, initial functional values $((s_i(x))_{i=1}^m : x \in \mathcal{X})$, desired number of iterations K

for $k = 1, \ldots, K$ **do**
 ▷ `Impute distributions`
 $\eta \leftarrow (\iota(s_1(x), \ldots, s_m(x)) : x \in \mathcal{X})$
 ▷ `Apply distributional Bellman operator`
 $\tilde{\eta} \leftarrow \mathcal{T}^\pi \eta$
 foreach state $x \in \mathcal{X}$ **do**
  for $i = 1, \ldots, m$ **do**
   ▷ `Update statistical functional values`
   $s_i(x) \leftarrow \psi_i(\tilde{\eta}(x))$
  end for
 end foreach
end for
return $((s_i(x))_{i=1}^m : x \in \mathcal{X})$

Example 8.14. For the quantile functionals $(\psi^Q_{\tau_i})_{i=1}^m$ with $\tau_i = \frac{2i-1}{2m}$ for $i = 1, \ldots, m$, an exact imputation strategy is

$$(q_1, \ldots, q_m) \mapsto \frac{1}{m} \sum_{i=1}^{m} \delta_{q_i} \,. \tag{8.9}$$

This follows because the $\frac{2i-1}{2m}$-quantile of $\frac{1}{m} \sum_{i=1}^{m} \delta_{q_i}$ is precisely q_i.

Note that when $\tau_1, \ldots, \tau_m \in (0, 1)$ are arbitrary levels with quantile values $(q_1, \ldots, q_m)$, however, it is generally not true that Equation 8.9 is an exact imputation strategy for the corresponding quantile functionals. △

Example 8.15. Categorical dynamic programming can be interpreted as an SFDP algorithm. Indeed, the parameters $p_1, \ldots, p_m$ found by the categorical projection correspond to the values of the following statistical functionals:

$$\psi_i^C(\nu) = \mathop{\mathbb{E}}_{Z\sim\nu}[h_i(\varsigma_m^{-1}(Z-\theta_i)], \quad i = 1, \ldots, m \tag{8.10}$$

where $(h_i)_{i=1}^m$ are the triangular and half-triangular kernels defining the categorical projection on $(\theta_i)_{i=1}^m$ (Section 5.6). An exact imputation strategy in this case is the function that returns the unique distribution supported on $(\theta_i)_{i=1}^m$ that matches the estimated functional values $p_i = \psi_i^C(\nu)$, $i = 1, \ldots, m$:

$$(p_1, \ldots, p_m) \mapsto \sum_{i=1}^m p_i \delta_{\theta_i}. \qquad \triangle$$

Mathematically, an exact imputation strategy always exists, because we defined imputation strategies in terms of valid functional values. However, there is no guarantee that an efficient algorithm exists to compute the application of this strategy to arbitrary functional values. In practice, we may favor *approximate strategies* with efficient implementations. For example, we may map functional values to probability distributions from a representation $\mathscr{F}$ by optimizing some notion of distance between functional values. The optimization process may not yield an exact match in $\mathscr{F}$ (one may not even exist) but can often be performed efficiently.

Example 8.16. Let $\psi_1^c, \ldots, \psi_m^c$ be the categorical functionals from Equation 8.10. Suppose we are given the corresponding functional values $p_1, \ldots, p_m$ of a probability distribution ν:

$$p_i = \psi_i^c(\nu), \quad i = 1, \ldots, m.$$

An approximate imputation strategy for these functionals is to find the n-quantile distribution (n possibly different from m)

$$\nu_\theta = \frac{1}{n}\sum_{i=1}^n \delta_{\theta_i}$$

that best fits p_i according to the loss

$$\mathcal{L}(\theta) = \sum_{i=1}^m \left| p_i - \psi_i^c(\nu_\theta) \right|. \tag{8.11}$$

Exercise 8.7 asks you to demonstrate that this strategy is approximate for $m > 2$. Although in this context, we know of an exact imputation strategy based on categorical distributions, this illustrates that it is possible to impute distributions from a different representation. △

8.5 Relationship to Distributional Dynamic Programming

In Chapter 5, we introduced distributional dynamic programming (DDP) as a class of methods that operates over return-distribution functions. In fact, every statistical functional dynamic programming is also a DDP algorithm (but not the other way around; see Exercise 8.8). This relationship is established by considering the implied representation

$$\mathscr{F} = \{\iota(s) : s \in I_\psi\} \subseteq \mathscr{P}(\mathbb{R})$$

and the projection $\Pi_{\mathscr{F}} = \iota \circ \psi$ (see Figure 8.3).

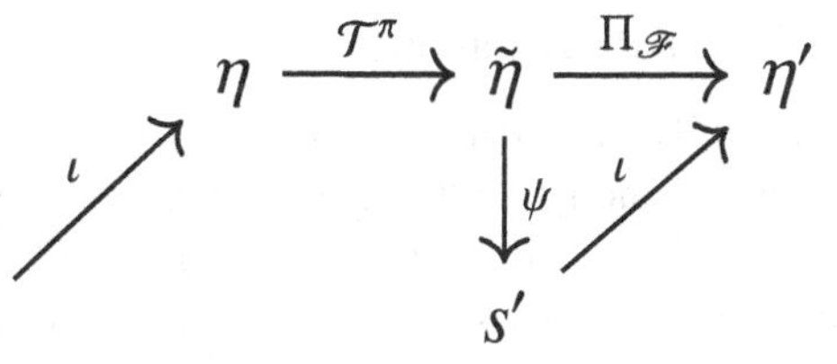

Figure 8.3
The interpretation of SFDP algorithms as distributional dynamic programming algorithms. Traversing along the diagram from η to η' corresponds to dynamic programming implementing a projected Bellman operator, while the path from s to s' corresponds to statistical functional dynamic programming (SFDP).

From this correspondence, we may establish the relationship between Bellman closedness and the notion of a diffusion-free projection developed in Chapter 5.

Proposition 8.17. Let ψ be a Bellman-closed sketch. Then for any choice of exact imputation strategy $\iota : I_\psi \to \mathscr{P}_\psi(\mathbb{R})$, the projection operator $\Pi_{\mathscr{F}} = \iota\psi$ is diffusion-free. △

Proof. We may directly check the diffusion-free property (omitting parentheses for conciseness):

$$\Pi_{\mathscr{F}}\mathcal{T}^\pi\Pi_{\mathscr{F}} = \iota\psi\mathcal{T}^\pi\iota\psi \overset{(a)}{=} \iota\mathcal{T}^\pi_\psi\psi\iota\psi \overset{(b)}{=} \iota\mathcal{T}^\pi_\psi\psi \overset{(a)}{=} \iota\psi\mathcal{T}^\pi = \Pi_{\mathscr{F}}\mathcal{T}^\pi\,,$$

where steps marked (a) follow from the identity $\psi\mathcal{T}^\pi = \mathcal{T}^\pi_\psi\psi$, and (b) follows from the identity $\psi\iota\psi = \psi$ for any exact imputation strategy ι for ψ. □

Imputation strategies formalize how one might interpret functional values as parameters of a probability distribution. Naturally, the chosen imputation strategy affects the approximation artifacts from distributional dynamic programming, the rate of convergence, and whether the algorithm converges at all.

Compared with representation-based algorithms of the style introduced in Chapter 5, working with statistical functionals allows us to design the projection $\Pi_{\mathscr{F}}$ in two separate pieces: a sketch ψ and an imputation strategy ι. In particular, this makes it possible to learn statistical functionals that would be difficult to directly capture in a probability distribution representation. As the next section demonstrates, this allows us to create new kinds of distributional reinforcement learning algorithms.

8.6 Expectile Dynamic Programming

Expectiles form a family of statistical functionals parameterized by a level $\tau \in (0, 1)$. They extend the notion of the mean of a distribution ($\tau = 0.5$) similar to how quantiles extend the notion of a median. Expectiles have classically found application in econometrics and finance as a form of risk measure (see the bibliographical remarks for further details). Based on the principles of statistical functional dynamic programming, *expectile dynamic programming*[64] uses an approximate imputation strategy in order to iteratively estimate the expectiles of the return function.

Definition 8.18. For a given $\tau \in (0, 1)$, the τ-*expectile* of a distribution $\nu \in \mathscr{P}_2(\mathbb{R})$ is

$$\psi^{\mathrm{E}}_{\tau}(\nu) = \operatorname*{arg\,min}_{z \in \mathbb{R}} \mathrm{ER}_{\tau}(z; \nu), \tag{8.12}$$

where

$$\mathrm{ER}_{\tau}(z; \nu) = \mathop{\mathbb{E}}_{Z \sim \nu}\left[|\mathbb{1}_{\{Z<z\}} - \tau| \times (Z - z)^2 \right] \tag{8.13}$$

is the *expectile loss*. △

The loss appearing in Definition 8.18 is strongly convex (Boyd and Vandenberghe 2004) and bounded below by 0. As a consequence, Equation 8.12 has a unique minimizer for a given τ; this verifies that the corresponding expectile is uniquely defined.

To understand the relationship to the mean functional and develop some intuition for the statistical property than an expectile encodes, observe that the mean of a distribution $\nu \in \mathscr{P}_2(\mathbb{R})$ can be expressed as

$$\mu_1(\nu) = \operatorname*{arg\,min}_{z \in \mathbb{R}} \mathop{\mathbb{E}}_{Z \sim \nu}\left[(Z - z)^2 \right].$$

64. The incremental analogue is called *expectile temporal-difference learning* (Rowland et al. 2019).

Similar to how a quantile is derived from a loss that weights errors asymmetrically (depending on whether the realization from Z is smaller or greater than z), the expectile loss for $\tau \in (0, 1)$ is the asymmetric version of the above. For τ greater than $1/2$, one can think of the expectile as an "optimistic" summary of the distribution – a value that emphasizes outcomes that are greater than the mean. Conversely, for τ smaller than $1/2$, the corresponding expectile is in a sense "pessimistic."

Expectile dynamic programming (EDP) estimates the values of a finite set of expectile functionals with values $0 < \tau_1 < \cdots < \tau_m < 1$. For a distribution $\nu \in \mathscr{P}_2(\mathbb{R})$, let us write

$$e_i = \psi^{\mathrm{E}}_{\tau_i}(\nu)\,.$$

Given the collection of expectile values $e_1, \ldots, e_m$, EDP uses an imputation strategy that outputs an n-quantile probability distribution that approximately has these expectile values.[65]

The imputation strategy finds a suitable reconstruction by finding a solution to a root-finding problem. To begin, this strategy outputs a n-quantile distribution $\hat{\nu}$, with n possibly different from m:

$$\hat{\nu} = \frac{1}{n}\sum_{j=1}^{n} \delta_{\theta_j}\,.$$

Following Definition 8.13, for this imputation to be exact, the expectiles of $\hat{\nu}$ at $\tau_1, \ldots, \tau_m$ should be equal to $e_1, \ldots, e_m$:

$$\psi^{\mathrm{E}}_{\tau_i}(\hat{\nu}) = e_i, \quad i = 1, \ldots, m\,.$$

This constraint implies that the derivatives of the expectile loss, instantiated with $\tau_1, \ldots, \tau_m$ and evaluated with $\hat{\nu}$, should all be 0:

$$\partial_z \,\mathrm{ER}_{\tau_i}\Big(z; \hat{\nu}\Big)\Big|_{z=e_i} = 0, \quad i = 1, \ldots, m\,. \tag{8.14}$$

Written out in full for the choice of $\hat{\nu}$ above, these derivatives take the form

$$\partial_z \,\mathrm{ER}_{\tau_i}\Big(z; \hat{\nu}\Big)\Big|_{z=e_i} = \frac{1}{n}\sum_{j=1}^{n} \frac{1}{2}(e_i - \theta_j)|\mathbb{1}_{\{\theta_j < e_i\}} - \tau_i|, \qquad i = 1, \ldots, m\,.$$

An alternative to the root-finding problem expressed in Equation 8.14 is the following optimization problem:

$$\text{minimise} \sum_{i=1}^{m} \left(\partial_z \,\mathrm{ER}_{\tau_i}\Big(z; \hat{\nu}\Big)\Big|_{z=e_i}\right)^2\,. \tag{8.15}$$

65. Of course, this particular form for the imputation strategy is a design choice; the reader is invited to consider what other imputation strategies might be sensible here.

A practical implementation of this imputation strategy therefore applies an optimization algorithm to the objective in Equation 8.15, or a root-finding method to Equation 8.14, viewed as functions of $\theta_1, \ldots, \theta_n$. Because the optimization algorithm may return a solution that does not exactly satisfy Equation 8.14, this method is an approximate (rather than exact) imputation strategy. It can be used in the *impute distributions* step of Algorithm 8.1, yielding a dynamic programming algorithm that aims to approximately learn return-distribution expectiles. If the root-finding algorithm is always able to find $\hat{\nu}$ exactly satisfying Equation 8.14, then the imputation strategy is exact in this instance; otherwise, it is approximate. A specific implementation is explored in detail in Exercise 8.10.

8.7 Infinite Collections of Statistical Functionals

Thus far, our treatment of statistical functionals has focused on finite collections of statistical functionals – what we call a sketch. From a computational standpoint, this is sensible since, to implement an SFDP algorithm, one needs to be able to operate on individual functional values. On the other hand, in Section 8.3, we saw that many sketches are not Bellman closed and must be combined with an imputation strategy in order to perform dynamic programming. An alternative, which we will study in greater detail in Chapter 10, is to implicitly parameterize an *infinite* family of statistical functionals.

Many (though not all) infinite families of functionals provide a lossless encoding of probability distributions and are consequently Bellman closed – that is, knowing the values taken on by these functionals is equivalent to knowing the distribution itself. We encode this property with the following definition.

Definition 8.19. Let Ψ be a set of statistical functionals. We say that Ψ *characterizes* probability distributions over the real numbers if, for each $\nu \in \mathscr{P}(\mathbb{R})$, there is a unique collection of functional values $(\psi(\nu) : \psi \in \Psi)$. △

The following families of statistical functionals all characterize probability distributions over $\mathbb{R}$.

The cumulative distribution function. The functionals mapping distributions ν to the probabilities $\mathbb{P}_{Z\sim\nu}(Z \leq z)$, indexed by $z \in \mathbb{R}$. Closely related are *upper-tail probabilities*,

$$\nu \mapsto \mathbb{P}_{Z\sim\nu}(Z \geq z)\,,$$

and the quantile functionals

$$\nu \mapsto F_\nu^{-1}(\tau)\,,$$

indexed by $\tau \in (0, 1)$.

The characteristic function. Functionals of the form

$$\nu \mapsto \underset{Z\sim\nu}{\mathbb{E}}[e^{\mathrm{i}uZ}] \in \mathbb{C},$$

indexed by $u \in \mathbb{R}$ (and where $\mathrm{i}^2 = -1$). The corresponding collection of statistical values is the *characteristic function* of ν, denoted χ_ν.

Moments and cumulants. The infinite collection of moment functionals $(\mu_p)_{p=1}^{\infty}$ does not unconditionally characterize the distribution ν: there are distinct distributions that have the same sequence of moments. However, if the sequence of moments does not grow too quickly, uniqueness is restored. In particular, a sufficient condition for uniqueness is that the underlying distribution ν has a *moment-generating function*

$$u \mapsto \underset{Z\sim\nu}{\mathbb{E}}[e^{uZ}],$$

which is finite in an open neighborhood of $u = 0$; see Remark 8.3 for further details. Under this condition, the moment-generating function itself also characterizes the distribution, as does the *cumulant-generating function*, defined as the logarithm of the moment-generating function,

$$u \mapsto \log\Big(\underset{Z\sim\nu}{\mathbb{E}}[e^{uZ}]\Big).$$

The *cumulants* $(\kappa_p)_{p=1}^{\infty}$ are defined through a power series expansion of the cumulant-generating function

$$\log\Big(\underset{Z\sim\nu}{\mathbb{E}}[e^{uZ}]\Big) = \sum_{p=1}^{\infty} \frac{\kappa_p u^p}{p!}.$$

Under the condition that the moment-generating function is finite in an open neighborhood of the origin, the sequences of cumulants and moments are determined by one another, and so the sequence of cumulants is another characterization of the distribution under this condition.

Example 8.20. Consider the return-variable Bellman equation

$$G^\pi(x,a) \overset{\mathcal{D}}{=} R + \gamma G^\pi(X',A'), \quad X = x, A = a.$$

If for each $u \in \mathbb{R}$ we apply the functional $\nu \mapsto \underset{Z\sim\nu}{\mathbb{E}}[e^{\mathrm{i}uZ}]$ to the distribution of the random variables on each side, we obtain the *characteristic function Bellman equation*:

$$\begin{aligned}
\chi_{\eta^\pi(x,a)}(u) &= \mathbb{E}_\pi[e^{\mathrm{i}u(R+\gamma G^\pi(X',A'))} \mid X = x, A = a] \\
&= \mathbb{E}_\pi[e^{\mathrm{i}uR} \mid X = x, A = a]\,\mathbb{E}_\pi[e^{\mathrm{i}\gamma u G^\pi(X',A')} \mid X = x, A = a] \\
&= \chi_{P_{\mathcal{R}}(\cdot\mid x,a)}(u)\,\mathbb{E}_\pi[\chi_{\eta^\pi(X',A')}(\gamma u) \mid X = x, A = a].
\end{aligned}$$

This is a different kind of distributional Bellman equation in which the addition of independent random variables corresponds to a multiplication of their characteristic functions. The equation highlights that the characteristic function of ν evaluated at u depends on the next-state characteristic functions evaluated at γu. This shows that for a set $S \subseteq \mathbb{R}$, the sketch $(\nu \mapsto \chi_\nu(u) : u \in S)$ cannot be Bellman closed unless S is infinite or $S = \{0\}$. Exercise 8.12 asks you to give a theoretical analysis of a dynamic programming approach based on characteristic functions. △

Another way to understand collections of statistical functionals that are characterizing (in the sense of Definition 8.19) is to interpret them in light of our definition of a probability distribution representation (Definition 5.2). Recall that a representation $\mathscr{F}$ is a collection of distributions indexed by a parameter θ:

$$\mathscr{F} = \{\nu_\theta \in \mathscr{P}(\mathbb{R}) : \theta \in \Theta\}.$$

Here, the functional values associated with the set of statistical functionals Ψ correspond to the (infinite-dimensional) parameter θ, so that

$$\mathscr{F}_\Psi = \mathscr{P}(\mathbb{R}).$$

This clearly implies that $\mathscr{F}_\Psi$ is closed under the distributional Bellman operator $\mathcal{T}^\pi$ (Section 5.3) and hence that approximation-free distributional dynamic programming is (mathematically) possible with $\mathscr{F}_\Psi$.

8.8 Moment Temporal-Difference Learning*

In Section 8.2, we introduced the m-moment Bellman operator, from which an exact dynamic programming algorithm can be derived. A natural follow-up is to apply the tools of Chapter 6 to derive an incremental algorithm for learning the moments of the return-distribution function from samples. Here, an algorithm that incrementally updates an estimate $M \in \mathbb{R}^{\mathcal{X}\times\mathcal{A}\times m}$ of the m first moments of the return function can be directly obtained through the unbiased estimation approach, as the corresponding operator can be written as an expectation. Given a sample transition (x, a, r, x', a'), the unbiased estimation approach yields the update rule (for $i = 1, \dots, m$)

$$M(x, a, i) \leftarrow (1-\alpha)M(x, a, i) + \alpha\left[\sum_{j=0}^{i} \gamma^{i-j}\binom{i}{j} r^j M(x', a', i-j)\right], \qquad (8.16)$$

where again we take $M(\cdot, \cdot, 0) = 1$ by convention.

Unlike the TD and CTD algorithms analyzed in Chapter 6, this algorithm is derived from an operator, $T^\pi_{(m)}$, which is not a contraction in a supremum-norm over states. As a result, the theory developed in Chapter 6 cannot immediately be applied to demonstrate convergence of this algorithm under appropriate conditions. With some care, however, a proof is possible; we now give an overview of what is needed.

The proof of Proposition 8.7 demonstrates that the behavior of $T^\pi_{(m)}$ is closely related to that of a contraction mapping. Specifically, the behavior of $T^\pi_{(m)}$ in updating the estimates of ith moments of returns is contractive if the lower moment estimates are sufficiently close to their correct values. To turn these observations into a proof of convergence, an inductive argument on the moments being learnt must be made, as in the proof of Proposition 8.7. Further, the approach of Chapter 6 needs to be extended to deal with a vanishing bias term in the update to account for this "near-contractivity" of $T^\pi_{(m)}$; to this end one may, for example, begin from the analysis of Bertsekas and Tsitsiklis (1996, Proposition 4.5).

Before moving on, let us remark that in practice, we are likely to be interested in centered moments such as the variance ($m = 2$); these take the form

$$\mathbb{E}_\pi\Big[\Big(\sum_{t=0}^{\infty}\gamma^t R_t - Q^\pi(x,a)\Big)^m \,\Big|\, X_0 = x, A_0 = a\Big],$$

These can be derived from their uncentered counterparts; for example, the variance of the return distribution $\eta^\pi(x,a)$ is obtained from the first two uncentered moments via Equation 8.2.

It is also possible to perform dynamic programming on centered moments directly, as was shown in the context of the mean and variance in Section 5.4 (Exercise 8.14 asks you to derive the Bellman operators for the more general case of the first m centered moments). Given in terms of state-action pairs, the Bellman equation for the return variances $\bar{M}^\pi(\cdot,\cdot,2) \in \mathbb{R}^{\mathcal{X}\times\mathcal{A}}$ is

$$\begin{aligned}\bar{M}^\pi(x,a,2) &= \mathrm{Var}_\pi(R \mid X = x, A = a) + \\ &\gamma^2\Big(\mathrm{Var}_\pi(Q^\pi(X',A') \mid X = x, A = a) + \mathbb{E}_\pi[\bar{M}^\pi(X',A',2) \mid X = x, A = a]\Big);\end{aligned} \tag{8.17}$$

contrast with Equation 5.20.

One challenge with deriving an incremental algorithm for learning the variance directly is that unbiasedly estimating some of the variance terms on the right-hand side requires multiple samples. For example, an unbiased estimator of

$$\mathrm{Var}_\pi(Q^\pi(X',A') \mid X = x, A = a)$$

in general requires two independent realizations of X', A' for a given source state-action pair x, a. Consequently, unbiased estimation of the corresponding operator application with a single transition is not feasible in this case. Despite the fact that the first m centered and uncentered moments of a probability distribution can be recovered from one another, there is a distinct advantage associated with working with uncentered moments when learning from samples.

8.9 Technical Remarks

Remark 8.1. Theorem 8.12 illustrates how dynamic programming over functional values must incur some approximation error, unless the underlying sketch is Bellman closed. One way to avoid this error is to augment the state space with additional information: for example, the return accumulated so far. We in fact took this approach when optimizing the conditional value-at-risk (CVaR) of the return in Chapter 7; in fact, risk measures are statistical functionals that may also take on the value $-\infty$ (see Definition 7.14). △

Remark 8.2 (Proof of Theorem 8.12). It is sufficient to consider a pair of states, x and y, such that x deterministically transitions to y with reward r. Because ψ is Bellman closed, we can identify an associated Bellman operator $\mathcal{T}^\pi_\psi$. For a given return function η whose state-indexed collection of functional values is $s = \psi(\eta)$, let us write $(\mathcal{T}^\pi_\psi s)_i(x)$ for the ith functional value at state x, for $i = 1, \dots, m$. By construction and definition of the operator $\mathcal{T}^\pi_\psi$, $(\mathcal{T}^\pi_\psi s)_i(x)$ is a function of the functional values at y as well as the reward r and discount factor γ, and so we may write

$$(\mathcal{T}^\pi_\psi s)_i(x) = g_i(r, \gamma, \psi_1(\eta(y)), \dots, \psi_m(\eta(y)))$$

for some function g_i. We next argue that g_i is affine[66] in the inputs $\psi_1(\eta(y)), \dots, \psi_m(\eta(y))$. This is readily observed as each functional $\psi_1, \dots, \psi_m$ is affine in its input distribution,

$$\begin{aligned}\psi_i(\alpha\nu + (1-\alpha)\nu') &= \mathop{\mathbb{E}}_{Z\sim\alpha\nu+(1-\alpha)\nu'}[f_i(Z)] \\ &= \alpha \mathop{\mathbb{E}}_{Z\sim\nu}[f_i(Z)] + (1-\alpha) \mathop{\mathbb{E}}_{Z\sim\nu'}[f_i(Z)] \\ &= \alpha\psi_i(\nu) + (1-\alpha)\psi_i(\nu'),\end{aligned}$$

and

$$(\mathcal{T}^\pi_\psi s)_i(x) = \mathop{\mathbb{E}}_{Z\sim\eta(y)}[f_i(r+\gamma Z)]$$

is also affine as a function of η. This affineness would be contradicted if g_i were not also affine. Hence, there exist functions $\beta_i : \mathbb{R} \times [0, 1) \to \mathbb{R}$ for $i = 1, \dots, m$

66. Recall that a function $h : M \to M'$ between vector spaces M and M' is affine if for $u_1, u_2 \in M$, $\lambda \in (0, 1)$, we have $h(\lambda u_1 + (1-\lambda)u_2) = \lambda h(u_1) + (1-\lambda)h(u_2)$.

such that

$$g_i(r, \gamma, \psi_1(\eta(y)), \dots, \psi_m(\eta(y))) = \beta_0(r, \gamma) + \sum_{i=1}^{m} \beta_i(r, \gamma)\psi_i(\eta(y)),$$

and therefore

$$\mathop{\mathbb{E}}_{Z\sim\eta(y)}[f_i(r + \gamma Z)] = \mathop{\mathbb{E}}_{Z\sim\eta(y)}\left[\sum_{j=0}^{m} \beta_i(r, \gamma) f_j(Z)\right],$$

where $f_0(z) = 1$. Taking $\eta(y)$ to be a Dirac delta δ_z then gives the following identity:

$$f_i(r + \gamma z) = \sum_{j=0}^{m} \beta_i(r, \gamma) f_j(z).$$

We therefore have that the finite-dimensional function space spanned by $f_0, f_1, \dots, f_m$ (where f_0 is the constant function equal to 1) is closed under translation (by $r \in \mathbb{R}$) and scaling (by $\gamma \in [0, 1)$). Engert (1970) shows that the only finite-dimensional subspaces of measurable functions closed under translation are contained in the span of finitely many functions of the form $z \mapsto z^\ell \exp(\lambda z)$, with $\ell \in \mathbb{N}$ and $\lambda \in \mathbb{C}$. Since we further require closure under scaling by $\gamma \in [0, 1)$, we deduce that we must have $\lambda = 0$ in any such function, and the subspace must be equal to the space spanned by the first n monomials (and the constant function).

To conclude, since each monomial $z \mapsto z^i$ for $i = 1, \dots, n$ is expressible as a linear combination of $f_0, \dots, f_m$, the corresponding expectations $\mathbb{E}_{Z\sim\nu}[Z^i]$ are expressible as linear combinations of the expectations $\mathbb{E}_{Z\sim\nu}[f_j(Z)]$, for any distribution ν. The converse also holds, and so we conclude that the sketch ψ encodes the same distributional information as the first n moments. △

Remark 8.3. The question of whether a distribution is characterized by its sequence of moments has been a subject of study in probability theory for over a century. The sufficient condition on the moment-generating function described in Section 8.8 means that the characteristic function of such a distribution can be written as a power series with scaled moments as coefficients, ensuring uniqueness of the distribution; see, for example, Billingsley (2012) for a detailed discussion. Lin (2017) gives a survey of known sufficient conditions for characterization, as well as examples where characterization does not hold. △

8.10 Bibliographical Remarks

8.1. Statistical functionals are a core notion in statistics; see, for example, the classic text by van der Vaart (2000). In reinforcement learning, specific functionals such as moments, quantiles, and CVaR have been of interest for risk-sensitive control (more on this in the bibliographical remarks of Chapter 7). Chandak et al. (2021) consider the problem of off-policy Monte Carlo policy evaluation of arbitrary statistical functionals of the return distribution.

8.2, 8.8. Sobel (1982) gives a Bellman equation for return-distribution moments for state-indexed value functions with deterministic policies. More recent work in this direction includes that of Lattimore and Hutter (2012), Azar et al. (2013), and Azar et al. (2017), who make use of variance estimates in combination with Bernstein's inequality to improve the efficiency of exploration algorithms, as well as the work of White and White (2016), who use estimated return variance to set trace coefficients in multistep TD learning methods. Sato et al. (2001), Tamar et al. (2012), Tamar et al. (2013), and Prashanth and Ghavamzadeh (2013) further develop methods for learning the variance of the return. Tamar et al. (2016) show that the operator $T^{\pi}_{(2)}$ is a contraction under a weighted norm (see Exercise 8.4), develop an incremental algorithm with a proof of convergence using the ODE method, and study both dynamic programming and incremental algorithms under linear function approximation (the topic of Chapter 9).

8.3–8.5. The notion of Bellman closedness is due to Rowland et al. (2019), although our presentation here is a revised take on the idea. The noted connection between Bellman closedness and diffusion-free representations and the term "statistical functional dynamic programming" are new to this book.

8.6. The expectile dynamic programming algorithm is new to this book but is directly derived from expectile temporal-difference learning (Rowland et al. 2019). Expectiles themselves were introduced by Newey and Powell (1987) in the context of testing in econometric regression models, with the asymmetric squared loss defining expectiles already appearing in Aigner et al. (1976). Expectiles have since found further application as risk measures, particularly within finance (Taylor 2008; Kuan et al. 2009; Bellini et al. 2014; Ziegel 2016; Bellini and Di Bernardino 2017). Our presentation here focuses on the asymmetric squared loss, requiring a finite second-moment assumption, but an equivalent definition allows expectiles to be defined for all distributions with a finite first moment (Newey and Powell 1987).

8.7. The study of characteristic functions in distributional reinforcement learning is due to Farahmand (2019), who additionally provides error propagation analysis for the *characteristic value iteration* algorithm, in which value iteration is carried out with characteristic function representations of return distributions. Earlier, Mandl (1971) studied the characteristic function of the return in Markov decision processes with deterministic immediate rewards and policies. Chow et al. (2015) combine a state augmentation method (see Chapter 7) with an infinite-dimensional Bellman equation for CVaR values to learn a CVaR-optimal policy. They develop an implementable version of the algorithm by tracking finitely many CVaR values and using linear interpolation for the remainder, an approach related to the imputation strategies described earlier in the chapter. Characterization via the quantile function has driven the success of several large-scale distributional reinforcement learning algorithms (Dabney et al. 2018a; Yang et al. 2019), and is the subject of further study in Chapter 10.

8.11 Exercises

Exercise 8.1. Consider the m-moment Bellman operator $T^\pi_{(m)}$ (Definition 8.6). For $M \in \mathbb{R}^{\mathcal{X}\times\mathcal{A}\times m}$, define the norm

$$\|M\|_{\infty,\text{MAX}} = \max_{i\in\{1,\dots,m\}} \sup_{\substack{x\in\mathcal{X}\\ a\in\mathcal{A}}} M(x,a,i).$$

By means of a counterexample, show that $T^\pi_{(m)}$ is not a contraction mapping in the metric induced by $\|\cdot\|_{\infty,\text{MAX}}$. △

Exercise 8.2. Let $\varepsilon > 0$. Determine a bound on the computational cost (in $O(\cdot)$ notation) of performing iterative policy evaluation with the m-moment Bellman operator to obtain an approximation $\hat{M}^\pi$ such that

$$\max_{i\in\{1,\dots,m\}} \sup_{\substack{x\in\mathcal{X}\\ a\in\mathcal{A}}} |\hat{M}^\pi(x,a,i) - M^\pi(x,a,i)| < \varepsilon.$$

You may find it convenient to refer to the proof of Proposition 8.7. △

Exercise 8.3. Equation 5.2 gives the value function V^π as the solution of the linear system of equations

$$V = r^\pi + \gamma P^\pi V.$$

Provide the analogous linear system for the moment function M^π. △

Exercise 8.4. The purpose of this exercise is to show that $T^\pi_{(2)}$ is a contraction mapping on $\mathbb{R}^{\mathcal{X}\times\mathcal{A}\times 2}$ in a *weighted* L^∞ norm, as shown by Tamar et al. (2016). Let $M \in \mathbb{R}^{\mathcal{X}\times\mathcal{A}\times 2}$ be a moment function estimate (specifically, for the first two moments). For each $\alpha \in (0,1)$, define the α-weighted norm on $\mathbb{R}^{\mathcal{X}\times\mathcal{A}\times 2}$ by

$$\|M\|_\alpha = \alpha\|M_{(1)}\|_\infty + (1-\alpha)\|M_{(2)}\|_\infty,$$

where $M_{(i)} = M(\cdot,\cdot,i) \in \mathbb{R}^{\mathcal{X}\times\mathcal{A}}$. For any $M, M' \in \mathbb{R}^{\mathcal{X}\times\mathcal{A}\times 2}$, show that

$$\|T^\pi_{(2)}(M-M')\|_\alpha \le \alpha\|\gamma P^\pi(M_{(1)} - M'_{(1)})\|_\infty$$
$$+ (1-\alpha)\|2\gamma C_r P^\pi(M_{(1)} - M'_{(1)}) + \gamma^2 P^\pi(M_{(2)} - M'_{(2)})\|_\infty,$$

where P^π is the state-action transition operator, defined by

$$(P^\pi Q)(x,a) = \sum_{(x',a')\in\mathcal{X}\times\mathcal{A}} P^\pi(x'|x,a)\pi(a'|x')Q(x',a'),$$

and C_r is the diagonal reward operator

$$(C_rQ)(x,a) = \mathbb{E}[R\,|\,X=x, A=a]Q(x,a).$$

Writing $\lambda \geq 0$ for the Lipschitz constant of C_rP^π with respect to the L^∞ metric, deduce that

$$\begin{aligned}&\|T^\pi_{(2)}(M-M')\|_\alpha \\ &\leq (\alpha\gamma + 2(1-\alpha)\gamma\lambda)\|M_{(1)} - M'_{(1)}\|_\infty + (1-\alpha)\gamma^2\|M_{(2)} - M'_{(2)}\|_\infty .\end{aligned}$$

Hence, deduce that there exist parameters $\alpha \in (0,1), \beta \in [0,1)$ such that

$$\|T^\pi_{(2)}(M-M')\|_\alpha \leq \beta\|M-M'\|_\alpha ,$$

as required. △

Exercise 8.5. Consider the median functional

$$\nu \mapsto F^{-1}_\nu(0.5).$$

Show that there does not exist a function $f : \mathbb{R} \to \mathbb{R}$ such that, for any $\nu \in \mathbb{R}$,

$$\mathop{\mathbb{E}}_{Z\sim\nu}[f(Z)] = F^{-1}_\nu(0.5).$$ △

Exercise 8.6. Consider the subset of probability distributions endowed with a probability density f_ν. Repeat the preceding exercise for the differential entropy functional

$$\nu \mapsto -\int_{z\in\mathbb{R}} f_\nu(z)\log(f_\nu(z))\mathrm{d}z .$$ △

Exercise 8.7. For the imputation strategy of Example 8.16:

(i) show that for $m = 2$, the imputation strategy is exact, for any $n \in \mathbb{N}^+$.

(ii) show that for $m > 2$, this imputation strategy is inexact. *Hint.* Find a distribution ν for which $\psi^{\mathrm{c}}_i(\iota(p_1,\ldots,p_m)) \neq p_i$, for some $i = 1,\ldots,m$. △

Exercise 8.8. In Section 8.4, we argued that every statistical functional dynamic programming algorithm is a distributional dynamic programming algorithm. Explain why the converse is false. Under what circumstances may we favor either an algorithm that operates on statistical functionals or one that operates on probability distribution representations? △

Exercise 8.9. Consider an imputation strategy ι for a sketch ψ. We say the (ψ,ι) pair is *mean-preserving* if, for any probability distribution $\nu \in \mathscr{P}_\psi(\mathbb{R})$,

$$\nu' = \iota\psi(\nu)$$

satisfies

$$\mathop{\mathbb{E}}_{Z\sim\nu'}[Z] = \mathop{\mathbb{E}}_{Z\sim\nu}[Z] .$$

Show that in this case, the operator

$$\psi \circ \mathcal{T}^\pi \circ \iota$$

is also mean-preserving. △

Exercise 8.10. Using your favorite numerical computation software, implement the expectile imputation strategy described in Section 8.6. Specifically:

(i) Implement a procedure for approximately determining the expectile values $e_1, \dots, e_m$ of a given distribution. *Hint.* An incremental approach in the style of quantile regression, or a binary search approach, will allow you to deal with continuous distributions.

(ii) Given a set of expectile values, $e_1, \dots, e_m$, implement a procedure that imputes an n-quantile distribution

$$\frac{1}{n}\sum_{i=1}^{n} \delta_{\theta_i}$$

by minimizing the objective given in Equation 8.15.

Test your implementation on discrete and continuous distributions, and compare it with the best m-quantile approximation of those distributions. Is one method better suited to discrete distributions than the other? More generally, when might one method be preferred over the other? △

Exercise 8.11. Formulate a variant of expectile dynamic programming that imputes n-quantile distributions and whose (possibly approximate) imputation strategy is mean-preserving in the sense of Exercise 8.9. △

Exercise 8.12 (*). This exercise applies the line of reasoning from Chapter 4 to characteristic functions and is based on Farahmand (2019). For a probability distribution ν, recall that its characteristic function χ_ν is

$$\chi_\nu(u) = \mathop{\mathbb{E}}_{Z\sim\nu}\left[e^{iuZ}\right] .$$

Now, for $p \in [1, \infty)$, define the probability metric

$$d_{1,p}(\nu, \nu') = \int_{u\in\mathbb{R}} \frac{|\chi_\nu(u) - \chi_{\nu'}(u)|}{|u|^p} \mathrm{d}u$$

and its supremum extension to return functions

$$\overline{d}_{1,p}(\eta, \eta') = \sup_{(x,a)\in\mathcal{X}\times\mathcal{A}} d_{1,p}(\eta(x,a), \eta'(x,a)) .$$

(i) Determine a subset $\mathscr{P}_{\chi,p}(\mathbb{R}) \subseteq \mathscr{P}(\mathbb{R})$ on which $d_{1,p}$ is a proper metric.

(ii) Provide assumption(s) under which the return function η^π lies in $\mathscr{P}_{\chi,p}(\mathbb{R})$.

(iii) Prove that for $p \geq 2$, the distributional Bellman operator is a contraction mapping in $d_{1,p}$, with modulus γ^{p-1}. △

Exercise 8.13 (*). Consider the probability metric

$$d_{2,2}(\nu, \nu') = \Big(\int_{u \in \mathbb{R}} \frac{(\chi_\nu(u) - \chi_{\nu'}(u))^2}{u^2} \mathrm{d}u \Big)^{1/2}.$$

Show that $d_{2,2}$ is the Cramér distance ℓ_2. *Hint.* Use the Parseval–Plancherel identity. △

Exercise 8.14. Let $m \in \mathbb{N}^+$. Derive a Bellman operator on $\mathbb{R}^{\mathcal{X} \times \mathcal{A} \times m}$ whose unique fixed point $\tilde{M}^\pi$ is the collection of centered moments:

$$\tilde{M}^\pi(x, a, i) = \mathbb{E}\Big[(G^\pi(x, a) - Q^\pi(x, a))^i\Big], \quad i = 1, \ldots, m.$$ △

9 Linear Function Approximation

A probability distribution representation is used to describe return functions in terms of a collection of numbers that can be stored in a computer's memory. With it, we can devise algorithms that operate on return distributions in a computationally efficient manner, including distributional dynamic programming algorithms and incremental algorithms such as CTD and QTD. *Function approximation* arises when our representation of the value or return function uses parameters that are shared across states. This allows reinforcement learning methods to be applied to domains where it is impractical or even impossible to keep in memory a table with a separate entry for each state, as we have done in preceding chapters. In addition, it makes it possible to make predictions about states that have not been encountered – in effect, to *generalize* a learned estimate to new states.

As a concrete example, consider the problem of determining an approximation to the optimal value function for the game of Go. In Go, players take turns placing white and black stones on a 19×19 grid. At any time, each location of the board is either occupied by a white or black stone, or unoccupied; consequently, there are astronomically many possible board states.[67] Any practical algorithm for this problem must therefore use a succinct representation of its value or return function.

Function approximation is also used to apply reinforcement learning algorithms to problems with continuous state variables. The classic Mountain Car domain, in which the agent must drive an underpowered car up a steep hill, is one such problem. Here, the state consists of the car's position and velocity (both bounded on some interval); learning to control the car requires being able

67. A naive estimate is $3^{19\times 19}$. The real figure is somewhat lower due to symmetries and the impossibility of certain states.

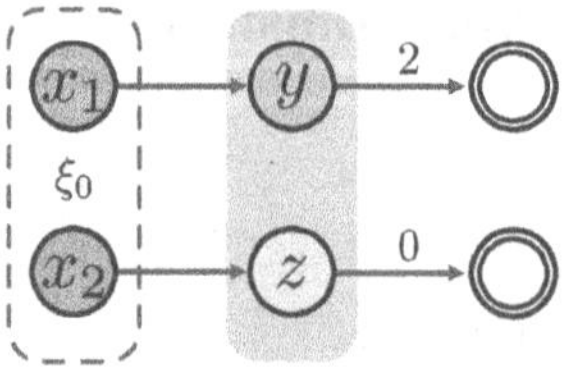

Figure 9.1
A Markov decision process in which aliasing due to function approximation may result in the wrong value function estimates even at the unaliased states x_1 and x_2.

to map two-dimensional points to a desired action, usually by predicting the return obtained following this action.[68]

While there are similarities between the use of probability distribution representations (parameterizing the output of a return function) and function approximation (parameterizing its input), the latter requires a different algorithmic treatment. When function approximation is required, it is usually because it is infeasible to exhaustively enumerate the state space and apply dynamic programming methods. One solution is to rely on samples (of the state space, transitions, trajectories, etc.), but this results in additional complexities in the algorithm's design. Combining incremental algorithms with function approximation may result in instability or even divergence; in the distributional setting, the analysis of these algorithms is complicated by two levels of approximation (one for probability distributions and one across states). With proper care, however, function approximation provides an effective way of dealing with large reinforcement learning problems.

9.1 Function Approximation and Aliasing

By necessity, when parameters are shared across states, a single parameter usually affects the predictions (value or distribution) at multiple states. In this case, we say that the states are *aliased*. State aliasing has surprising consequences in the context of reinforcement learning, including the unwanted propagation of errors and potential instability in the learning process.

Example 9.1. Consider the Markov decision process in Figure 9.1, with four nonterminal states x_1, x_2, y, and z, a single action, a deterministic reward function, an initial state distribution ξ_0, and no discounting. Consider an

68. Domains such as Mountain Car – which have a single initial state and a deterministic transition function – can be solved without function approximation: for example, by means of a search algorithm. However, function approximation allows us to learn a control policy that can in theory be applied to any given state and has a low run-time cost.

approximation based on three parameters w_{x_1}, w_{x_2}, and w_{yz}, such that

$$\hat{V}(x_1) = w_{x_1} \quad \hat{V}(x_2) = w_{x_2} \quad \hat{V}(y) = \hat{V}(z) = w_{yz} .$$

Because the rewards from y and z are different, no choice of w_{yz} can yield $\hat{V} = V^\pi$. As such, any particular choice of w_{yz} trades off approximation error at y and z. When a reinforcement learning algorithm is combined with function approximation, this trade-off is made (implicitly or explicitly) based on the algorithm's characteristics and the parameters of the learning process. For example, the best approximation obtained by the incremental Monte Carlo algorithm (Section 3.2) correctly learns the value of x_1 and x_2:

$$\hat{V}(x_1) = 2 , \quad \hat{V}(x_2) = 0 ,$$

but learns a parameter w_{yz} that depends on the frequency at which states y and z are visited. This is because w_{yz} is updated toward 2 whenever the estimate $\hat{V}(y)$ is updated and toward 0 whenever $\hat{V}(z)$ is updated. In our example, the frequency at which this occurs is directly implied by the initial state distribution ξ_0, and we have

$$\begin{aligned} w_{yz} &= 2 \times \mathbb{P}_\pi(X_1 = y) + 0 \times \mathbb{P}_\pi(X_1 = z) \\ &= 2\xi_0(x_1) . \end{aligned} \tag{9.1}$$

When the approximation is learned using a bootstrapping procedure, aliasing can also result in incorrect estimates at states that are not themselves aliased. The solution found by temporal-difference learning, w_{yz}, is as per Equation 9.1, but the algorithm also learns the incorrect value at x_1 and x_2:

$$\begin{aligned} \hat{V}(x_1) = \hat{V}(x_2) &= 0 + \gamma \times \hat{V}(z) \\ &= 2\xi_0(x_1) . \end{aligned}$$

Thus, errors due to function approximation can compound in unexpected ways; we will study this phenomenon in greater detail in Section 9.3. △

In a *linear value function approximation*, the value estimate at a state x is given by a weighted combination of features of x. This is in opposition to a *tabular* representation, where value estimates are stored in a table with one entry per state.[69] As we will see, linear approximation is simple to implement and relatively easy to analyze.

Definition 9.2. Let $n \in \mathbb{N}^+$. A *state representation* is a mapping $\phi : \mathcal{X} \to \mathbb{R}^n$. A linear value function approximation $V_w \in \mathbb{R}^{\mathcal{X}}$ is parameterized by a weight

69. Technically, a tabular representation can also be expressed using the trivial collection of indicator features. In practice, the two are used in distinct problem settings.

vector $w \in \mathbb{R}^n$ and maps states to their expected return estimates according to

$$V_w(x) = \phi(x)^\top w .$$

A *feature* $\phi_i(x) \in \mathbb{R}$, $i = 1, \dots, n$ is an individual element of $\phi(x)$. We call the vectors $\phi_i \in \mathbb{R}^{\mathcal{X}}$ *basis functions.* △

As its name implies, a linear value function approximation is linear in the weight vector w. That is, for any $w_1, w_2 \in \mathbb{R}^n$, $\alpha, \beta \in \mathbb{R}$, we have

$$V_{\alpha w_1 + \beta w_2} = \alpha V_{w_1} + \beta V_{w_2} .$$

In addition, the gradient of $V_w(x)$ with respect to w is given by

$$\nabla_w V_w(x) = \phi(x) .$$

As we will see, these properties affect the learning dynamics of algorithms that use linear value function approximation.

We extend linear value function approximation to state-action values in the usual way. For a state representation $\phi : \mathcal{X} \times \mathcal{A} \to \mathbb{R}^n$, we define

$$Q_w(x, a) = \phi(x, a)^\top w .$$

A practical alternative is to use a distinct set of weights for each action and a common representation $\phi(x)$ across actions. In this case, we use a collection of weight vectors $(w_a : a \in \mathcal{A})$, with $w_a \in \mathbb{R}^n$, and write

$$Q_w(x, a) = \phi(x)^\top w_a .$$

Remark 9.1 discusses the relationship between these two methods.

9.2 Optimal Linear Value Function Approximations

In this chapter, we will assume that there is a finite (but very large) number of states. In this case, the state representation $\phi : \mathcal{X} \to \mathbb{R}^n$ can expressed as a *feature matrix* $\Phi \in \mathbb{R}^{\mathcal{X} \times n}$ whose rows are the vectors $\phi(x)$, $x \in \mathcal{X}$. This yields the approximation

$$V_w = \Phi w .$$

The state representation determines a space of value function approximations that are constructed from linear combinations of features. Expressed in terms of the feature matrix, this space is

$$\mathcal{F}_\phi = \{\Phi w : w \in \mathbb{R}^n\} .$$

We first consider the problem of finding the best linear approximation to a value function V^π. Because $\mathcal{F}_\phi$ is a n-dimensional linear subspace of the space of value functions $\mathbb{R}^{\mathcal{X}}$, there are necessarily some value functions that cannot be represented with a given state representation (unless $n = N_{\mathcal{X}}$). We measure the discrepancy between a value function V^π and an approximation $V_w = \Phi w$ in ξ-weighted L^2 norm, for $\xi \in \mathscr{P}(\mathcal{X})$:

$$\|V^\pi - V_w\|_{\xi,2} = \Big(\sum_{x \in \mathcal{X}} \xi(x)(V^\pi(x) - V_w(x))^2\Big)^{1/2}.$$

The weighting ξ reflects the relative importance given to different states. For example, we may weigh states according to the frequency at which they are visited, or we may put greater importance on initial states. Provided that $\xi(x) > 0$ for all $x \in \mathcal{X}$, the norm $\|\cdot\|_{\xi,2}$ induces the ξ-weighted L^2 metric on $\mathbb{R}^{\mathcal{X}}$:[70]

$$d_{\xi,2}(V, V') = \|V - V'\|_{\xi,2}.$$

The best linear approximation under this metric is the solution to the minimization problem

$$\min_{w \in \mathbb{R}^n} \|V^\pi - V_w\|_{\xi,2}. \tag{9.2}$$

One advantage of measuring approximation error in a weighted L^2 norm, rather than the L^∞ norm used in the analyses of previous chapters, is that a solution w^* to Equation 9.2 can be easily determined by solving a least-squares system.

Proposition 9.3. Suppose that the columns of the feature matrix Φ are linearly independent and $\xi(x) > 0$ for all $x \in \mathcal{X}$. Then, Equation 9.2 has a unique solution w^* given by

$$w^* = (\Phi^\top \Xi \Phi)^{-1} \Phi^\top \Xi V^\pi, \tag{9.3}$$

where $\Xi \in \mathbb{R}^{\mathcal{X} \times \mathcal{X}}$ is a diagonal matrix with entries $(\xi(x) : x \in \mathcal{X})$. △

Proof. By a standard calculus argument, any optimum w must satisfy

$$\nabla_w \sum_{x \in \mathcal{X}} \xi(x)(V^\pi(x) - \phi(x)^\top w)^2 = 0$$

$$\implies \sum_{x \in \mathcal{X}} \xi(x)(V^\pi(x) - \phi(x)^\top w)\phi(x) = 0.$$

Written in matrix form, this is

$$\Phi^\top \Xi(\Phi w - V^\pi) = 0$$

$$\implies \Phi^\top \Xi \Phi w = \Phi^\top \Xi V^\pi.$$

70. Technically, $\|\cdot\|_{\xi,2}$ is only a proper norm if ξ is strictly positive for all x; otherwise, it is a semi-norm. Under the same condition, $d_{\xi,2}$ is proper metric; otherwise, it is a pseudo-metric. Assuming that $\xi(x) > 0$ for all x addresses uniqueness issues and simplifies the analysis.

Because Φ has rank n, then so does $\Phi^\top \Xi \Phi$: for any $u \in \mathbb{R}^n$ with $u \neq 0$, we have $\Phi u = v \neq 0$ and so

$$\begin{aligned} u^\top \Phi^\top \Xi \Phi u &= v^\top \Xi v \\ &= \sum_{x \in \mathcal{X}} \xi(x) v(x)^2 > 0 \,, \end{aligned}$$

as $\xi(x) > 0$ and $v(x)^2 \geq 0$ for all $x \in \mathcal{X}$, and $\sum_{x \in \mathcal{X}} v(x)^2 > 0$. Hence, $\Phi^\top \Xi \Phi$ is invertible and the only solution w^* to the above satisfies Equation 9.3. □

9.3 A Projected Bellman Operator for Linear Value Function Approximation

Dynamic programming finds an approximation to the value function V^π by successively computing the iterates

$$V_{k+1} = T^\pi V_k \,.$$

As we saw in preceding chapters, dynamic programming makes it easy to derive incremental algorithms for learning the value from samples and also allows us to find an approximation to the optimal value function Q^*. Often, it is the de facto approach for finding an approximation of the return-distribution function. It is also particularly useful when using function approximation, where it enables algorithms that learn by extrapolating to unseen states.

When dynamic programming is combined with function approximation, we obtain a range of methods called *approximate dynamic programming*. In the case of linear value function approximation, the iterates $(V_k)_{k \geq 0}$ are given by linear combinations of features, which allows us to apply dynamic programming to problems with larger state spaces than can be described in memory. In general, however, the space of approximations $\mathcal{F}_\phi$ is not closed under the Bellman operator, in the sense that

$$V \in \mathcal{F}_\phi \not\Rightarrow T^\pi V \in \mathcal{F}_\phi \,.$$

Similar to the notion of a distributional projection introduced in Chapter 5, we address the issue by projecting, for $V \in \mathbb{R}^{\mathcal{X}}$, the value function $T^\pi V$ back onto $\mathcal{F}_\phi$. Let us define the projection operator $\Pi_{\phi,\xi} : \mathbb{R}^{\mathcal{X}} \to \mathbb{R}^{\mathcal{X}}$ as

$$(\Pi_{\phi,\xi} V)(x) = \phi(x)^\top w^* \quad \text{such that} \quad w^* \in \arg\min_{w \in \mathbb{R}^n} \|V - V_w\|_{\xi,2} \,.$$

This operator returns the approximation $V_{w^*} = \Phi w^*$ that is closest to $V \in \mathbb{R}^{\mathcal{X}}$ in the ξ-weighted L^2 norm. As established by Proposition 9.3, when ξ is fully supported on $\mathcal{X}$ and the basis functions $(\phi_i)_{i=1}^n$ are linearly independent, then this projection is unique.[71] By repeatedly applying the projected Bellman operator $\Pi_{\phi,\xi}T^\pi$ from an initial condition $V_0 \in \mathcal{F}_\phi$, we obtain the iterates

$$V_{k+1} = \Pi_{\phi,\xi}T^\pi V_k\,. \tag{9.4}$$

Unlike the approach taken in Chapter 5, however, it is usually impractical to implement Equation 9.4 as is, as there are too many states to enumerate. A simple solution is to rely on a sampling procedure that approximates the operator itself. For example, one may sample a batch of K states and find the best linear fit to $T^\pi V_k$ at these states. In the next section, we will study the related approach of using an incremental algorithm to learn the linear value function approximation from sample transitions. Understanding the behavior of the exact projected operator $\Pi_{\phi,\xi}T^\pi$ informs us about the behavior of these approximations, as it describes in some sense the ideal behavior that one expects from both of these approaches.

Also different from the setting of Chapter 5 is the presence of aliasing across states. As a consequence of this aliasing, we have limited freedom in the choice of projection if we wish to guarantee the convergence of the iterates of Equation 9.4. To obtain such a guarantee, in general, we need to impose a condition on the distribution ξ that defines the projection $\Pi_{\phi,\xi}$. We will demonstrate that the projected Bellman operator is a contraction mapping with modulus γ with respect to the ξ-weighted L^2 norm, for a specific choice of ξ. Historically, this approach predates the analysis of distributional dynamic programming and is in fact a key inspiration for our analysis of distributional reinforcement learning algorithms as approximating projected Bellman operators (see bibliographical remarks).

To begin, let us introduce the convention that the Lipschitz constant of an operator with respect to a norm (such as L^∞) follows Definition 5.20, applied to the metric associated with the norm. In the case of L^∞, this metric defines the distance between $u, u' \in \mathbb{R}^\mathcal{X}$ by

$$\|u - u'\|_\infty\,.$$

Now recall that the Bellman operator T^π is a contraction mapping in L^∞ norm, with modulus γ. One reason this holds is because the transition matrix P^π satisfies

$$\|P^\pi u\|_\infty \le \|u\|_\infty, \qquad \text{for all } u \in \mathbb{R}^\mathcal{X}\,;$$

71. If only the first of those two conditions hold, then there may be multiple optimal weight vectors. However, they all result in the same value function, and the projection remains unique.

we made use of this fact in the proof of Proposition 4.4. This is equivalent to requiring that the Lipschitz constant of P^π satisfy

$$\|P^\pi\|_\infty \leq 1 .$$

Unfortunately, the Lipschitz constant of $\Pi_{\phi,\xi}$ in the L^∞ norm may be greater than 1, precluding a direct analysis in that norm (see Exercise 9.5). We instead prove that the Lipschitz constant of P^π in the ξ-weighted L^2 norm satisfies the same condition when ξ is taken to be a *steady-state distribution* under policy π.

Definition 9.4. Consider a Markov decision process and let π be a policy defining the probability distribution $\mathbb{P}_\pi$ over the random transition (X, A, R, X'). We say that $\xi \in \mathscr{P}(\mathcal{X})$ is a *steady-state distribution* for π if for all $x' \in \mathcal{X}$,

$$\xi(x') = \sum_{x \in \mathcal{X}} \xi(x)\mathbb{P}_\pi(X' = x' \mid X = x) . \qquad \triangle$$

Assumption 9.5. There is a unique steady-state distribution ξ_π, and it satisfies $\xi_\pi(x) > 0$ for all $x \in \mathcal{X}$. $\triangle$

Qualitatively, Assumption 9.5 ensures that approximation error at any state is reflected in the norm $\|\cdot\|_{\xi_\pi,2}$; contrast with the setting in which ξ_π is nonzero only at a handful of states. Uniqueness is not strictly necessary but simplifies the exposition. There are a number of practical scenarios in which the assumption does not hold, most importantly when there is a terminal state. We discuss how to address such a situation in Remark 9.2.

Lemma 9.6. Let $\pi : \mathcal{X} \to \mathscr{P}(\mathcal{A})$ be a policy and let ξ_π be a steady-state distribution for this policy. The transition matrix is a nonexpansion with respect to the ξ_π-weighted L^2 metric. That is,

$$\|P^\pi\|_{\xi_\pi,2} = 1 . \qquad \triangle$$

Proof. A simple algebraic argument shows that if $U \in \mathbb{R}^{\mathcal{X}}$ is such that $U(x) = 1$ for all x, then

$$P^\pi U = U .$$

This shows that $\|P^\pi\|_{\xi_\pi,2} \geq 1$. Now for an arbitrary $U \in \mathbb{R}^{\mathcal{X}}$, write

$$\begin{aligned}
\|P^\pi U\|^2_{\xi_\pi,2} &= \sum_{x \in \mathcal{X}} \xi_\pi(x)((P^\pi U)(x))^2 \\
&= \sum_{x \in \mathcal{X}} \xi_\pi(x)\Big(\sum_{x' \in \mathcal{X}} P^\pi(x' \mid x)U(x')\Big)^2 \\
&\overset{(a)}{\leq} \sum_{x \in \mathcal{X}} \xi_\pi(x) \sum_{x' \in \mathcal{X}} P^\pi(x' \mid x)(U(x'))^2
\end{aligned}$$

$$= \sum_{x' \in \mathcal{X}} (U(x'))^2 \sum_{x \in \mathcal{X}} \xi_\pi(x) P^\pi(x' \mid x)$$

$$\stackrel{(b)}{=} \sum_{x' \in \mathcal{X}} \xi_\pi(x')(U(x'))^2$$

$$= \|U\|^2_{\xi_\pi,2} ,$$

where (*a*) follows by Jensen's inequality and (*b*) by Definition 9.4. Since

$$\|P^\pi\|_{\xi_\pi,2} = \sup_{U \in \mathbb{R}^{\mathcal{X}}} \frac{\|P^\pi U\|_{\xi_\pi,2}}{\|U\|_{\xi_\pi,2}} \leq 1 ,$$

this concludes the proof. □

Lemma 9.7. For any $\xi \in \mathscr{P}(\mathcal{X})$ with $\xi(x) > 0$, the projection operator $\Pi_{\phi,\xi}$ is a nonexpansion in the ξ-weighted L^2 metric, in the sense that

$$\|\Pi_{\phi,\xi}\|_{\xi,2} = 1 .$$

△

The proof constitutes Exercise 9.4.

Theorem 9.8. Let π be a policy and suppose that Assumption 9.5 holds; let ξ_π be the corresponding steady-state distribution. The projected Bellman operator $\Pi_{\phi,\xi_\pi} T^\pi$ is a contraction with respect to the ξ_π-weighted L^2 norm with modulus γ, in the sense that for any $V, V' \in \mathbb{R}^{\mathcal{X}}$,

$$\|\Pi_{\phi,\xi_\pi} T^\pi V - \Pi_{\phi,\xi_\pi} T^\pi V'\|_{\xi_\pi,2} \leq \gamma \|V - V'\|_{\xi_\pi,2} .$$

As a consequence, this operator has a unique fixed point

$$\hat{V}^\pi = \Pi_{\phi,\xi_\pi} T^\pi \hat{V}^\pi , \tag{9.5}$$

which satisfies

$$\|\hat{V}^\pi - V^\pi\|_{\xi_\pi,2} \leq \frac{1}{\sqrt{1-\gamma^2}} \|\Pi_{\phi,\xi_\pi} V^\pi - V^\pi\|_{\xi_\pi,2} . \tag{9.6}$$

In addition, for an initial value function $V_0 \in \mathbb{R}^{\mathcal{X}}$, the sequence of iterates

$$V_{k+1} = \Pi_{\phi,\xi_\pi} T^\pi V_k \tag{9.7}$$

converges to this fixed point. △

Proof. The contraction result and consequent convergence of the iterates in Equation 9.7 to a unique fixed point follow from Lemmas 9.6 and 9.7, which, combined with Lemma 5.21, allow us to deduce that

$$\|\Pi_{\phi,\xi_\pi} T^\pi\|_{\xi_\pi,2} \leq \gamma .$$

Because Assumption 9.5 guarantees that $\|\cdot\|_{\xi_\pi,2}$ induces a proper metric, we may then apply Banach's fixed-point theorem. For the error bound of Equation 9.6, we use Pythagoras's theorem to write

$$\begin{aligned}\|\hat{V}^\pi - V^\pi\|^2_{\xi_\pi,2} &= \|\hat{V}^\pi - \Pi_{\phi,\xi_\pi} V^\pi\|^2_{\xi_\pi,2} + \|\Pi_{\phi,\xi_\pi} V^\pi - V^\pi\|^2_{\xi_\pi,2} \\ &= \|\Pi_{\phi,\xi_\pi} T^\pi \hat{V}^\pi - \Pi_{\phi,\xi_\pi} T^\pi V^\pi\|^2_{\xi_\pi,2} + \|\Pi_{\phi,\xi_\pi} V^\pi - V^\pi\|^2_{\xi_\pi,2} \\ &\leq \gamma^2 \|\hat{V}^\pi - V^\pi\|^2_{\xi_\pi,2} + \|\Pi_{\phi,\xi_\pi} V^\pi - V^\pi\|^2_{\xi_\pi,2} ,\end{aligned}$$

since $\|\Pi_{\phi,\xi_\pi} T^\pi\|_{\xi_\pi,2} \leq \gamma$. The result follows by rearranging terms and taking the square root of both sides. □

Theorem 9.8 implies that the iterates $(V_k)_{k\geq 0}$ are guaranteed to converge when the projection is performed in ξ_π-weighted L^2 norm. Of course, this does not imply that a projection in a different norm may not result in a convergent algorithm (see Exercise 9.8), but divergence is a practical concern (we return to this point in the next section). A sound alternative to imposing a condition on the distribution ξ is to instead impose a condition on the feature matrix Φ; this is explored in Exercise 9.10.

When the feature matrix Φ forms a basis of $\mathbb{R}^{\mathcal{X}}$ (i.e., it has rank $N_{\mathcal{X}}$), it is always possible to find a weight vector w^* for which

$$\Phi w^* = T^\pi V ,$$

for any given $V \in \mathbb{R}^{\mathcal{X}}$. As a consequence, for any $\xi \in \mathscr{P}(\mathcal{X})$ we have

$$\Pi_{\phi,\xi} T^\pi = T^\pi ,$$

and Theorem 9.8 reduces to the analysis of the (unprojected) Bellman operator given in Section 4.2. On the other hand, when $n < N_{\mathcal{X}}$, the fixed point of Equation 9.5 is in general different from the minimum-error solution $\Pi_{\phi,\xi_\pi} V^\pi$ and is called the *temporal-difference learning fixed point*. Similar to the diffusion effect studied in Section 5.8, successive applications of the projected Bellman operator result in compounding approximation errors. The nature of this fixed point is by now well studied in the literature (see bibliographical remarks).

9.4 Semi-Gradient Temporal-Difference Learning

We now consider the design of a sample-based, incremental algorithm for learning the linear approximation of a value function V^π. In the context of domains with large state spaces, algorithms that learn from samples have an advantage over dynamic programming approaches: whereas the latter require some form of enumeration and hence have a computational cost that depends on the size of $\mathcal{X}$, the computational cost of the former instead depends on the size of the function approximation (in the linear case, on the number of features n).

To begin, let us consider learning a linear value function approximation using an incremental Monte Carlo algorithm. We are presented with a sequence of state-return pairs $(x_k, g_k)_{k\geq 0}$, with the assumption that the source states x_k are realizations of independent draws from the distribution ξ and that each g_k is a corresponding independent realization of the random return $G^\pi(x_k)$. As before, we are interested in the optimal weight vector w^* for the problem

$$\min_{w\in\mathbb{R}^n} \|V^\pi - V_w\|_{\xi,2}, \quad V_w(x) = \phi(x)^\top w\,. \tag{9.8}$$

Of note, the optimal approximation V_{w^*} is also the solution to the problem

$$\min_{w\in\mathbb{R}^n} \mathbb{E}[\|G^\pi - V_w\|_{\xi,2}]\,. \tag{9.9}$$

Consequently, a simple approach for finding w_* is to perform *stochastic gradient descent* with a loss function that reflects Equation 9.9. For $x \in \mathcal{X}$ and $z \in \mathbb{R}$, let us define the sample loss

$$\mathcal{L}(w) = (z - \phi(x)^\top w)^2$$

whose gradient with respect to w is

$$\nabla_w \mathcal{L}(w) = -2(z - \phi(x)^\top w)\phi(x)\,.$$

Stochastic gradient descent updates the weight vector w by following the (negative) gradient of the sample loss constructed from each sample. Instantiating the sample loss with $x = x_k$ and $z = g_k$, this results in the update rule

$$w \leftarrow w + \alpha_k(g_k - \phi(x_k)^\top w)\phi(x_k)\,, \tag{9.10}$$

where $\alpha_k \in [0, 1)$ is a time-varying step size that also subsumes the constant from the loss. Under appropriate conditions, this update rule finds the optimal weight vector w^* (see, e.g., Bottou 1998). Exercise 9.9 asks you to verify that the optimal weight vector w^* is a fixed point of the expected update.

Let us now consider the problem of learning the value function from a sequence of sample transitions $(x_k, a_k, r_k, x'_k)_{k\geq 0}$, again assumed to be independent realizations from the appropriate distributions. Given a weight vector w, the temporal-difference learning target for linear value function approximation is

$$r_k + \gamma V_w(x'_k) = r_k + \gamma\phi(x'_k)^\top w\,.$$

We use this target in lieu of g_k in Equation 9.10 to obtain the *semi-gradient temporal-difference learning* update rule

$$w \leftarrow w + \alpha_k \underbrace{(r_k + \gamma\phi(x'_k)^\top w - \phi(x_k)^\top w)}_{\text{TD error}} \phi(x_k)\,, \tag{9.11}$$

in which the temporal-difference (TD) error appears, now with value function estimates constructed from linear approximation. By substituting the Monte Carlo target g_k for the temporal-difference target, the intent is to learn an approximation to V^π by a bootstrapping process, as in the tabular setting. The term "semi-gradient" reflects the fact that the update rule does not actually follow the gradient of the sample loss

$$(r_k + \gamma\phi(x_k')^\top w - \phi(x_k)^\top w)^2 ,$$

which contains additional terms related to $\phi(x')$ (see bibliographical remarks).

We can understand the relationship between semi-gradient temporal-difference learning and the projected Bellman operator $\Pi_{\phi,\xi}T^\pi$ by way of an update rule defined in terms of a second set of weights $\tilde{w}$, the *target weights*. This update rule is

$$w \leftarrow w + \alpha_k(r_k + \gamma\phi(x_k')^\top \tilde{w} - \phi(x_k)^\top w)\phi(x_k) . \tag{9.12}$$

When $\tilde{w} = w$, this is Equation 9.11. However, if $\tilde{w}$ is a separate weight vector, this is the update rule of stochastic gradient descent on the sample loss

$$(r_k + \gamma\phi(x_k')^\top \tilde{w} - \phi(x_k)^\top w)^2 .$$

Consequently, this update rule finds a weight vector w^* that approximately minimizes

$$\|T^\pi V_{\tilde{w}} - V_w\|_{\xi,2} ,$$

and Equation 9.12 describes an incremental algorithm for computing a single step of the projected Bellman operator applied to $V_{\tilde{w}}$; its solution w^* satisfies

$$\Phi w^* = \Pi_{\phi,\xi}T^\pi V_{\tilde{w}} .$$

This argument suggests that semi-gradient temporal-difference learning tracks the behavior of the projected Bellman operator. In particular, at the fixed point $\hat{V}^\pi = \Phi\hat{w}$ of this operator, semi-gradient TD learning (applied to realizations from the sample transition model (X, A, R, X'), with $X \sim \xi$) leaves the weight vector unchanged, in expectation.

$$\mathbb{E}_\pi \left[(R + \gamma\phi(X')^\top \hat{w} - \phi(X)^\top \hat{w})\phi(X)\right] = 0 .$$

In semi-gradient temporal-difference learning, however, the sample target $r_k + \gamma\phi(x_k')^\top w$ depends on w and is used to update w itself. This establishes a feedback loop that, combined with function approximation, can result in divergence – even when the projected Bellman operator is well behaved. The following example illustrates this phenomenon.

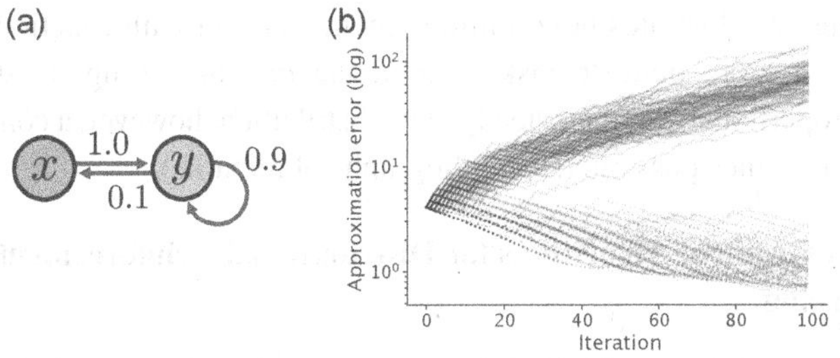

Figure 9.2
(a) A Markov decision process for which semi-gradient temporal-difference learning can diverge. **(b)** Approximation error (measured in unweighted L^2 norm) over the course of 100 runs of the algorithm with $\alpha = 0.01$ and the same initial condition $w_0 = (1, 0, 0)$ but different draws of the sample transitions. Dashed (top) and dotted (bottom) lines indicate runs with $\xi(x) = 1/2$ and $1/11$, respectively.

Example 9.9 (Baird's counterexample (Baird 1995))**.** Consider the Markov decision process depicted in Figure 9.2a and the state representation

$$\phi(x) = (1, 3, 2) \quad \phi(y) = (4, 3, 3).$$

Since the reward is zero everywhere, the value function for this MDP is $V^\pi = 0$. However, if $X_k \sim \xi$ with $\xi(x) = \xi(y) = 1/2$, the semi-gradient update

$$w_{k+1} = w_k + \alpha(0 + \gamma\phi(X'_k)^\top w_k - \phi(X_k)^\top w_k)\phi(X_k)$$

diverges unless $w_0 = 0$. Figure 9.2b depicts the effect of this divergence on the approximation error: on average, the distance to the fixed point $\|\Phi w_k - 0\|_{\xi,2}$ grows exponentially with each update. Also shown is the approximation error over time if we take ξ to be the steady-state distribution

$$\xi(x) = 1/11 \quad \xi(y) = 10/11,$$

in which case the approximation error becomes close to zero as $k \to \infty$. This demonstrates the impact of the relative update frequency of different states on the behavior of the algorithm.

On the other hand, note that the basis functions implied by ϕ span $\mathbb{R}^{\mathcal{X}}$ and thus

$$\Pi_{\phi,\xi} T^\pi V = T^\pi V$$

for any $V \in \mathbb{R}^2$. Hence, the iterates $V_{k+1} = \Pi_{\phi,\xi} T^\pi V_k$ in this case converge to 0 for any initial $V_0 \in \mathbb{R}^2$. This illustrates that requiring the convergence that the sequence of iterates derived from the projected Bellman operator is not sufficient to guarantee the convergence of the semi-gradient iterates. △

Example 9.9 illustrates how reinforcement learning with function approximation is a more delicate task than in the tabular setting. If states are updated proportionally to their steady-state distribution, however, a convergence guarantee becomes possible (see bibliographical remarks).

9.5 Semi-Gradient Algorithms for Distributional Reinforcement Learning

With the right modeling tools, function approximation can also be used to tractably represent the return functions of large problems. One difference with the expected-value setting is that it is typically more challenging to construct an approximation that is linear in the true sense of the word. With linear value function approximations, adding weight vectors is equivalent to adding approximations:

$$V_{w_1+w_2}(x) = V_{w_1}(x) + V_{w_2}(x) .$$

In the distributional setting, the same cannot apply because probability distributions do not form a vector space. This means that we cannot expect a return-distribution function representation to satisfy

$$\eta_{w_1+w_2}(x) \stackrel{?}{=} \eta_{w_1}(x) + \eta_{w_2}(x) ; \tag{9.13}$$

the right-hand side is not a probability distribution (it is, however, a *signed distribution*: more on this in Section 9.6). An alternative is to take a slightly broader view and consider distributions whose *parameters* depend linearly on w. There are now two sources of approximation: one due to the finite parameterization of probability distributions in $\mathscr{F}$, another because those parameters are themselves aliased. This is an expressive framework, albeit one under which the analysis of algorithms is significantly more complex.

Linear QTD. Let us first derive a linear approximation of quantile temporal-difference learning. *Linear QTD* represents the locations of quantile distributions using linear combinations of features. If we write $w \in \mathbb{R}^{m \times n}$ for the matrix whose columns are $w_1, \ldots, w_m \in \mathbb{R}^n$, then the linear QTD return function estimate takes the form

$$\eta_w(x) = \frac{1}{m} \sum_{i=1}^{m} \delta_{\phi(x)^\top w_i} .$$

One can verify that $\eta_w(x)$ is not a linear combination of features, even though its parameters are. We construct the linear QTD update rule by following the negative gradient of the quantile loss (Equation 6.12), taken with respect to the parameters $w_1, \ldots, w_m$. We first rewrite this loss in terms of a function ρ_τ:

$$\rho_\tau(\Delta) = |\mathbb{1}_{\{\Delta < 0\}} - \tau| \times |\Delta|$$

so that for a sample $z \in \mathbb{R}$ and estimate $\theta \in \mathbb{R}$, the loss of Equation 6.12 can be expressed as

$$|\mathbb{1}_{\{z<\theta\}} - \tau| \times |z - \theta| = \rho_\tau(z - \theta)\,.$$

We instantiate the quantile loss with $\theta = \phi(x)^\top w_i$ and $\tau = \tau_i = \frac{2i-1}{2m}$ to obtain the loss

$$\mathcal{L}_{\tau_i}(w) = \rho_{\tau_i}(z - \phi(x)^\top w_i)\,. \tag{9.14}$$

By the chain rule, the gradient of this loss with respect to w_i is

$$\nabla_{w_i}\rho_{\tau_i}(z - \phi(x)^\top w_i) = -(\tau_i - \mathbb{1}\{z < \phi(x)^\top w_i\})\phi(x)\,. \tag{9.15}$$

As in our derivation of QTD, from a sample transition (x, a, r, x'), we construct m sample targets:

$$g_j = r + \gamma\phi(x')^\top w_j, \quad j = 1, \ldots, m\,.$$

By instantiating the gradient expression in Equation 9.15 with $z = g_j$ and taking the average over the m sample targets, we obtain the update rule

$$w_i \leftarrow w_i - \alpha\frac{1}{m}\sum_{j=1}^{m}\nabla_{w_i}\rho_{\tau_i}\big(g_j - \phi(x)^\top w_i\big),$$

which is more explicitly

$$w_i \leftarrow w_i + \alpha\frac{1}{m}\sum_{j=1}^{m}\underbrace{\big(\tau_i - \mathbb{1}\{g_j < \phi(x)^\top w_i\}\big)}_{\text{quantile TD error}}\phi(x). \tag{9.16}$$

Note that, by plugging g_j into the expression for the gradient (Equation 9.15), we obtain a semi-gradient update rule. That is, analogous to the value-based case, Equation 9.16 is not equivalent to the gradient update

$$w_i \leftarrow w_i - \alpha\frac{1}{m}\sum_{j=1}^{m}\nabla_{w_i}\rho_{\tau_i}\big(r + \gamma\phi(x')^\top w_j - \phi(x)^\top w_i\big),$$

because in general

$$\nabla_{w_i}\rho_{\tau_i}\big(r + \gamma\phi(x')^\top w_i - \phi(x)^\top w_i\big) \neq \big(\tau_i - \mathbb{1}\{g_i < \phi(x)^\top w_i\}\big)\phi(x)\,.$$

Linear CTD. To derive a linear approximation of categorical temporal-difference learning, we represent the probabilities of categorical distributions using linear combinations of features. Specifically, we apply the softmax function to transform the parameters $(\phi(x)^\top w_i)_{i=1}^m$ into a probability distribution. We write

$$p_i(x; w) = \frac{e^{\phi(x)^\top w_i}}{\sum_{j=1}^{m} e^{\phi(x)^\top w_j}}\,. \tag{9.17}$$

Recall that the probabilities $(p_i(x; w))_{i=1}^m$ correspond to m locations $(\theta_i)_{i=1}^m$. The softmax transformation guarantees that the expression

$$\eta_w(x) = \sum_{i=1}^m p_i(x; w)\delta_{\theta_i}$$

describes a bona fide probability distribution. We thus construct the sample target $\bar{\eta}(x)$:

$$\bar{\eta}(x) = \Pi_c(\mathrm{b}_{r,\gamma})_\# \eta_w(x') = \Pi_c\Big[\sum_{i=1}^m p_i(x'; w)\delta_{r+\gamma\theta_i}\Big]$$
$$= \sum_{i=1}^m \bar{p}_i \delta_{\theta_i},$$

where $\bar{p}_i$ denotes the probability assigned to the location θ_i by the sample target $\bar{\eta}(x)$. Expressed in terms of the CTD coefficients (Equation 6.10), the probability $\bar{p}_i$ is

$$\bar{p}_i = \sum_{j=1}^m \zeta_{i,j}(r) p_j(x'; w).$$

As with quantile regression, we adjust the weights w by means of a gradient descent procedure. Here, we use the cross-entropy loss between $\eta_w(x)$ and $\bar{\eta}(x)$:[72]

$$\mathcal{L}(w) = -\sum_{i=1}^m \bar{p}_i \log p_i(x; w). \tag{9.18}$$

Combined with the softmax function, Equation 9.18 becomes

$$\mathcal{L}(w) = -\sum_{i=1}^m \bar{p}_i\Big(\phi(x)^\top w_i - \log \sum_{j=1}^m e^{\phi(x)^\top w_j}\Big).$$

With some algebra and again invoking the chain rule, we obtain that the gradient with respect to the weights w_i is

$$\nabla_{w_i}\mathcal{L}(w) = -(\bar{p}_i - p_i(x; w))\phi(x).$$

By adjusting the weights in the opposite direction of this gradient, this results in the update rule

$$w_i \leftarrow w_i + \alpha \underbrace{(\bar{p}_i - p_i(x; w))}_{\text{CTD error}} \phi(x). \tag{9.19}$$

72. The choice of cross-entropy loss is justified because it is the *matching loss* for the softmax function, and their combination results in a convex objective (Auer et al. 1995; see also Bishop 2006, Section 4.3.6).

While interesting in their own right, linear CTD and QTD are particularly important in that they can be straightforwardly adapted to learn return-distribution functions with nonlinear function approximation schemes such as deep neural networks; we return to this point in Chapter 10. For now, it is worth noting that the update rules of linear TD, linear QTD, and linear CTD can all be expressed as

$$w_i \leftarrow w_i + \alpha \varepsilon_i \phi(x) \,,$$

where ε_i is an error term. One interesting difference is that for both linear CTD and linear QTD, ε_i lies in the interval $[-1, 1]$, while for linear TD, it is unbounded. This gives evidence that we should expect different learning dynamics from these algorithms. In addition, combining linear TD or linear QTD to a tabular state representation recovers the corresponding incremental algorithms from Chapter 6. For linear CTD, the update corresponds to a tabular representation of the softmax parameters rather than the probabilities themselves, and the correspondence is not as straightforward.

Analyzing linear QTD and CTD is complicated by the fact that the return functions themselves are not linear in $w_1, \ldots, w_m$. One solution is to relax the requirement that the approximation $\eta_w(x)$ be a probability distribution; as we will see in the next section, in this case the distributional approximation behaves much like the value function approximation, and a theoretical guarantee can be obtained.

9.6 An Algorithm Based on Signed Distributions*

So far, we made sure that the outputs of our distributional algorithms were valid probability distributions (or could be as interpreted as such: for example, when working with statistical functionals). This was done explicitly when using the softmax parameterization in defining linear CTD and implicitly in the mixture update rule of CTD in Chapter 3. In this section, we consider an algorithm that is similar to linear CTD but omits the softmax function. As a consequence of this change, this modified algorithm's outputs are *signed distributions*, which we briefly encountered in Chapter 6 in the course of analyzing categorical temporal-difference learning.

Compared to linear CTD, this approach has the advantage of being both closer to the tabular algorithm (it finds a best fit in ℓ_2 distance, like tabular CTD) and closer to linear value function approximation (making it amenable to analysis). Although the learned predictions lack some aspects of probability distributions – such as well-defined quantiles – the learned signed distributions can be used to estimate expectations of functions, including expected values.

To begin, let us define a (finite) signed distribution ν as a weighted sum of two probability distributions:

$$\nu = \lambda_1 \nu_1 + \lambda_2 \nu_2 \quad \lambda_1, \lambda_2 \in \mathbb{R}, \nu_1, \nu_2 \in \mathscr{P}(\mathbb{R}). \tag{9.20}$$

We write $\mathscr{M}(\mathbb{R})$ for the space of finite signed distributions. For $\nu \in \mathscr{M}(\mathbb{R})$ decomposed as in Equation 9.20, we define its cumulative distribution function F_ν and the expectation of a function $f : \mathbb{R} \to \mathbb{R}$ as

$$F_\nu(z) = \lambda_1 F_{\nu_1}(z) + \lambda_2 F_{\nu_2}(z),\ z \in \mathbb{R}$$
$$\mathop{\mathbb{E}}_{Z\sim\nu}[f(Z)] = \lambda_1 \mathop{\mathbb{E}}_{Z\sim\nu_1}[f(Z)] + \lambda_2 \mathop{\mathbb{E}}_{Z\sim\nu_2}[f(Z)]. \tag{9.21}$$

Exercise 9.14 asks you to verify that these definitions are independent of the decomposition of ν into a sum of probability distributions. The *total mass* of ν is given by

$$\kappa(\nu) = \lambda_1 + \lambda_2.$$

We make these definitions explicit because signed distributions are not probability distributions; in particular, we cannot draw samples from ν. In that sense, the notation $Z \sim \nu$ in the definition of expectation is technically incorrect, but we use it here for convenience.

Definition 9.10. The *signed m-categorical representation* parameterizes the mass of m particles at fixed locations $(\theta_i)_{i=1}^m$:

$$\mathscr{F}_{\text{S},m} = \Big\{ \sum_{i=1}^m p_i \delta_{\theta_i} : p_i \in \mathbb{R} \text{ for } i = 1, \ldots, m, \sum_{i=1}^m p_i = 1 \Big\}. \qquad \triangle$$

Compared to the usual m-categorical representation, its signed analogue adds a degree of freedom: it allows the mass of its particles to be negative and of magnitude greater than 1 (we reserve "probability" for values strictly in [0, 1]). However, in our definition, we still require that signed m-categorical distributions have unit total mass; as we will see, this avoids a number of technical difficulties. Exercise 9.15 asks you to verify that $\nu \in \mathscr{F}_{\text{S},m}$ is a signed distribution in the sense of Equation 9.20.

Recall from Section 5.6 that the categorical projection $\Pi_c : \mathscr{P}(\mathbb{R}) \to \mathscr{F}_{\text{C},m}$ is defined in terms of the triangular and half-triangular kernels $h_i : \mathbb{R} \to [0, 1]$, $i = 1, \ldots, m$. We use Equation 9.21 to extend this projection to signed distributions, written $\Pi_c : \mathscr{M}(\mathbb{R}) \to \mathscr{F}_{\text{S},m}$. Given $\nu \in \mathscr{M}(\mathbb{R})$, the masses of $\Pi_c \nu = \sum_{i=1}^m p_i \delta_{\theta_i}$ are given by

$$p_i = \mathop{\mathbb{E}}_{Z\sim\nu}\big[h_i(\varsigma_m^{-1}(Z - \theta_i))\big],$$

where as before, $\varsigma_m^{-1} = \theta_{i+1} - \theta_i$ ($i < m$), and we write $\mathop{\mathbb{E}}_{Z\sim\nu}$ in the sense of Equation 9.21. We also extend the notation to signed return-distribution functions in the usual way. Observe that if ν is a probability distribution, then $\Pi_c \nu$ matches our earlier definition. The distributional Bellman operator, too, can be extended to signed distributions. Let $\eta \in \mathscr{M}(\mathbb{R})^{\mathcal{X}}$ be a *signed return function*. We define $\mathcal{T}^\pi : \mathscr{M}(\mathbb{R})^{\mathcal{X}} \to \mathscr{M}(\mathbb{R})^{\mathcal{X}}$:

$$(\mathcal{T}^\pi \eta)(x) = \mathbb{E}_\pi[(\mathrm{b}_{R,\gamma})_\# \eta(X') \mid X = x] .$$

This is the same equation as before, except that now the operator constructs convex combinations of signed distributions.

Definition 9.11. Given a state representation $\phi : \mathcal{X} \to \mathbb{R}^n$ and evenly spaced particle locations $\theta_1, \ldots, \theta_m$, a *signed linear return function approximation* $\eta_w \in \mathscr{F}_{\mathrm{S},m}^{\mathcal{X}}$ is parameterized by a weight matrix $w \in \mathbb{R}^{n \times m}$ and maps states to signed return function estimates according to

$$\eta_w(x) = \sum_{i=1}^m p_i(x; w) \delta_{\theta_i}, \quad p_i(x; w) = \phi(x)^\top w_i + \frac{1}{m}\Big(1 - \sum_{j=1}^m \phi(x)^\top w_j\Big), \quad (9.22)$$

where w_i is the ith column of w. We denote the space of signed return functions that can be represented in this form by $\mathscr{F}_{\phi,\mathrm{S},m}$. △

Equation 9.22 can be understood as approximating the mass of each particle linearly and then adding mass to all particles uniformly to normalize the signed distribution to have unit total mass. It defines a subset of signed m-categorical distributions that are constructed from linear combinations of features. Because of the normalization, and unlike the space of linear value function approximations constructed from ϕ, $\mathscr{F}_{\phi,\mathrm{S},m}$ is not a linear subspace of $\mathscr{M}(\mathbb{R})$. However, for each $x \in \mathcal{X}$, the mapping

$$w \mapsto \eta_w(x)$$

is said to be *affine*, a property that is sufficient to permit theoretical analysis (see Remark 9.3).

Using a linearly parameterized representation of the form given in Equation 9.22, we seek to build a distributional dynamic programming algorithm based on the signed categorical representation. The complication now is that the distributional Bellman operator takes the return function approximation away from our linear parameterization in two separate ways: first, the distributions themselves may move away from the categorical representation, and second, the distributions may no longer be representable by a linear combination of features. We address these issues with a doubly projected distributional operator:

$$\Pi_{\phi,\xi,\ell_2} \Pi_c \mathcal{T}^\pi ,$$

where the outer projection finds the best approximation to $\Pi_c \mathcal{T}^\pi$ in $\mathscr{F}_{\phi,\mathrm{S},m}$. By analogy with the value-based setting, we define "best approximation" in terms of a ξ-weighted Cramér distance, denoted $\ell_{\xi,2}$:

$$\ell_2^2(\nu,\nu') = \int_{\mathbb{R}} (F_\nu(z) - F_{\nu'}(z))^2 \mathrm{d}z\,,$$

$$\ell_{\xi,2}^2(\eta,\eta') = \sum_{x\in\mathcal{X}} \xi(x)\ell_2^2(\eta(x),\eta'(x))\,,$$

$$\Pi_{\phi,\xi,\ell_2}\eta = \arg\min_{\eta'\in\mathscr{F}_{\phi,\mathrm{S},m}} \ell_{\xi,2}^2(\eta,\eta')\,.$$

Because we are now dealing with signed distributions, the Cramér distance $\ell_2^2(\nu,\nu')$ is infinite if the two input signed distributions ν,ν' do not have the same total mass. This justifies our restriction to signed distributions with unit total mass in defining both the signed m-categorical representation and the linear approximation. The following lemma shows that the distributional Bellman operator preserves this property (see Exercise 9.16).

Lemma 9.12. Suppose that $\eta \in \mathscr{M}(\mathbb{R})^{\mathcal{X}}$ is a signed return-distribution function. Then,

$$\kappa((\mathcal{T}^\pi\eta)(x)) = \sum_{x'\in\mathcal{X}} \mathbb{P}_\pi(X'=x' \mid X=x)\kappa(\eta(x')).$$

In particular, if all distributions of η have unit total mass – that is, $\kappa(\eta(x))=1$ for all x – then

$$\kappa((\mathcal{T}^\pi\eta)(x)) = 1\,. \qquad \triangle$$

While it is possible to derive algorithms that are not restricted to outputting signed distributions with unit total mass, one must then deal with added complexity due to the distributional Bellman operator moving total mass from state to state.

We next derive a semi-gradient update rule based on the doubly projected Bellman operator. Following the unbiased estimation method for designing incremental algorithms (Section 6.2), we construct the signed target

$$\Pi_c\tilde{\eta}(x) = \Pi_c(\mathrm{b}_{r,\gamma})_\# \eta_w(x')\,.$$

Compared to the sample target of Equation 6.9, here η_w is a signed return function in $\mathscr{F}_{\mathrm{S},m}^{\mathcal{X}}$.

The semi-gradient update rule adjusts the weight vectors w_i to minimize the ℓ_2 distance between $\Pi_c\tilde{\eta}(x)$ and the predicted distribution $\eta_w(x)$. It does so by taking a step in direction of the negative gradient of the squared ℓ_2 distance[73]

$$w_i \leftarrow w_i - \alpha\nabla_{w_i}\ell_2^2(\eta_w(x), \Pi_c\tilde{\eta}(x))\,.$$

73. In this expression, $\tilde{\eta}(x)$ is used as a target estimate and treated as constant with regards to w.

We obtain an implementable version of this update rule by expressing distributions in $\mathscr{F}_{\text{S},m}$ in terms of m-dimensional vectors of masses. For a signed categorical distribution

$$\nu = \sum_{i=1}^{m} p_i \delta_{\theta_i} ,$$

let us denote by $p \in \mathbb{R}^m$ the vector $(p_1, \ldots, p_m)$; similarly, denote by p' the vector of masses for a signed distribution ν'. Because the cumulative functions of signed categorical distributions with unit total mass are constant on the intervals $[\theta_i, \theta_{i+1}]$ and equal outside of the $[\theta_1, \theta_m]$ interval, we can write

$$\begin{aligned} \ell_2^2(\nu, \nu') &= \int_{\mathbb{R}} (F_\nu(z) - F_{\nu'}(z))^2 \mathrm{d}z \\ &= \varsigma_m \sum_{i=1}^{m-1} (F_\nu(\theta_i) - F_{\nu'}(\theta_i))^2 \\ &= \varsigma_m \|Cp - Cp'\|_2^2 , \end{aligned} \tag{9.23}$$

where $\|\cdot\|_2$ is the usual Euclidean norm and $C \in \mathbb{R}^{m \times m}$ is the lower-triangular matrix

$$C = \begin{bmatrix} 1 & 0 & \cdots & 0 & 0 \\ 1 & 1 & \cdots & 0 & 0 \\ \vdots & & \ddots & & \vdots \\ 1 & 1 & \cdots & 1 & 0 \\ 1 & 1 & \cdots & 1 & 1 \end{bmatrix} .$$

Letting $p(x)$ and $\tilde{p}(x)$ denote the vector of masses for the signed approximation $\eta_w(x)$ and the signed target $\Pi_c \tilde{\eta}(x)$, respectively, we rewrite the above in terms of matrix-vector operations (Exercise 9.17 asks you to derive the gradient of ℓ_2^2 with respect to w_i):

$$w_i \leftarrow w_i + \alpha \varsigma_m (\tilde{p}(x) - p(x))^\top C^\top C \tilde{e}_i \phi(x) , \tag{9.24}$$

where $\tilde{e}_i \in \mathbb{R}^m$ is a vector whose entries are

$$\tilde{e}_{ij} = \mathbb{1}_{\{i=j\}} - \frac{1}{m} .$$

By precomputing the vectors $C^\top C \tilde{e}_i$, this update rule can be applied in $O(m^2 + mn)$ operations per sample transition.

9.7 Convergence of the Signed Algorithm*

We now turn our attention to establishing a contraction result for the doubly projected Bellman operator, by analogy with Theorem 9.8. We write

$$\mathscr{M}_{\ell_2,1}(\mathbb{R}) = \{\lambda_1 \nu_1 + \lambda_2 \nu_2 : \lambda_1, \lambda_2 \in \mathbb{R}, \lambda_1 + \lambda_2 = 1, \nu_1, \nu_2 \in \mathscr{P}_{\ell_2}(\mathbb{R})\} ,$$

for finite signed distributions with unit total mass and finite Cramér distance from one another. In particular, note that $\mathscr{F}_{S,m} \subseteq \mathscr{M}_{\ell_2,1}(\mathbb{R})$.

Theorem 9.13. Suppose Assumption 4.29(1) holds and that Assumption 9.5 holds with unique stationary distribution ξ_π. Then the projected operator $\Pi_{\phi,\xi_\pi,\ell_2}\Pi_c\mathcal{T}^\pi : \mathscr{M}_{\ell_2,1}(\mathbb{R})^{\mathcal{X}} \to \mathscr{M}_{\ell_2,1}(\mathbb{R})^{\mathcal{X}}$ is a contraction with respect to the metric $\ell_{\xi_\pi,2}$, with contraction modulus $\gamma^{1/2}$. △

This result is established by analyzing separately each of the three operators that composed together constitute the doubly projected operator $\Pi_{\phi,\xi_\pi,\ell_2}\Pi_c\mathcal{T}^\pi$.

Lemma 9.14. Under the assumptions of Theorem 9.13, $\mathcal{T}^\pi : \mathscr{M}_{\ell_2,1}(\mathbb{R}) \to \mathscr{M}_{\ell_2,1}(\mathbb{R})$ is a $\gamma^{1/2}$-contraction with respect to $\ell_{\xi_\pi,2}$. △

Lemma 9.15. The categorical projection operator $\Pi_c : \mathscr{M}_{\ell_2,1}(\mathbb{R}) \to \mathscr{F}_{S,m}$ is a nonexpansion with respect to ℓ_2. △

Lemma 9.16. Under the assumptions of Theorem 9.13, the function approximation projection operator $\Pi_{\phi,\xi_\pi,\ell_2} : \mathscr{F}_{S,m}^{\mathcal{X}} \to \mathscr{F}_{S,m}^{\mathcal{X}}$ is a nonexpansion with respect to $\ell_{\xi_\pi,2}$. △

The proof of Lemma 9.14 essentially combines the reasoning of Proposition 4.20 and Lemma 9.6 and so is left as Exercise 9.18. Similarly, the proof of Lemma 9.15 follows according to exactly the same calculations as in the proof of Lemma 5.23, and so it is left as Exercise 9.19. The central observation is that the corresponding arguments made for the Cramér distance in Chapter 5 under the assumption of probability distributions do not actually make use of the monotonicity of the cumulative distribution functions and so extend to the signed distributions under consideration here.

Proof of Lemma 9.16. Let $\eta, \eta' \in \mathscr{F}_{S,m}^{\mathcal{X}}$ and write $p(x)$, $p'(x)$ for the vector of masses of $\eta(x)$ and $\eta'(x)$, respectively. From Equation 9.23, we have

$$\begin{aligned}\ell^2_{\xi_\pi,2}(\eta,\eta') &= \sum_{x\in\mathcal{X}} \xi_\pi(x)\ell_2^2(\eta(x),\eta'(x))\\ &= \sum_{x\in\mathcal{X}} \xi_\pi(x)\varsigma_m\|Cp(x) - Cp'(x)\|_2^2\\ &= \varsigma_m\|\boldsymbol{p} - \boldsymbol{p}'\|^2_{\Xi\otimes C^\top C},\end{aligned}$$

where $\Xi \otimes C^\top C$ is a positive semi-definite matrix in $\mathbb{R}^{(\mathcal{X}\times m)\times(\mathcal{X}\times m)}$, with $\Xi = \text{diag}(\xi_\pi) \in \mathbb{R}^{\mathcal{X}\times\mathcal{X}}$, and $\boldsymbol{p}, \boldsymbol{p}' \in \mathbb{R}^{\mathcal{X}\times m}$ are the vectorized probabilities associated with η, η'. Therefore, $\Pi_{\phi,\xi_\pi,\ell_2}$ can be interpreted as a Euclidean projection under the norm $\|\cdot\|_{\Xi\otimes C^\top C}$ and hence is a nonexpansion with respect to the corresponding norm; the result follows as this norm precisely induces the metric $\ell_{\xi_\pi,2}$. □

Proof of Theorem 9.13. We use similar logic to that of Lemma 5.21 to combine the individual contraction results established above, taking care that the operators to be composed have several different domains between them. Let $\eta, \eta' \in \mathscr{M}_{\ell_2,1}(\mathbb{R})^{\mathcal{X}}$. Then note that

$$\begin{aligned}\ell_{\xi_\pi,2}(\Pi_{\phi,\xi_\pi,\ell_2}\Pi_{\mathrm{c}}\mathcal{T}^\pi\eta, \Pi_{\phi,\xi_\pi,\ell_2}\Pi_{\mathrm{c}}\mathcal{T}^\pi\eta') &\overset{(a)}{\leq} \ell_{\xi_\pi,2}(\Pi_{\mathrm{c}}\mathcal{T}^\pi\eta, \Pi_{\mathrm{c}}\mathcal{T}^\pi\eta') \\ &\overset{(b)}{\leq} \ell_{\xi_\pi,2}(\mathcal{T}^\pi\eta, \mathcal{T}^\pi\eta') \\ &\overset{(c)}{\leq} \gamma^{1/2}\ell_{\xi_\pi,2}(\eta, \eta'),\end{aligned}$$

as required. Here, (a) follows from Lemma 9.16, since $\Pi_{\mathrm{c}}\mathcal{T}^\pi\eta, \Pi_{\mathrm{c}}\mathcal{T}^\pi\eta' \in \mathscr{F}^{\mathcal{X}}_{\mathrm{S},m}$. (b) follows from Lemma 9.15 and the straightforward corollary that $\ell_{\xi_\pi,2}(\Pi_{\mathrm{c}}\eta, \Pi_{\mathrm{c}}\eta') \leq \ell_{\xi_\pi,2}(\eta, \eta')$ for $\eta, \eta' \in \mathscr{M}_{\ell_2,1}(\mathbb{R})$. Finally, (c) follows from Lemma 9.14. □

The categorical projection is a useful computational device as it allows Π_{ϕ,ξ,ℓ_2} to be implemented strictly in terms of signed m-categorical representations, and we rely on it in our analysis. Mathematically, however, it is not strictly necessary as it is implied by the projection onto the $\mathscr{F}_{\phi,\mathrm{S},m}$; this is demonstrated by the following Pythagorean lemma.

Lemma 9.17. For any signed m-categorical distribution $\nu \in \mathscr{F}_{\mathrm{S},m}$ (for which $\nu = \Pi_{\mathrm{c}}\nu$) and any signed distribution $\nu' \in \mathscr{M}_{\ell_2,1}(\mathbb{R})$,

$$\ell_2^2(\nu, \nu') = \ell_2^2(\nu, \Pi_{\mathrm{c}}\nu') + \ell_2^2(\Pi_{\mathrm{c}}\nu', \nu'). \tag{9.25}$$

Consequently, for any $\eta \in \mathscr{M}_{\ell_2,1}(\mathbb{R})$,

$$\Pi_{\phi,\xi,\ell_2}\Pi_{\mathrm{c}}\mathcal{T}^\pi\eta = \Pi_{\phi,\xi,\ell_2}\mathcal{T}^\pi\eta. \qquad \triangle$$

Equation 9.25 is obtained by a similar derivation to the one given in Remark 5.4, which proves the identity when ν and ν' are the usual m-categorical distributions.

Just as in the case studied in Chapter 5 without function approximation, it is now possible to establish a guarantee on the quality of the fixed point of the operator $\Pi_{\phi,\xi_\pi,\ell_2}\Pi_C\mathcal{T}^\pi$.

Proposition 9.18. Suppose that the conditions of Theorem 9.13 hold, and let $\hat{\eta}^\pi_s$ be the resulting fixed point of the projected operator $\Pi_{\phi,\xi_\pi,\ell_2}\Pi_C\mathcal{T}^\pi$. We have the following guarantee on the quality of $\hat{\eta}^\pi_s$ compared with the $(\ell_{2,\xi_\pi}, \mathscr{F}_{\phi,S,m})$-optimal approximation of η^π: namely, $\Pi_{\phi,\xi_\pi,\ell_2}\eta^\pi$:

$$\ell_{\xi_\pi,2}(\eta^\pi, \hat{\eta}^\pi_s) \leq \frac{\ell_{\xi_\pi,2}(\eta^\pi, \Pi_{\phi,\xi_\pi,\ell_2}\eta^\pi)}{\sqrt{1-\gamma}}. \qquad \triangle$$

Proof. We calculate directly:

$$\begin{aligned}
&\ell^2_{\xi_\pi,2}(\eta^\pi, \hat{\eta}^\pi_s) \\
&\quad\overset{(a)}{=} \ell^2_{\xi_\pi,2}(\eta^\pi, \Pi_{\phi,\xi_\pi,\ell_2}\Pi_c\eta^\pi) + \ell^2_{\xi_\pi,2}(\Pi_{\phi,\xi_\pi,\ell_2}\Pi_c\eta^\pi, \Pi_{\phi,\xi_\pi,\ell_2}\Pi_c\hat{\eta}^\pi_s) \\
&\quad\overset{(b)}{=} \ell^2_{\xi_\pi,2}(\eta^\pi, \Pi_{\phi,\xi_\pi,\ell_2}\Pi_c\eta^\pi) + \ell^2_{\xi_\pi,2}(\Pi_{\phi,\xi_\pi,\ell_2}\Pi_c\mathcal{T}^\pi\eta^\pi, \Pi_{\phi,\xi_\pi,\ell_2}\Pi_c\mathcal{T}^\pi\hat{\eta}^\pi_s) \\
&\quad\overset{(c)}{\leq} \ell^2_{\xi_\pi,2}(\eta^\pi, \Pi_{\phi,\xi_\pi,\ell_2}\Pi_c\eta^\pi) + \gamma\ell^2_{\xi_\pi,2}(\eta^\pi, \hat{\eta}^\pi_s) \\
&\qquad\qquad \implies \ell^2_{\xi_\pi,2}(\eta^\pi, \hat{\eta}^\pi_s) \leq \frac{1}{1-\gamma}\ell^2_{\xi_\pi,2}(\eta^\pi, \Pi_{\phi,\xi_\pi,\ell_2}\Pi_c\eta^\pi),
\end{aligned}$$

where (a) follows from the Pythagorean identity in Lemma 9.17 and a similar identity concerning $\Pi_{\phi,\xi_\pi,\ell_2}$, (b) follows since η^π is fixed by $\mathcal{T}^\pi$ and $\hat{\eta}^\pi_s$ is fixed by $\Pi_{\phi,\xi_\pi,\ell_2}\Pi_C\mathcal{T}^\pi$, and (c) follows from $\gamma^{1/2}$-contractivity of $\Pi_{\phi,\xi_\pi,\ell_2}\Pi_C\mathcal{T}^\pi$ in ℓ_{2,ξ_π}. □

This result provides a quantitative guarantee on how much the approximation error compounds when $\hat{\eta}^\pi_s$ is computed by approximate dynamic programming. Of note, here there are two sources of error: one due to the use of a finite number m of distributional parameters and another due to the use of function approximation.

Example 9.19. Consider linearly approximating the return-distribution function of the safe policy in the Cliffs domain (Example 2.9). If we use a three-dimensional state representation $\phi(x) = [1, x_r, r_c]$ where x_r, x_c are the row and column indices of a given state, then we observe aliasing in the approximated return distribution (Figure 9.3). In addition, the use of a signed distribution representation results in negative mass being assigned to some locations in the optimal approximation. This is by design; given the limited capacity of the approximation, the algorithms find a solution that mitigates errors at some locations by introducing negative mass at other locations. As usual, the use of a bootstrapping procedure introduces diffusion error, here quite significant due to the low-dimensional state representation. △

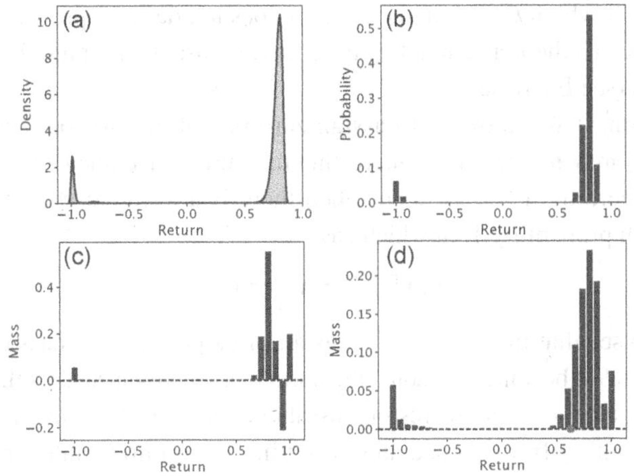

Figure 9.3
Signed linear approximations of the return distribution of the initial state in Example 2.9. **(a–b)** Ground-truth return-distribution function and categorical Monte Carlo approximation, for reference. **(c)** Best approximation from $\mathscr{F}_{\phi,\mathrm{S},m}$ based on the state representation $\phi(x) = [1, x_r, r_c]$ where x_r, x_c are the row and column indices of a given state, with $(0, 0)$ denoting the top-left corner. **(d)** Fixed point of the signed categorical algorithm.

9.8 Technical Remarks

Remark 9.1. For finite action spaces, we can easily convert a state representation $\phi(x)$ to a state-action representation $\phi(x, a)$ by repeating a basic feature matrix $\Phi \in \mathbb{R}^{\mathcal{X}\times n}$. Let $N_{\mathcal{A}}$ be the number of actions. We build a block-diagonal feature matrix $\Phi_{\mathcal{X},\mathcal{A}} \in \mathbb{R}^{(\mathcal{X}\times\mathcal{A})\times(nN_{\mathcal{A}})}$ that contains $N_{\mathcal{A}}$ copies of Φ:

$$\Phi_{\mathcal{X},\mathcal{A}} := \begin{bmatrix} \Phi & 0 & \cdots & 0 \\ 0 & \Phi & \cdots & 0 \\ \vdots & \vdots & \ddots & \vdots \\ 0 & 0 & \cdots & \Phi \end{bmatrix}.$$

The weight vector w is also extended to be of dimension $nN_{\mathcal{A}}$, so that

$$Q_w(x, a) = (\Phi_{\mathcal{X},\mathcal{A}} w)(x, a)$$

as before. This is equivalent to but somewhat more verbose than the use of per-action weight vectors. △

Remark 9.2. Assumption 9.5 enabled us to demonstrate that the projected Bellman operator $\Pi_{\phi,\xi} T^\pi$ has a unique fixed point, by invoking Banach's fixed point theorem. The first part of the assumption, on the uniqueness of ξ_π, is relatively mild and is only used to simplify the exposition. However, if there is a state x for which $\xi_\pi(x) = 0$, then $\|\cdot\|_{\xi_\pi,2}$ does not define a proper metric on $\mathbb{R}^n$ and Banach's theorem cannot be used. In this case, there might be multiple fixed points (see Exercise 9.7).

In addition, if we allow ξ_π to assign zero probability to some states, the norm $\|\cdot\|_{\xi_\pi,2}$ may not be a very interesting measure of accuracy. One common situation where the issue arises is when there is a terminal state $x_\varnothing$ that is reached with probability 1, in which case

$$\lim_{t\to\infty} \mathbb{P}_\pi(X_t = x_\varnothing) = 1 .$$

It is easy to see that in this case, ξ_π puts all of its probability mass on $x_\varnothing$, so that Theorem 9.8 becomes vacuous: the norm $\|\cdot\|_{\xi_\pi,2}$ only measures the error at the terminal state. A more interesting distribution to consider is the distribution of states resulting from immediately resetting to an initial state when $x_\varnothing$ is reached. In particular, this corresponds to the distribution used in many practical applications. Let ξ_0 be the initial state distribution; without loss of generality, let us assume that $\xi_0(x_\varnothing) = 0$. Define the substochastic transition operator

$$\mathcal{P}_{\mathcal{X},\varnothing}(x' \mid x, a) = \begin{cases} 0 & \text{if } x' = x_\varnothing, \\ P_{\mathcal{X}}(x' \mid x, a) & \text{otherwise.} \end{cases}$$

In addition, define a transition operator that replaces transitions to the terminal state by transition to one of the initial states, according to the initial distribution ξ_0:

$$\mathcal{P}_{\mathcal{X},\xi_0}(x' \mid x, a) = \mathcal{P}_{\mathcal{X},\varnothing}(x' \mid x, a) + \mathbb{1}_{\{x' \neq x_\varnothing\}} P_{\mathcal{X}}(x_\varnothing \mid x, a)\xi_0(x') .$$

One can show that the Bellman operator $T^\pi_\varnothing$ (defined from $\mathcal{P}_{\mathcal{X},\varnothing}$) satisfies

$$T^\pi_\varnothing V = T^\pi V$$

for any $V \in \mathbb{R}^{\mathcal{X}}$ for which $V(x_\varnothing) = 0$ and that the steady-state distribution $\xi_\varnothing$ induced by $\mathcal{P}_{\mathcal{X},\xi_0}$ and a policy π is such that

$$\text{there exists } t \in \mathbb{N},\ \mathbb{P}_\pi(X_t = x) > 0 \implies \xi_\varnothing(x) > 0 .$$

Let $P^\pi_\varnothing$ be the transition matrix corresponding to $\mathcal{P}_{\mathcal{X},\varnothing}(x' \mid x, a)$ and the policy π. Exercise 9.21 asks you to prove that

$$\|P^\pi_\varnothing\|_{\xi_\varnothing,2} \leq 1 ,$$

from which a modified version of Theorem 9.8 can be obtained. △

Remark 9.3. Let M and M' be vector spaces. A mapping $O : M \to M'$ is said to be affine if, for any $U, U' \in M$ and $\alpha \in [0, 1]$,

$$O(\alpha U + (1-\alpha)U') = \alpha OU + (1-\alpha)OU' .$$

When η_w is a signed linear return function approximation parameterized by w, the map

$$w \mapsto \eta_w(x)$$

is affine for each $x \in \mathcal{X}$, as we now show. It is this property that allows us to express the ℓ_2 distance between $\nu, \nu' \in \mathscr{F}_{\mathrm{S},m}$ in terms of a difference of vectors of probabilities (Equation 9.23), needed in the proof of Lemma 9.16.

Let $w, w' \in \mathbb{R}^{n \times m}$ be the parameters of signed return functions of the form of Equation 9.22. For these, write $p_w(x)$ for the vector of masses determined by w:

$$p_w(x) = w^\top \phi(x) + \frac{1}{m} e(1 - e^\top w^\top \phi(x)),$$

where e is the m-dimensional vector of ones. We then have

$$\sum_{i=1}^{m} \phi(x)^\top w_i = e^\top w^\top \phi(x),$$

and similarly for w'. Hence,

$$\begin{aligned} p_{\alpha w+(1-\alpha)w'}(x) &= (\alpha w + (1-\alpha)w')^\top \phi(x) - \frac{1}{m} ee^\top (\alpha w + (1-\alpha)w')^\top \phi(x) \\ &= \alpha p_w(x) + (1-\alpha) p_{w'}(x). \end{aligned}$$

We conclude that $w \mapsto \eta_w(x)$ is indeed affine in w. △

9.9 Bibliographical Remarks

9.1–9.2. Linear value function approximation as described in this book is in effect a special case of linear regression where the inputs are taken from a finite set $(\phi(x))_{x \in \mathcal{X}}$ and noiseless labels; see Strang (1993), Bishop (2006), and Murphy (2012) for a discussion on linear regression. Early in the history of reinforcement learning research, function approximation was provided by connectionist systems (Barto et al. 1983; Barto et al. 1995) and used to deal with infinite state spaces (Boyan and Moore 1995; Sutton 1996). An earlier-still form of linear function approximation and temporal-difference learning was used by Samuel (1959) to train a strong player for the game of checkers.

9.3. Bertsekas and Tsitsiklis (1996, Chapter 6) studies linear value function approximation and its combination with various reinforcement learning methods, including TD learning (see also Bertsekas 2011, 2012). Tsitsiklis and Van Roy (1997) establish the contractive nature of the projected Bellman operator $\Pi_{\phi,\xi} T^\pi$ under the steady-state distribution (Theorem 9.8 in this book). The temporal-difference fixed point can be determined directly by solving a least-squares problem, yielding the least-squares TD algorithm (LSTD; Bradtke and Barto 1996), which is extended to the control setting by least-squares policy iteration (LSPI; Lagoudakis and Parr 2003). In the control setting, the method is also called fitted Q-iteration (Gordon 1995; Ernst et al. 2005; Riedmiller 2005).

Bertsekas (1995) gives an early demonstration that the TD fixed point is in general different from (and has greater error than) the best approximation to V^π, measured in ξ-weighted L^2 norm. However, it is possible to interpret the temporal-difference fixed point as the solution to an oblique projection problem that minimizes errors between consecutive states (Harmon and Baird 1996; Scherrer 2010).

9.4. Barnard (1993) provides a proof that temporal-difference learning is not a true gradient-descent method, justifying the term *semi-gradient* (Sutton and Barto 2018). The situation in which ξ differs from the steady-state (or sampling) distribution of the induced Markov chain is called *off-policy learning*; Example 9.9 is a simplified version of the original counterexample due to Baird (1995). Baird argued for the direct minimization of the *Bellman residual* by gradient descent as a replacement to temporal-difference learning, but this method suffers from other issues and can produce undesirable solutions even in mild scenarios (Sutton et al. 2008a). The GTD line of work (Sutton et al. 2009; Maei 2011) is a direct attempt at handling the issue by using a pair of approximations (see Qu et al. 2019 for applications of this idea in a distributional context); more recent work directly considers a corrected version of the Bellman residual (Dai et al. 2018; Chen and Jiang 2019). The convergence of temporal-difference learning with linear function approximation was proven under fairly general conditions by Tsitsiklis and Van Roy (1997), using the ODE method from stochastic approximation theory (Benveniste et al. 2012; Kushner and Yin 2003; Ljung 1977).

Parr et al. (2007), Parr et al. (2008) and Sutton et al. (2008b) provide a domain-dependent analysis of linear value function approximation, which is extended by Ghosh and Bellemare (2020) to establish the existence of representations that are stable under a greater array of conditions (Exercises 9.10 and 9.11). Kolter (2011) studies the space of temporal-difference fixed point as a function of the distribution ξ.

9.5. Linear CTD and QTD were implicitly introduced by Bellemare et al. (2017a) and Dabney et al. (2018b), respectively, in the design of deep reinforcement learning agents (see Chapter 10). Their presentation as given here is new.

9.6–9.7. The algorithm based on signed distributions is new to this book. It improves on an algorithm proposed by Bellemare et al. (2019b) in that it adjusts the total mass of return distributions to always be 1. Lyle et al. (2019) give evidence that the original algorithm generally underperforms the categorical algorithm. They further establish that, in the risk-neutral control setting, distributional algorithms cannot do better than value-based algorithms when using linear approximation. The reader interested in the theory of signed measures is referred to Doob (1994).

9.10 Exercises

Exercise 9.1. Use the update rules of Equation 9.10 and 9.11 to prove the results from Example 9.1. △

Exercise 9.2. Prove that for a linear value function approximation $V_w = \Phi w$, we have, for any $w, w' \in \mathbb{R}^n$ and $\alpha, \beta \in \mathbb{R}$,

$$V_{\alpha w + \beta w'} = \alpha V_w + \beta V_{w'}, \nabla_w V_w(x) = \phi(x).$$ △

Exercise 9.3. In the statement of Proposition 9.3, we required that

(i) the columns of the feature matrix Φ be linearly independent;
(ii) $\xi(x) > 0$ for all $x \in \mathcal{X}$.

Explain how the result is affected when either of these requirements is omitted. △

Exercise 9.4. Prove Lemma 9.7. *Hint.* Apply Pythagoras's theorem to a well-chosen inner product. △

Exercise 9.5. The purpose of this exercise is to empirically study the contractive and expansive properties of the projection in L^2 norm. Consider an integer-valued state space $x \in \mathcal{X} = \{1, \dots, 10\}$ with two-dimensional state representation $\phi(x) = (1, x)$, and write Π_ϕ for the L^2 projection of a vector $v \in \mathbb{R}^{\mathcal{X}}$ onto the linear subspace defined by ϕ. That is,

$$\Pi_\phi v = \Phi(\Phi^\top \Phi)^{-1} \Phi^\top v,$$

where Φ is the feature matrix.

With this representation, we can represent the vector $u(x) = 0$ exactly, and hence $\Pi_\phi u = u$. Now consider the vector u' defined by

$$u'(x) = \log x.$$

With a numerical experiment, show that

$$\|\Pi_\phi u' - \Pi_\phi u\|_2 \leq \|u' - u\|_2 \quad \text{but} \quad \|\Pi_\phi u' - \Pi_\phi u\|_\infty > \|u' - u\|_\infty.$$ △

Exercise 9.6. Provide an example Markov decision process and state representation that result in the left- and right-hand sides of Equation 9.6 being equal. *Hint.* A diagram might prove useful. △

Exercise 9.7. Following Remark 9.2, suppose that the steady-state distribution ξ_π is such that $\xi_\pi(x) = 0$ for some state x. Discuss the implications on the analysis performed in this chapter, in particular on the set of solutions to Equation 9.8 and the behavior of semi-gradient temporal-difference learning when source states are drawn from this distribution. △

Exercise 9.8. Let ξ_0 be some initial distribution and π a policy. Define the discounted state-visitation distribution

$$\bar{\xi}(x') = (1-\gamma)\xi_0(x') + \gamma \sum_{x\in\mathcal{X}} \mathbb{P}_\pi(X' = x' \mid X = x)\bar{\xi}(x), \quad \text{for all } x' \in \mathcal{X}.$$

Following the line of reasoning leading to Lemma 9.6, show that the projected Bellman operator $\Pi_{\phi,\bar{\xi}}T^\pi$ is a $\gamma^{1/2}$ contraction in the $\bar{\xi}$-weighted L^2 metric:

$$\|T^\pi\|_{\bar{\xi},2} \leq \gamma^{1/2}.$$

△

Exercise 9.9. Let $\xi \in \mathscr{P}(\mathcal{X})$. Suppose that the feature matrix $\Phi \in \mathbb{R}^{\mathcal{X}\times n}$ has linearly independent columns and $\xi(x) > 0$ for all $x \in \mathcal{X}$. Show that the unique optimal weight vector w^* that is a solution to Equation 9.2 satisfies

$$\mathbb{E}\Big[(G^\pi(X) - \phi(X)^\top w^*)\phi(X)\Big] = 0, \quad X \sim \xi.$$

△

Exercise 9.10 (*). This exercise studies the divergence of semi-gradient temporal-difference learning from a dynamical systems perspective. Recall that $\Xi \in \mathbb{R}^{\mathcal{X}\times\mathcal{X}}$ is the diagonal matrix whose entries are $\xi(x)$.

(i) Show that in expectation and in matrix form, Equation 9.11 produces the sequence of weight vectors $(w_k)_{k\geq 0}$ given by

$$w_{k+1} = w_k + \alpha_k \Phi^\top \Xi(r_\pi + \gamma P^\pi \Phi w_k - \Phi w_k).$$

(ii) Assume that $\alpha_k = \alpha > 0$. Express the above as an update of the form

$$w_{k+1} = Aw_k + b, \tag{9.26}$$

where $A \in \mathbb{R}^{n\times n}$ and $b \in \mathbb{R}^n$.

(iii) Suppose that the matrix $C = \Phi^\top \Xi(\gamma P^\pi - I)\Phi$ has eigenvalues that are all real. Show that if one of these is positive, then A has at least one eigenvalue greater than 1.

(iv) Use the preceding result to conclude that under those conditions, there exists a w_0 such that $\|w_k\|_2 \to \infty$.

(v) Suppose now that all of the matrix C's eigenvalues are real and nonpositive. Show that in this case, there exists an $\alpha \in (0, 1)$ such that taking $\alpha_k = \alpha$, the sequence $(\Phi w_k)_{k\geq 0}$ converges to the temporal-difference fixed point. △

Exercise 9.11 (*). Suppose that the state representation is such that

(i) the matrix $P^\pi\Phi \in \mathbb{R}^{\mathcal{X}\times n}$ has columns that lie in the column span of Φ, and

(ii) the matrix has columns that are orthogonal with regards to the ξ-weighted inner product. That is,

$$\Phi^\top \Xi \Phi = I.$$

Show that in this case, the eigenvalues of the matrix

$$\Phi^\top \Xi(\gamma P^\pi - I)\Phi$$

are all nonpositive, for any choice of $\xi \in \mathscr{P}(\mathcal{X})$. Based on the previous exercise, conclude on the dynamical properties of semi-gradient temporal-difference learning with this representation. △

Exercise 9.12. Using your favorite numerical computation software, implement the Markov decision process from Baird's counterexample (Example 9.9) and the semi-gradient update

$$w_{k+1} = w_k + \alpha(\gamma\phi(x'_k)w_k - \phi(x_k)w_k)\phi(x_k),$$

for the state representation of the example and a small, constant value of α. Here, it is assumed that X_k has distribution ξ and X'_k is drawn from the transition kernel depicted in Figure 9.2.

(i) Vary $\xi(x)$ from 0 to 1, and plot the norm of the value function estimate, $\|V_k\|_{\xi,2}$, as a function of k.

(ii) Now replace $\phi(X'_k)w_k$ by $\phi(X'_k)\tilde{w}_k$, where $\tilde{w}_k = w_{k-k \bmod L}$ for some integer $L \geq 1$. That is, $\tilde{w}_k$ is kept fixed for L iterations. Plot the evolution of $\|V_k\|_{\xi,2}$ for different values of L. What do you observe? △

Exercise 9.13. Repeat Exercise 3.3, replacing the uniform grid encoding the state with a representation ϕ defined by

$$\phi(x) = (1, \sin(x_1), \cos(x_1), \ldots, \sin(x_4), \cos(x_4)),$$

where (x_1, x_2) denote the Cart–Pole state variables. Implement linear CTD and QTD and use these with the state representation ϕ to learn a return-distribution function approximation for the uniform and forward-leaning policies. Visualize the learned approximations at selected states, including the initial state, and compare them to ground-truth return distributions estimated by the nonparametric Monte Carlo algorithm (Remark 3.1). △

Exercise 9.14. Recall that given a finite signed distribution ν expressed as a sum of probability distributions $\nu = \lambda_1\nu_1 + \lambda_2\nu_2$ (with $\lambda_1, \lambda_2 \in \mathbb{R}$, $\nu_1, \nu_2 \in \mathscr{P}(\mathbb{R})$), we defined expectations under ν in terms of sums of expectations under ν_1 and ν_2. Verify that this definition is not dependent on the choice of decomposition of ν into a sum of probability distributions. That is, suppose that there also exist $\lambda'_1, \lambda'_2 \in \mathbb{R}$ and $\nu'_1, \nu'_2 \in \mathscr{P}(\mathbb{R})$ with $\nu = \lambda'_1\nu'_1 + \lambda'_2\nu'_2$. Show that for any (measurable) function $f : \mathbb{R} \to \mathbb{R}$ with

$$\mathop{\mathbb{E}}_{Z\sim\nu_1}[|f(Z)|] < \infty, \mathop{\mathbb{E}}_{Z\sim\nu_2}[|f(Z)|] < \infty, \mathop{\mathbb{E}}_{Z\sim\nu'_1}[|f(Z)|] < \infty, \mathop{\mathbb{E}}_{Z\sim\nu'_2}[|f(Z)|] < \infty,$$

we have

$$\lambda_1 \mathop{\mathbb{E}}_{Z\sim\nu_1}[f(Z)] + \lambda_2 \mathop{\mathbb{E}}_{X\sim\nu_2}[f(Z)] = \lambda_1 \mathop{\mathbb{E}}_{Z\sim\nu'_1}[f(Z)] + \lambda_2 \mathop{\mathbb{E}}_{Z\sim\nu'_2}[f(Z)]. \qquad \triangle$$

Exercise 9.15. Show that any m-categorical signed distribution $\nu \in \mathscr{F}_{\mathrm{S},m}$ can be written as the weighted sum of two m-categorical (probability) distributions $\nu_1, \nu_2 \in \mathscr{F}_{\mathrm{C},m}$. △

Exercise 9.16. Suppose that we consider a return function $\eta \in \mathscr{M}(\mathbb{R})^{\mathcal{X}}$ defined over signed distributions (not necessarily with unit total mass). Show that

$$\kappa((\mathcal{T}^\pi\eta)(x)) = \sum_{x'\in\mathcal{X}} \mathbb{P}_\pi(X' = x \mid X = x)\kappa(\eta(x')).$$

Conclude that if $\eta \in \mathscr{F}^{\mathcal{X}}_{\mathrm{S},m}$, then

$$\kappa((\mathcal{T}^\pi\eta)(x)) = 1. \qquad \triangle$$

Exercise 9.17. Consider two signed m-categorical distributions $\nu, \nu' \in \mathscr{F}_{\mathrm{S},m}$. Denote their respective vectors of masses by $p, p' \in \mathbb{R}^m$. Prove the correctness of Equation 9.24. That is, show that

$$\nabla_{w_i}\ell_2^2(\nu,\nu') = -2\varsigma_m(p' - p)^\top C^\top C\tilde{e}_i\phi(x). \qquad \triangle$$

Exercise 9.18. Prove Lemma 9.14. △

Exercise 9.19. Prove Lemma 9.15. △

Exercise 9.20. Prove Lemma 9.17. △

Exercise 9.21. Following the discussion in Remark 9.2, show that

$$\|P^\pi_\varnothing\|_{\xi_\varnothing,2} \le 1. \qquad \triangle$$

10 Deep Reinforcement Learning

As a computational description of how an organism learns to predict, temporal-difference learning reduces the act of learning, prediction, and control to numerical operations and mostly ignores how the inputs and outputs to these operations come to be. By contrast, real learning systems are but a component within a larger organism or machine, which in some sense provides the interface between the world and the learning component. To understand and design learning agents, we must also situate their computational operations within a larger system.

In recent years, *deep reinforcement learning* has emerged as a framework that more directly studies how characteristics of the environment and the architecture of an artificial agent affect learning and behavior. The name itself comes from the combination of reinforcement learning techniques with deep neural networks, which are used to make sense of low-level perceptual inputs, such as images, and also structure complex outputs (such as the many degrees of freedom of a robotic arm). Many of reinforcement learning's recent applied successes can be attributed to this combination, including superhuman performance in the game of Go, robots that learn to grasp a wide variety of objects, champion-level backgammon play, helicopter control, and autonomous balloon navigation.

Much of the recent research in distributional reinforcement learning, including our own, is rooted in deep reinforcement learning. This is no coincidence, since return distributions are naturally complex objects, whose learning is facilitated by the use of deep neural networks. Conversely, predicting the return also translates into practical benefits, often in the guise of accelerated and more stable learning. In this context, the distributional method helps the system organize its inputs into features that are useful for further learning, a process known as *representation learning*. This chapter surveys some of these ideas and discusses practical considerations in introducing distributional reinforcement learning to a larger-scale system.

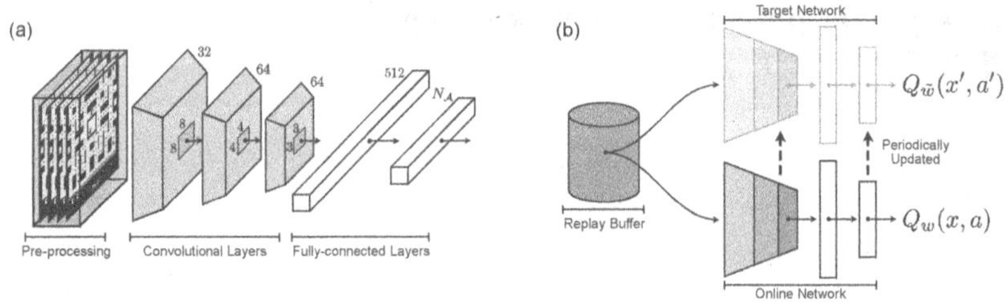

Figure 10.1
(a) The deep neural network at the heart of the DQN agent architecture. The network takes as inputs four preprocessed images and outputs the predicted state-action value estimates. **(b)** The replay buffer provides the sample transitions used to update the main (online) network (Equation 10.2). A second *target* network is used to construct sample targets and is periodically updated with the weights of the online network. Example input to the network is visualized with frames from the Atari game Ms. PAC-MAN (published by Midway Manufacturing).

10.1 Learning with a Deep Neural Network

The term *deep neural network* (or simply deep network) designates a fairly broad class of functions built from a composition of atomic elements (called neurons or hidden units), typically organized in layers and parameterized by weight vectors and one or many activation functions. The function's inputs are transformed by the first layer, whose outputs become the inputs of the second layer, and so on. Canonically, the weights of the network are adjusted to minimize a given loss function by gradient descent. In that respect, one may view linear function approximation as a simple neural network architecture that maps inputs to outputs by means of a single, nonadjustable layer implementing the chosen state representation.

More sophisticated architectures include: convolutional neural networks (LeCun and Bengio 1995), particularly suited to learning functions that exhibit some degree of translation invariance; recurrent networks (Hochreiter and Schmidhuber 1997), designed to deal with sequences; and attention mechanisms (Bahdanau et al. 2015; Vaswani et al. 2017). The reader interested in further details is invited to consult the work of Goodfellow et al. (2016).

By virtue of their parametric flexibility, deep neural networks have proven particularly effective at dealing with reinforcement learning problems with high-dimensional structured inputs. As an example, the DQN algorithm (Deep Q-Networks; Mnih et al. 2015) applies the tools of deep reinforcement learning to achieve expert-level game play in Atari 2600 video games. More properly speaking, DQN is a system or *agent architecture* that combines a deep neural network with a semi-gradient Q-learning update rule and a few algorithmic components that improve learning stability and efficiency. DQN also contains subroutines to transform received observations into a useful notion of state and to select actions to be enacted in the environment. We divide the computation performed by DQN into four main subtasks:

(a) *preprocessing* observations (Atari 2600 images) into an input vector,
(b) *predicting* future outcomes given such an input vector,
(c) *behaving* on the basis of these predictions, and finally
(d) *learning* to improve the agent's predictions from its experience.

Preprocessing. The Arcade Learning Environment or *ALE* (Naddaf 2010; Bellemare et al. 2013a) provides a reinforcement learning interface to the Stella Atari 2600 emulator (Mott et al. 1995–2023). This interface produces 210×160, 7-bit pixel images in response to one of eighteen possible joystick motions (combinations of an up-down motion, a left-right motion, and a button press). A single emulation step or *frame* lasts 1/60th of a second, but the agent only selects a new action every four frames, what might be more appropriately called a time step (that is, fifteen times per emulated second; for further implementation details, see Machado et al. (2018)).

In order to simplify the learning process and make it more suitable for reinforcement learning, DQN transforms or *preprocesses* the images and rewards it receives from the ALE. Every time step, the DQN agent converts the most recent frame to 7-bit grayscale (preserving luminance). It combines this frame with the immediately preceding one, similarly converted, by means of a *max-pooling* operation that takes the pixel-wise maximum of the two images. The result is then *downscaled* to a smaller 84×84 size, and pixel values are normalized from 0 (black) to 1 (white). The four most recent images (spanning a total of 16 frames, a little over 1/4th of a second) are concatenated together to form the input vector, of size $84 \times 84 \times 4$.[74] In essence, DQN uses this input vector as a surrogate for the state x.

The Arcade Learning Environment also provides an integer reward indicating the change in score (positive or negative) between consecutive time steps. DQN applies an operation called *reward clipping*, which keeps only the sign of the provided reward (−1, 0, or 1). The result is what we denote by r.

Prediction. DQN uses a five layer neural network to predict the expected return from the agent's current state (see Figure 10.1a). The network's input is the stack of frames produced by the preprocessor. This input is successively

74. In many games, these images do not provide a complete picture of the game state. In this case, the problem is said to be *partially observable*. See, for example, Kaelbling et al. (1998).

transformed by three convolutional layers that extract features from the image, and then by a *fully connected* layer that applies a linear transformation and a nonlinearity to the output of the final convolutional layer. The result is a 512-dimensional vector, which is then linearly transformed into $N_{\mathcal{A}} = 18$ values corresponding to the predicted expected returns for each action. If we denote by w the entire collection of weights of the neural network, then the approximation $Q_w(x, a)$ denotes the resulting prediction made by DQN. Here, x is described by the input image stack and a indexes into the network's $N_{\mathcal{A}}$-dimensional output vector.

Behavior and experience. DQN acts according to an ε-greedy policy derived from its approximation Q_w. More precisely, this policy assigns $1 - \varepsilon$ of its probability mass to a greedy selection rule that breaks ties uniformly at random. Provided that Q_w assigns a higher value to better actions, acting greedily results in good performance. The remaining ε probability mass is divided equally among all actions, allowing for some *exploration* of unseen situations and eventually improving the network's value estimates from experience.

Similar to the online setting described by Algorithms 3.1 and 3.2, the sample produced by the agent's interactions with the Arcade Learning Environment is used to improve the network's value predictions (see **Learning** below). Here, however, the transitions are not provided to the learning algorithm as they are experienced but are instead stored in a *circular replay buffer*. The replay buffer stores the last 1 million transitions experienced by the agent; as the name implies, these are typically replayed multiple times as part of the learning process. Compared to learning from each piece of experience once and then discarding it, using a replay buffer has been empirically demonstrated to result in greater sample efficiency and stability.

Learning. Every four time steps, DQN samples a *minibatch* of thirty-two sample transitions from its replay buffer, uniformly at random. As with linear approximation, these sample transitions are used to adjust the network weights w with a semi-gradient update rule. Given a second *target network* with weights $\tilde{w}$ and a sample transition (x, a, r, x'), the corresponding sample target is

$$r + \gamma \max_{a' \in \mathcal{A}} Q_{\tilde{w}}(x', a') . \tag{10.1}$$

Canonically, the discount factor is chosen to be $\gamma = 0.99$.

In deep reinforcement learning, the update to the network's parameters is typically expressed first and foremost in terms of a loss function $\mathcal{L}$ to be minimized. This is because automatic differentiation can be used to efficiently compute the gradient $\nabla_w \mathcal{L}(w)$ by means of the backpropagation algorithm (Werbos 1982; Rumelhart et al. 1986). In the case of DQN, the sample target is used to construct the squared loss

$$\mathcal{L}(w) = (r + \gamma \max_{a' \in \mathcal{A}} Q_{\tilde{w}}(x', a') - Q_w(x, a))^2 .$$

Taking the gradient of $\mathcal{L}$ with respect to w, we obtain the semi-gradient update rule:

$$w \leftarrow w + \alpha(r + \gamma \max_{a' \in \mathcal{A}} Q_{\tilde{w}}(x', a') - Q_w(x, a))\nabla_w Q_w(x, a) . \tag{10.2}$$

In practice, the actual loss to be minimized is the average loss over the sampled minibatch. Consequently, the semi-gradient update rule adjusts the weight vector on the basis of all sampled transitions simultaneously. After a specified number of updates have been performed (10,000 in the original implementation), the weights w of the main network (also called *online network*) are copied over to the target network; the process is illustrated in Figure 10.1b. As discussed in Section 9.4, the use of a target network allows the semi-gradient update rule to behave more like the projected Bellman operator and mitigates some of the convergence issues of semi-gradient methods.

Equation 10.2 generalizes the semi-gradient rule for linear value function approximation (Equation 9.11) to nonlinear schemes; observe that if $Q_w(x, a) = \phi(x, a)^\top w$, then

$$\nabla_w Q_w(x, a) = \phi(x, a) .$$

In practice, more sophisticated adaptive gradient descent methods (see, e.g., Tieleman and Hinton 2012; Kingma and Ba 2015) are used to accelerate convergence. Additionally, the *Huber loss* is a popular alternative to the squared loss. For $\kappa \geq 0$, the Huber error function is defined as

$$\mathcal{H}_\kappa(u) = \begin{cases} \frac{1}{2}u^2, & \text{if } |u| \leq \kappa \\ \kappa(|u| - \frac{1}{2}\kappa), & \text{otherwise} . \end{cases}$$

DQN's Huber loss instantiates this error function with $\kappa = 1$:

$$\mathcal{L}(w) = \mathcal{H}_1\big(r + \gamma \max_{a' \in \mathcal{A}} Q_{\tilde{w}}(x', a') - Q_w(x, a)\big) .$$

The Huber loss corresponds to the squared loss for small errors but behaves like the L^1 loss for larger errors. In effect, this puts a limit on the magnitude of the updates to the weight vector, which in turn reduces instability in the learning process.

Viewed as an agent interacting with its environment, DQN operates by acting in the world according to its state-action value function estimates, collecting new experience along the way. It then uses this experience to produce an improved policy that is used to collect better experience and so on. Given sufficient training time, this approach can learn fairly complex behavior from pixels and rewards alone, including behavior that exploits computer bugs in some of Atari

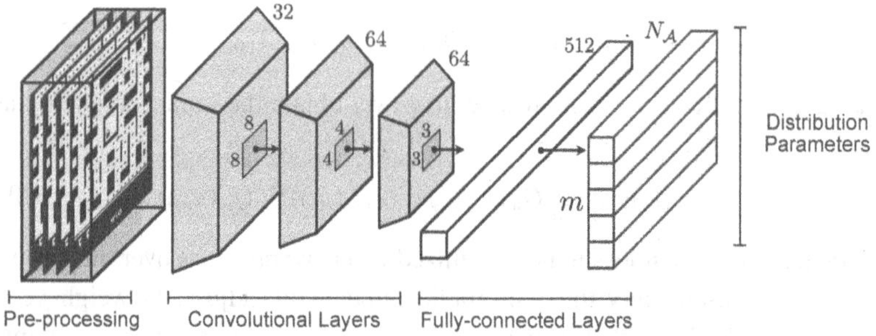

Figure 10.2
The extension of DQN to the distributional setting uses a deep neural network that outputs m distribution parameters per action. The same network architecture is used for both C51 (categorical representation) and QR-DQN (quantile representation). Example input to the network is visualized with frames from the Atari game Ms. PAC-MAN (published by Midway Manufacturing).

2600 games. Although it is possible to achieve the same behavior by using a fixed state representation and learning a linear value approximation (Machado et al. 2018), in practice, doing so without a priori knowledge of the specific game being played is computationally demanding and eventually does not scale as effectively.

10.2 Distributional Reinforcement Learning with Deep Neural Networks

We extend DQN to the distributional setting by mapping state-action pairs to return distributions rather than to scalar values. A simple approach is to change the network to output one m-dimensional parameter vector per action (equivalently, one $N_{\mathcal{A}} \times m$-dimensional parameter matrix, as illustrated in Figure 10.2). These vectors are then interpreted as the parameters of return distributions, by means of a probability distribution representation (Section 5.2). The target network is modified in the same way.

By making different choices of distribution representation or rule for updating the parameters of these distributions, we obtain different distributional algorithms. The *C51* algorithm, in particular, is based on the categorical representation, while the *QR-DQN* algorithm is based on the quantile representation. Their respective update rules follow those of linear CTD and linear QTD in Section 9.5, and here we emphasize the differences from the linear approach. We begin with a description of the C51 algorithm.

Prediction. C51 owes its name to its use of a 51-categorical representation in its original implementation (Bellemare et al. 2017a), but the name generally applies for any such agent with $m \geq 2$. Given an input state x, its network outputs a collection of softmax parameters $((\varphi_i(x, a))_{i=1}^m : a \in \mathcal{A})$. These parameters are used to define the state-action categorical distributions:

$$\eta_w(x, a) = \sum_{i=1}^{m} p_i(x, a; w)\delta_{\theta_i} ,$$

where $p_i(x, a; w)$ is a probability given by the softmax distribution:

$$p_i(x, a; w) = \frac{e^{\varphi_i(x,a;w)}}{\sum_{j=1}^{m} e^{\varphi_j(x,a;w)}} .$$

The standard implementation of C51 parameterizes the locations $\theta_1, \ldots, \theta_m$ in two specific ways: first, θ_1 and $-\theta_m$ are chosen to be negative of each other, so that the support of the distribution is symmetric around 0. Second, it is customary to take m to be an odd number so that the central location $\theta_{\frac{m-1}{2}}$ is zero. The canonical choice is $\theta_1 = -10$ and $\theta_m = 10$; we will study the effect of this choice in Section 10.4. Even though the largest theoretically achievable return is $V_{\text{MAX}} = \frac{R_{\text{MAX}}}{1-\gamma} = 100$, the choice of $\theta_m = 10$ is sensible because in most Atari 2600 video games, the player's score only changes infrequently. Consequently, the reward is zero on most time steps and the largest return actually observed is typically much smaller than the analytical maximum (the same argument applies to V_{MIN} and θ_1).

Behavior. C51 is designed to optimize for the risk-neutral objective. The value function induced by its distributional predictions is

$$Q_w(x, a) = \mathop{\mathbb{E}}_{Z \sim \eta_w(x,a)}[Z] = \sum_{i=1}^{m} p_i(x, a; w)\theta_i ,$$

and similarly for $Q_{\tilde{w}}$. The agent acts according to an ε-greedy policy defined from the main network's state-action value estimates $Q_w(x, a)$.

Learning. The sample target is obtained from the combination of an update and projection step. Following Section 9.5, given a transition (x, a, r, x'), the sample target is given by

$$\bar{\eta}(x, a) = \sum_{j=1}^{m} \bar{p}_j \delta_{\theta_j} = \Pi_c\big((\mathrm{b}_{r,\gamma})_\# \eta_{\tilde{w}}(x', \mathrm{a}_{\tilde{w}}(x'))\big) ,$$

where

$$\mathrm{a}_{\tilde{w}}(x') = \arg\max_{a' \in \mathcal{A}} Q_{\tilde{w}}(x', a')$$

is the greedy action for the induced state-action value function $Q_{\tilde{w}}$. As in the linear setting, this sample target is used to formulate a cross-entropy loss to be minimized (Equation 9.18):

$$\mathcal{L}(w) = -\sum_{j=1}^{m} \bar{p}_j \log p_j(x, a; w) .$$

Updating the parameters w in the direction of the negative gradient of this loss yields the (first-order) semi-gradient update rule

$$w \leftarrow w + \alpha \sum_{i=1}^{m} (\bar{p}_i - p_i(x, a; w)) \nabla_w \theta_i(x, a; w) .$$

It is useful to contrast the above with the linear CTD semi-gradient update rule (Equation 9.19):

$$w_i \leftarrow w_i + \alpha(\bar{p}_i - p_i(x, a; w))\phi(x, a) .$$

With linear CTD, the update rule takes a simple form where one computes the per-particle categorical TD error $(\bar{p}_i - p_i(x, a; w))$ and moves the weight vector w_i in the direction $\phi(x, a)$ in proportion to this error. This is possible because the softmax parameters $\varphi_i(x, a; w) = \phi(x, a)^\top w_i$ are linear in ϕ and consequently

$$\nabla_{w_i} \varphi_i(x, a; w) = \phi(x, a) ,$$

similar to the value-based setting. When using a deep network, however, the weights at earlier layers affect the entire predicted distribution and it is not possible to perform per-particle weight updates independently.

QR-DQN. QR-DQN uses the same neural network as C51 but interprets the output vectors as parameters of a m-quantile representation, rather than a categorical representation:

$$\eta_w(x, a) = \frac{1}{m} \sum_{i=1}^{m} \delta_{\theta_i(x,a;w)} .$$

Its induced value function is simply

$$Q_w(x, a) = \frac{1}{m} \sum_{i=1}^{m} \theta_i(x, a; w) .$$

Given a sample transition (x, a, r, x') and levels $\tau_1, \ldots, \tau_m$, the parameters are updated by performing gradient descent on the *Huber quantile loss*

$$\mathcal{L}(w) = \frac{1}{m} \sum_{i,j=1}^{m} \rho^{\mathcal{H}}_{\tau_i}(r + \gamma \theta_j(x', \mathrm{a}_{\tilde{w}}(x'); \tilde{w}) - \theta_i(x, a; w)) ,$$

where $\rho^{\mathcal{H}}_{\tau}(u) = |\mathbb{1}_{\{u<0\}} - \tau| \mathcal{H}_1(u)$.

The Huber quantile loss behaves like the quantile loss (Equation 6.12) for large errors but like the expectile loss (the single-sample equivalent of Equation 8.13) for sufficiently small errors. This loss is minimized by a statistical functional between a quantile and expectile. The Huber quantile loss can be interpreted as a smoothed version of the quantile regression loss (Sections 6.4 and 9.5), which leads to more stable behavior when combined with the nonlinear function approximation and adaptive gradient descent methods used in deep reinforcement learning.

10.3 Implicit Parameterizations

The value function Q^π can be viewed as a mapping from state-action pairs to real values. When there are finitely many actions and the state space is small and finite, a tabular representation of this mapping is usually sufficient – in this case, each entry in the table corresponds to the value at a given state and action. Much like linear function approximation, neural networks improve on the tabular representation by parameterizing the relationship between similar states, allowing us to generalize the learned function to unseen states. It is therefore useful to think of the inputs of DQN's neural network as arguments to a function and its outputs as the evaluation of this function. Under this perspective, the network implements a function mapping $\mathcal{X}$ to $\mathbb{R}^{\mathcal{A}}$, which can be interpreted as the function $Q_w : \mathcal{X} \times \mathcal{A} \to \mathbb{R}$ by a further indexing into the output vector.

Similarly, we can represent probability distributions as different kinds of functions. For example, the probability density function f_ν of a suitable distribution $\nu \in \mathscr{P}(\mathbb{R})$ is a mapping from $\mathbb{R}$ to $[0, \infty)$ and its cumulative distribution is a monotonically increasing function from $\mathbb{R}$ to $[0, 1]$. A distribution can also be represented by its inverse cumulative distribution function $F_\nu^{-1} : (0, 1) \to \mathbb{R}$, which we call *quantile function* in the context of this section. By extension, a state-action return function can be viewed as a function of three arguments: $x \in \mathcal{X}$, $a \in \mathcal{A}$, and a distribution parameter $\tau \in (0, 1)$. That is, we may represent η by the mapping

$$(x, a, \tau) \mapsto F^{-1}_{\eta(x,a)}(\tau) .$$

Doing so gives rise to an *implicit* approach to distributional reinforcement learning, which makes the distribution parameter an additional input to the network.

To understand this idea, it is useful to contrast it with the two algorithms of the previous section. Both C51 and QR-DQN take a half-and-half approach: the state x is provided as input to the network, but the parameters of the distribution (the probabilities p_i or locations θ_i, accordingly) are given by the network's

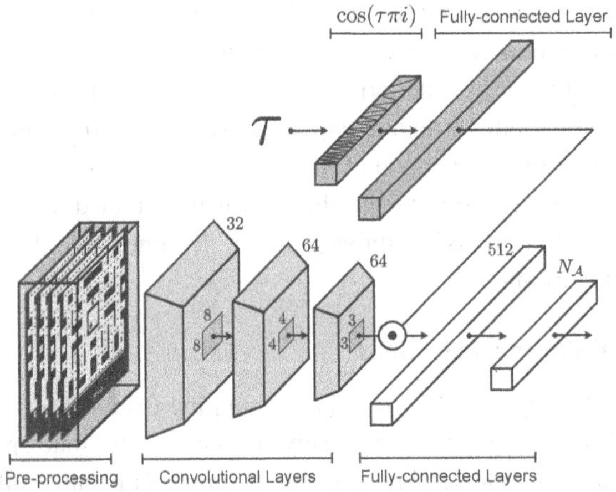

Figure 10.3
The deep neural network used by IQN. IQN outputs a single scalar per action, corresponding to the queried τth quantile. The input τ to the network is encoded using a cosine embedding and multiplicatively combined with one of the network's inner layers (see Remark 10.1). Example input to the network is visualized with frames from the Atari game Ms. PAC-MAN (published by Midway Manufacturing).

outputs, one m-dimensional vector per action. This is conceptually simple and has the advantage of leveraging probability distribution representations whose behavior is theoretically well understood in the tabular setting. On the other hand, the implied discretization of the return distributions incurs some cost, for example, due to diffusion (Section 5.8). If instead we turn the distribution parameter into an input to the network, we gain the benefits of generalization, and in principle we can learn an approximation to the return function that is only limited by the capacity of the neural network.

Prediction. The *implicit quantile network* (IQN) algorithm instantiates the argument-as-inputs principle by parameterizing the quantile function within the neural network (Figure 10.3). As before, one of the network's inputs is a description of the state x: in the case of a DQN-type architecture, the stack of preprocessed Atari 2600 images. In addition, the network also receives as input a desired level $\tau \in (0, 1)$. This level is encoded by the *cosine embedding* $\varphi(\tau) \in \mathbb{R}^M$:

$$\varphi(\tau) = (\cos(\pi\tau i))_{i=0}^{M-1} .$$

In effect, the cosine embedding represents τ as a real-valued vector, making it easier for the neural network to work with. The network's output is an $N_{\mathcal{A}}$-dimensional vector describing the evaluation of the quantile function $F^{-1}_{\eta(x,a)}(\cdot)$

at this τ. With this construction, the neural network outputs the approximation

$$\theta_w(x, a, \tau) = \phi_w(x, \tau)^\top w_a \approx F^{-1}_{\eta(x,a)}(\tau)\,,$$

where $\phi_w(x, \tau) \in \mathbb{R}^n$ are learned features for the (x, τ) pair and $w_a \in \mathbb{R}^n$ are per-action weights. Remark 10.1 provides details on how $\phi_w(x, \tau)$ is implemented in the network.

Behavior. Because IQN represents the quantile function implicitly, the induced state-action value functions Q_w and $Q_{\tilde{w}}$ are approximated by a sampling procedure, where as before $\tilde{w}$ denotes the weights of the target network. This approximation is obtained by drawing m levels $\tau_1, \dots, \tau_m$ uniformly and independently from the $(0, 1)$ interval and averaging the output of the network at these levels:

$$Q_w(x, a) \approx \frac{1}{m} \sum_{i=1}^{m} \theta_w(x, a, \tau_i)\,. \tag{10.3}$$

As with the other algorithms presented in this chapter, IQN acts according to an ε-greedy policy derived from Q_w, with greedy action $\mathrm{a}_w(x)$.

Learning. Where QR-DQN aims to learn the quantiles of the return distribution at a finite number of fixed levels, IQN aims to approximate the entire quantile function of this distribution.[75] Because each query to the network returns the network's prediction evaluated for a single level, learning proceeds somewhat differently. First, a pair of levels τ and τ' is sampled uniformly from $(0, 1)$. These determine the level at which the quantile function is to be updated and a level from which a sample target is constructed. For a given sample transition (x, a, r, x') and levels τ, τ', the two-sample IQN loss is

$$\mathcal{L}_{\tau,\tau'}(w) = \rho^{\mathcal{H}}_{\tau}\big(r + \gamma\theta_{\tilde{w}}(x', \mathrm{a}_{\tilde{w}}(x'), \tau') - \theta_w(x, a, \tau)\big)\,.$$

The variance of the sample gradient of this loss is reduced by averaging the two-sample loss over many pairs of levels $\tau_1, \dots, \tau_{m_1}$ and $\tau'_1, \dots, \tau'_{m_2}$:

$$\mathcal{L}(w) = \frac{1}{m_2} \sum_{i=1}^{m_1} \sum_{j=1}^{m_2} \mathcal{L}_{\tau_i,\tau'_j}(w)\,.$$

Risk-sensitive control. An appealing side effect of using an implicit parameterization is that many risk-sensitive objectives can be computed simply by changing the sampling distribution for Equation 10.3. For example, given a predicted quantile function $\theta_w(x, a, \cdot)$ with instantiated random variable $G_w(x, a)$, recall that the CVaR of $G_w(x, a)$ for a given level $\bar{\tau} \in (0, 1)$ is, in the integral

75. Of course, in practice, the predictions made by IQN might not actually correspond to the return distribution of any fixed policy, because of approximation, bootstrapping, and issues arising in the control setting (Chapter 7).

form of Equation 7.20,

$$\mathrm{CVaR}_{\bar{\tau}}(G_w(x,a)) = \frac{1}{\bar{\tau}} \int_0^{\bar{\tau}} \theta_w(x,a,\tau)\mathrm{d}\tau\,.$$

Similar to the procedure for estimating the expected value of $G_w(x, a)$ (Equation 10.3), this integral can be approximated by sampling m levels $\tau_1, \ldots, \tau_m$, but now from the $(0, \bar{\tau})$ interval:

$$\mathrm{CVaR}_{\bar{\tau}}(G_w(x,a)) \approx \frac{1}{m}\sum_{i=1}^{m} \theta_w(x,a,\tau_i), \quad \tau_i \sim \mathcal{U}([0,\bar{\tau}])\,.$$

Treating the distribution parameter as a network input opens up the possibility for a number of different algorithms, of which IQN is but one instantiation. In particular, in problems where actions are real-valued (for example, when controlling a robotic arm), it is common to also make the action a an input to the network.

10.4 Evaluation of Deep Reinforcement Learning Agents

To illustrate the practical value of deep reinforcement learning and the added benefits from predicting return distributions, let us take a closer look at how well the algorithms presented in this chapter can learn to play Atari 2600 video games. As the Arcade Learning Environment provides an interface to more than sixty different games, a standard evaluation procedure is to apply the same algorithm across a large set of games and report its performance on a per-game basis, as well as aggregated across games (see Machado et al. 2018 for a more complete discussion). Here, performance is measured in terms of the in-game score achieved during one play-through (i.e., an episode). One particularly attractive feature of Atari 2600 games, from a benchmarking perspective, is that almost all games explicitly provide such a score. Measuring performance in terms of game score has the additional advantage that it allows us to numerically compare the playing skill of learning agents to that of human players.

The goal of the learning agent is to improve its performance at a given game by repeatedly playing that game – in more formal terms, to optimize the risk-neutral control objective from sample interactions.[76] The agent interacts with the environment for a total of 200 million frames per game (about 925 hours of game-play). Experimentally, we repeat this process multiple times in order to evaluate the performance of the agent across different initial network weights

76. Note, however, that the in-game score differs from the agent's actual learning objective, which involves a discount factor (canonically, $\gamma = 0.99$) and clipped rewards. This metric-objective mismatch is well studied in the literature (e.g., van Hasselt et al. 2016a), and exists in part because optimizing for the undiscounted, unclipped return with DQN produces unstable behavior.

and realizations of the various sources of randomness. Following common usage, we call each repetition a *trial* and the process itself *training* the agent. Figure 10.4 (left) illustrates the overall progress made by DQN, C51, QR-DQN, and IQN over the course these interactions, evaluated every 1 million frames. Quantitatively, we measure an agent's performance in terms of an aggregate metric called *human-normalized interquartile mean score* (HNIQM), which we now define.

Definition 10.1. Denote by g the score achieved by a learning agent on a particular game and by g^{H} and g^{R} the score achieved by two reference agents (a human expert and the random policy). Assuming that $g^{\text{H}} > g^{\text{R}}$, we define the *human-normalized score* as

$$\text{HNS}(g, g^{\text{H}}, g^{\text{R}}) = \frac{g - g_{\text{R}}}{g_{\text{H}} - g_{\text{R}}} . \qquad \triangle$$

In the standard evaluation protocol described here, a learning agent's performance during training corresponds to its average per-episode score, measured from 500,000 frames of additional, evaluation-only interactions with the environment.

Definition 10.2. Let $\mathcal{E}$ be a set of E games on which we have evaluated an agent for K trials. For $1 \leq k \leq K$ and $e \in \mathcal{E}$, denote by s_e^k the human-normalized score achieved on trial k:

$$s_e^k = \text{HNS}(g_e^k, g_e^{\text{H}}, g_e^{\text{R}}) ,$$

where g_e^k, g_e^{H}, and g_e^{R} are respectively the agent's, human player's, and random policy's score on game e. Suppose that $(\hat{s}_i)_{i=1}^{K \times E}$ denotes the vector of these human-normalized scores, sorted so that $\hat{s}_i \leq \hat{s}_{i+1}$, for all $i = 1, \ldots, K \times E - 1$. The *human-normalized interquartile mean score* is given by

$$\frac{1}{E_1 - E_0} \sum_{i=E_0}^{E_1 - 1} \hat{s}_i , \qquad (10.4)$$

where E_0 is the integer nearest to $\frac{KE}{4}$ and $E_1 = KE - E_0$. $\triangle$

The normalization to human performance makes it possible to compare scores across games; although it is also possible to evaluate agents in terms of their mean or median normalized scores, these aggregates tend to be less representative of the full distribution of scores. As shown in Figure 10.4, it is accepted practice (see, e.g., Agarwal et al. (2021)) to measure the degree of variability in performance across trials and games using some empirically-determined confidence interval. One should be mindful that such an interval, while informative, does not typically guarantee statistical significance – for example, because there are too few samples to aggregate or because these samples are not identically

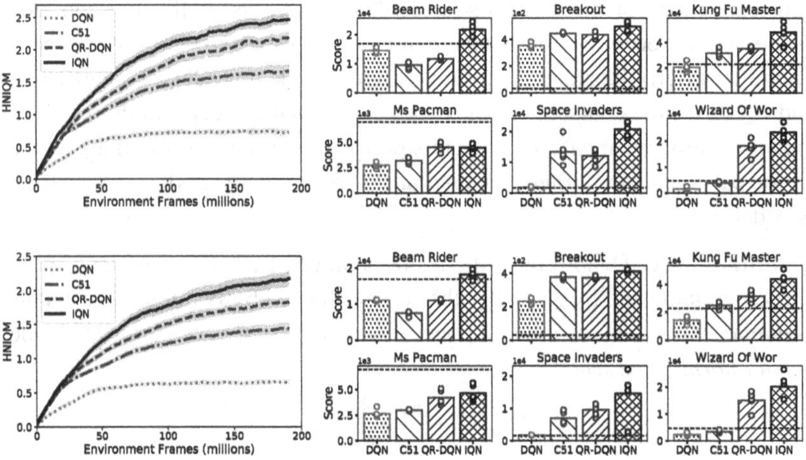

Figure 10.4
Top and bottom: Evaluation using the deterministic-action and sticky-action versions of the Arcade Learning Environment, respectively. **Left**: Human-normalized interquartile mean score (HNIQM; see main text) across fifty-seven Atari 2600 games during the course of learning. A normalized score of 1.0 indicates human-level performance, on aggregate. Per-game scores were obtained from five independent trials for each algorithm × game configuration. Shading indicates bootstrapped 95 percent confidence intervals (see main text). **Right**: Game scores obtained by different algorithms and a human expert (dashed line); the scale of these scores is indicated at the top left of each graph; dots indicate individual per-trial scores. The reported scores are measured at the end of training.

distributed. Consequently, it is also generally recommended to report individual game scores in addition to aggregate performance. Figure 10.4 (right) illustrates this for a selected subset of games.

These results show that reinforcement learning, combined with deep neural networks, can achieve a high level of performance on a wide variety of Atari 2600 games. Additionally, this performance improves with more training. This is particularly remarkable given that the learning algorithm is only provided game images as input, rather than game-specific features such as the location of objects or the number of remaining lives. It is worth noting that although each agent uses the same network architecture and hyperparameters[77] across Atari 2600 games, there are a few differences between the hyperparameters used by

77. Following common usage, we use the term "hyperparameter" to distinguish the learned parameters (e.g., w) from the parameters selected by the user (e.g., m and $\theta_1, \dots, \theta_m$).

these agents; these differences match what is reported in the literature and are the default choices in the code used for these experiments (Quan and Ostrovski 2020). The question of how to account for hyperparameters in scientific studies is an active area of research (see, e.g., Henderson et al. 2018; Ceron and Castro 2021; Madeira Auraújo et al. 2021).

Historically, all four agents whose performance is reported here were trained on what is called the *deterministic-action version* of the Arcade Learning Environment, in which arbitrarily complicated joystick motions can be performed. For example, nothing prevents the agent from alternating between the "left" and "right" actions every four frames (fifteen times per emulated second). This makes the comparison with human players somewhat unrealistic, as human play involves a minimum reaction time and interaction with a mechanical device that may not support such high-frequency decisions. In addition, some of the policies found by agents in the deterministic setting exploit quirks of the emulator in ways that were clearly not intended by the designer.

To address this issue, more recent versions of the Arcade Learning Environment implement what is called *sticky actions* – a procedure that introduces a variable delay in the environment's response to the agent's actions. Figure 10.4 (bottom panels) shows the results of the same experiment as above, but now with sticky actions. The performance of the various algorithms considered here generally remains similar, with some per-game differences (e.g., for the game Space Invaders).

Although Atari 2600 games are fundamentally deterministic, randomness is introduced in the learning process by a number of phenomena, including side effects of distributional value iteration (Section 7.4), state aliasing (Section 9.1), the use of a stochastic ε-greedy policy, and the sticky-actions delay added by the Arcade Learning Environment. In many situations, this results in distributional agents making surprisingly complex predictions (Figure 10.5). A common theme is the appearance of bimodal or skewed distributions when the outcome is uncertain – for example, when the agent's behavior in the next few time steps is critical to its eventual success or failure. Informally, we can imagine that because the agent predicts such outcomes, it in some sense "knows" something more about the state than, say, an agent that only predicts the expected return. We will see some evidence to this effect in the next section.

Furthermore, incorporating distributional predictions in a deep reinforcement learning agent provides an additional degree of freedom in defining the number and type of predictions that an agent makes at any given point in time. C51, for example, is parameterized by the number of particles m used to represent probability distributions as well as the range of its support (described by θ_m). Figure 10.6 illustrates the change in human-normalized interquartile mean (measured

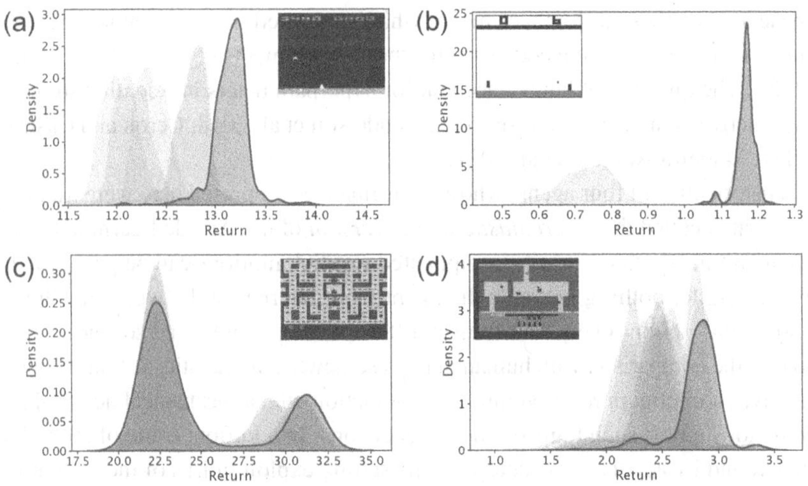

Figure 10.5
Example return distributions predicted by IQN agents in four Atari 2600 games: **(a)** SPACE INVADERS (published by Atari, Inc.), **(b)** PONG (from VIDEO OLYMPICS, published by Atari, Inc.), **(c)** Ms. PAC-MAN (published by Midway Manufacturing), and **(d)** H.E.R.O. (published by Atari, Inc.). In each panel, action-return distributions for the game state shown are estimated for each action using kernel density estimation (over 1000 samples τ). The outlined distribution corresponds to the action with the highest expected return estimate (chosen by the greedy selection rule).

at the end of training) that results from varying both of these hyperparameters. These results illustrate the commonly reported finding that predicting distributions leads to improved performance, with more accurate predictions generally helping. More generally, this illustrates that an agent's performance depends to a good degree on the chosen distribution representation; this is an example of what is called an *inductive bias*. In addition, as illustrated by the results for $m = 201$ and larger values of θ_m, there is clearly a complex relationship between an agent's parameterization and its aggregate performance across Atari 2600 games.

The development of deep distributional reinforcement learning agent architectures continues to be an active area of research. Recent advances have provided further improvements to the game-playing performance of such agents in Atari 2600 video games, including fully parameterized quantile networks (FQF; Yang et al. 2019), which extend the ideas underlying QR-DQN and IQN, and moment-matching DQN (MM-DQN; Nguyen et al. 2021), which defines a training loss

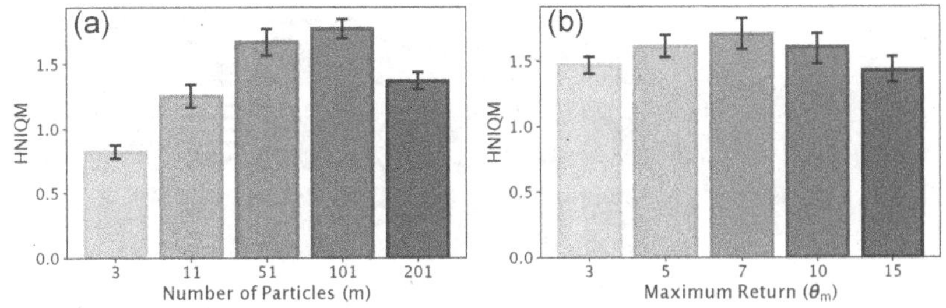

Figure 10.6
Aggregate performance (HNIQM) of C51 as a function of (**a**) the number of particles m and (**b**) the largest return that can be predicted (θ_m), for $m = 51$. Performance is averaged over the last 10 million frames of training, and error bars indicate bootstrapped 95 percent confidence intervals (see main text).

via the MMD metric described in Chapter 4. Distributional reinforcement learning has also been combined with a variety of other deep reinforcement learning techniques, as well as being used to improve exploration; we discuss some of these techniques in the bibliographical remarks.

10.5 How Predictions Shape State Representations

Deep reinforcement learning algorithms adjust the weights w of their neural network in order to minimize the error in their predictions. For example, DQN's semi-gradient update rule

$$w \leftarrow w + \alpha(r + \gamma \max_{a' \in \mathcal{A}} Q_{\tilde{w}}(x', a') - Q_w(x, a))\nabla_w Q_w(x, a)$$

adjusts the network weights w in proportion to the temporal-difference error and the first-order relationship between each weight and the action-value prediction $Q_w(x, a)$. One way to understand how predictions influence the agent's behavior is to consider how these predictions affect the parameters w.

In all agent architectures studied in this chapter, the algorithm's predictions are formed from linear combinations of the outputs of the hidden units at the penultimate layer. Consequently, we may separate the network weights into two sets: the weights that map inputs to the penultimate layer and the weights that map this layer to predictions (Figures 10.1 and 10.2). In fact, we may think of the output of the penultimate layer as a state representation $\phi(x)$, where the mapping $\phi : \mathcal{X} \rightarrow \mathbb{R}^n$ is implemented by the earlier layers. Viewed this way, the learning process simultaneously modifies the parameterized mapping ϕ and the weights of the final layer in order to make better predictions.

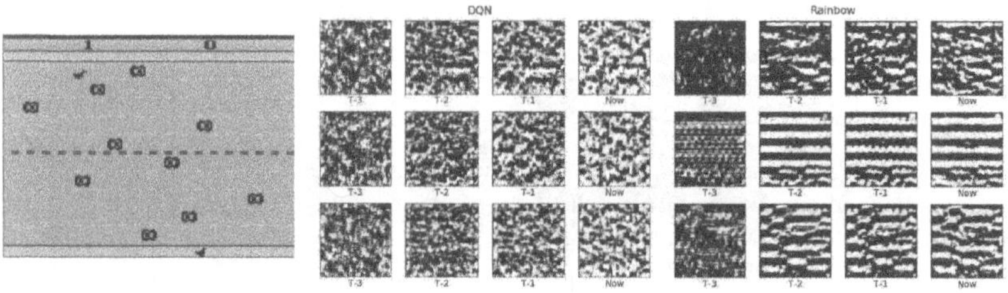

Figure 10.7
Left: Example frame from the game FREEWAY (published by Activision, Inc.). **Center and right**: Reproduced from Such et al. (2019) with permission. Input images synthesized to maximally activate individual hidden units in either a DQN or Rainbow (C51) network. Each row represents a different hidden unit; individual columns correspond to one of the four images provided as input to the networks.

We should also expect that the type and number of predictions made by the agent should influence the nature of the state representation. To test this hypothesis, Such et al. (2019) studied the patterns detected by individual hidden units in the penultimate layer of trained neural networks. In their experiments, input images were synthesized to maximize the output of a given hidden unit (see, e.g., Olah et al. 2018, for a discussion of how this can be done). For example, Figure 10.7 compares the result of this process applied to networks that either predict action-value functions (DQN) or categorical return distributions (Rainbow (Hessel et al. 2018), a variant of C51 enhanced with additional algorithmic components). Across multiple Atari 2600 games, Such et al. found that making distributional predictions resulted in state representations that better reflected the structure of the Atari 2600 game being played: for example, identifying horizontal elements (car lanes) in the game FREEWAY (shown in the figure).

Understanding how distributional reinforcement learning affects the state representation implied by the network and how that representation affects performance is an active area of research. One challenge is that deep reinforcement learning architectures have a great deal of moving parts, many of which are tuned by means of hyperparameters, and obtaining relevant empirical evidence tends to be computationally demanding. In addition, changing the state representation has a number of downstream effects on the learning process, including changing the optimization landscape, reducing or amplifying the parameter noise due to the use of an incremental update rule, and of course affecting the quality of the best achievable value function approximation. Yet, the hope is that understanding why making distributional predictions improves performance in domains such as the Arcade Learning Environment should shed light on the design of more efficient deep reinforcement learning agents.

10.6 Technical Remarks

Remark 10.1. Figure 10.1 shows that the DQN neural network is composed of three convolutional layers, a fully connected layer, and finally a linear mapping to the output $Q_w(x, a)$. One appeal of neural networks is that these operations can be composed fairly easily to add capacity to the learning algorithm or combine different inputs: for example, the cosine embedding discussed in Section 10.3.

Let us denote the input to the network by $x \in \mathbb{R}^{n_0}$, where for simplicity of exposition, we omit the structured nature of the input (in the case of DQN, it is a $84 \times 84 \times 4$ array of values; four square images). Thus, for $i \geq 0$, we denote by $n_i \in \mathbb{N}$ the size of the "flattened" input vector to layer $i+1$. For $i > 0$, the ith layer transforms its inputs using a function $f_i : \mathbb{R}^{n_{i-1}} \to \mathbb{R}^{n_i}$ parameterized by a weight vector w_i:

$$x_i = f_i(x_{i-1}; w_i), \quad i > 0 .$$

In the case of DQN, f_1, f_2, and f_3 are convolutional, while f_4 is fully connected. The latter is defined by a weight matrix W_4 and bias vector b_4 (which in our notation are part of w_4):

$$f_4(x; w_4) = [W_4^\top x + b_4]^+ ;$$

recall that for $z \in \mathbb{R}$, $[z]^+ = \max(0, z)$. In the field of deep learning, this function is called a rectified linear transformation (ReLU; Nair and Hinton 2010).

IQN augments this network by transforming the cosine embedding $\varphi(\tau)$ with another fully connected layer with the same number of dimensions as the output of the last convolutional layer (n_3):

$$x_{3b} = f_{3b}(\varphi(\tau); w_{3b}) .$$

The output of these two layers is then composed by element-wise multiplication:

$$x_{3c} = x_3 \odot x_{3b} ,$$

which becomes the input to the original network's final layer:

$$x_4 = f_4(x_{3c}; w_4) ,$$

itself linearly transformed into the quantile function estimate for (x, a, τ). The reader interested in further details should consult the work of Dabney et al. (2018a). △

10.7 Bibliographical Remarks

Beyond Atari 2600 games, the general effectiveness of deep reinforcement learning is well established in the literature. Perhaps the most famous example to date is AlphaGo (Silver et al. 2016), a computer Go program that learned to play better than the world's best Go players. Further evidence that distributional predictions improve the performance of deep reinforcement learning agents can be found in experiments on a variety of domains, including the cooperative card game Hanabi (Bard et al. 2020), stratospheric balloon flight (Bellemare et al. 2020), robotic manipulation (Bodnar et al. 2020; Cabi et al. 2020; Vecerik et al. 2019), simulated race car control (Wurman et al. 2022), and magnetic control of plasma (Degrave et al. 2022).

10.1. Early reinforcement learning research has close ties with the study of connectionist systems; see, for example, Barto et al. (1983) and Bertsekas and Tsitsiklis (1996). Tesauro (1995) combined temporal-difference learning with a single-layer neural network to produce TD-Gammon, a grandmaster-level Backgammon player and early deep reinforcement learning algorithm. Neural Fitted Q-iteration (Riedmiller 2005) implements many of the ingredients later found in DQN, including replaying past experience (Lin 1992) and the use of a target network; the method has been successfully used to control a soccer-playing robot (Riedmiller et al. 2009).

The Arcade Learning Environment (Bellemare et al. 2013a), itself based on the Stella emulator (Mott et al. 1995–2023), introduced Atari 2600 game-playing as a challenge domain for artificial intelligence. Early results on the Arcade Learning Environment included both reinforcement learning (Bellemare et al. 2012a, 2012b) and planning (Bellemare et al. 2013b; Lipovetzky et al. 2015) solutions. The DQN algorithm demonstrated the ability of deep neural networks to effectively tackle this domain (Mnih et al. 2015). Since then, deep reinforcement learning has been applied to produce high-performing policies for a variety of video games and image-based control problems (e.g., Beattie et al. 2016; Levine et al. 2016; Kempka et al. 2016; Bhonker et al. 2017; Cobbe et al. 2020). Machado et al. (2018) study the relative performance of linear and deep methods in the context of the Arcade Learning Environment. See François-Lavet et al. (2018) and Arulkumaran et al. (2017) for reviews of deep reinforcement learning and Graesser and Keng (2019) for a practical overview. Montfort and Bogost (2009) give an excellent history of the Atari 2600 video game console itself.

10.2–10.3. The C51, QR-DQN, and IQN agent architectures and algorithms were respectively introduced by Bellemare et al. (2017a), Dabney et al. (2018b), and Dabney et al. (2018a). Open-source implementations of these three algorithms are available in the Dopamine framework (Castro et al. 2018) and the DQN Zoo (Quan and Ostrovski 2020). The idea of implicitly parameterizing other arguments of the prediction function has been used extensively to deal with continuous actions; see, for example, Lillicrap et al. (2016b) and Barth-Maron et al. (2018).

There is by now a wide variety of deep distributional reinforcement learning algorithms, many of which outperform IQN. FQF (Yang et al. 2019) approximates the return distribution with a weighted combination of Diracs by combining the IQN architecture with a method for selecting which values of $\tau \in (0, 1)$ to feed into the network. MM-DQN (Nguyen et al. 2021) use an architecture based on QR-DQN in combination with an MMD-based loss as described in Chapter 4; typically, the Gaussian kernel has been found to provide the best empirical performance, despite a lack of theoretical guarantees. Both Freirich et al. (2019) and Doan et al. (2018) propose the use of generative adversarial networks (Goodfellow et al. 2014) to model the reward distribution. Freirich et al. also extend this approach to the case of multivariate rewards. There are also several recent modifications to the QR-DQN architecture that seek to address the *quantile-crossing* problem – namely, that the outputs of the QR-DQN network need not satisfy the natural monotonicity constraints of distribution quantiles. Yue et al. (2020) propose to use deep generative models combined with a postprocessing sorting step to obtain monotonic quantile estimates. Zhou et al. (2021) parameterize the difference between successive quantiles, rather than the quantile locations themselves, to enforce monotonicity; this approach was extended by Luo et al. (2021), who directly parameterize the quantile function via rational-quadratic splines. Developing and improving deep distributional reinforcement learning agents continues to be an exciting direction of research.

Several agents also incorporate a distributional loss in combination with a variety of other deep reinforcement learning techniques. Munchausen-IQN (Vieillard et al. 2020) combines IQN with a form of entropy regularization, and Rainbow (Hessel et al. 2018) combines C51 with a variety of modifications to DQN, including double Q-networks (van Hasselt et al. 2016b), prioritized experience replay (Schaul et al. 2016), a value-advantage dueling architecture (Wang et al. 2016), parameter noise for exploration (Fortunato et al. 2018), and multistep returns (Sutton 1988). There has also been a wide variety of work combining distributional RL with the actor-critic framework, typically by modifying the critic to include distributional predictions; see, for example, Tessler et al. (2019), Kuznetsov et al. (2020), and Duan et al. (2021) and Nam et al. (2021).

Several recent deep reinforcement learning agents have also leveraged return distribution estimates to improve exploration. Nikolov et al. (2019) combine C51 with information-directed exploration to obtain an agent that outperforms IQN, as judged by human-normalized mean and median performance on Atari 2600. Mavrin et al. (2019) extract an estimate of parametric uncertainty from distributions learned under QR-DQN and use this information to specify an exploration bonus. Clements et al. (2020) similarly build on QR-DQN to estimate both aleatoric and epistemic uncertainties and use these for exploratory action selection. Zhang and Yao (2019) use the quantile representation learned by QR-DQN to form a set of options, with policies that vary in their risk-sensitivity, to improve exploration. Further use cases continue to be developed, with Lin et al. (2019) decomposing the reward signal into independent streams, all of which are then predicted.

10.4. The training and evaluation protocol presented here, including the idea of a human-normalized score, is due to Mnih et al. (2015); more generally, Bellemare et al. (2013a) propose the use of a normalization scheme in order to compare agents across Atari 2600 games. The sticky-actions mechanism was proposed by Machado et al. (2018), who give evidence that a naive trajectory optimization algorithm (see also Bellemare et al. 2015) can achieve scores comparable to DQN when evaluated with the deterministic version of the ALE. Our use of the interquartile mean to compare score follows the recommendations of Agarwal et al. (2021), who also highlight reproducibility concerns when evaluating across multiple-domain games. See also Henderson et al. (2018). The experiments reported here were performed using the DQN Zoo (Quan and Ostrovski 2020).

10.5. The success of distributional reinforcement learning algorithms on large benchmarks such as the Arcade Learning Environment is often attributed to their use as *auxiliary tasks* (Jaderberg et al. 2017). Auxiliary tasks are ancillary predictions made by the neural network that stabilize learning, shape the state representation implied by the neural network, and improve end performance (Lample and Chaplot 2017; Kartal et al. 2019; Agarwal et al. 2020; Laskin et al. 2020; Guo et al. 2020; Lyle et al. 2021). Their effect on the penultimate layer of the network is discussed more formally by Chung et al. (2018), Bellemare et al. (2019a), and Le Lan et al. (2022). Dabney et al. (2020a) argue that representation learning plays a particularly acute role in deep reinforcement learning when considering the control problem, in which the policy under evaluation changes over time. General value functions (GVFs) provide a language for describing auxiliary tasks and expressing an agent's knowledge (Sutton et al. 2011). Schlegel et al. (2021) extend some of these ideas to the deep learning setting, in particular considering how GVFs are useful features in the context of sequential prediction.

As an alternative explanation for the empirical successes of distributional reinforcement learning, Imani and White (2018) study the effect of distributional predictions on optimisation landscapes.

10.8 Exercises

Many of the exercises in this chapter are hands-on and open-ended in nature and require programming or applying published learning algorithms to standard reinforcement learning domains. More generally, these exercises are designed to give the reader some experience in performing deep reinforcement learning experiments. As a starting point, we recommend that the reader use open-source implementations of the algorithms covered in this chapter. The Dopamine framework[78] (Castro et al. 2018) provides implementations of the DQN, C51, QR-DQN, and IQN agents and support for the Acrobot domain. DQN Zoo[79] (Quan and Ostrovski 2020) is a collection of open-source reference implementations that were used to generate the results of this chapter.

Exercise 10.1. In this exercise, you will apply deep reinforcement learning methods to the Acrobot domain (Sutton 1996). Acrobot is a two-link pendulum where the state gives the joint angles of the two links and their corresponding angular velocities. Because the inputs are vectors rather than images, the DQN agent described in this chapter must be adapted to this domain by replacing the convolutional layers by one or multiple fully connected layers and substituting a simple state encoding for the image preprocessing (in Dopamine, this encoding is given by what is called *Fourier features* (Konidaris et al. 2011)). Of course, the reader is encouraged to think of possible alternatives to both of these choices.

(i) Train a DQN agent, varying the frequency at which the target network is updated. Plot the value function over time, at the initial starting state and for different update frequencies.

(ii) Train a DQN agent, but now varying the size of the replay buffer. Describe the effect of the replay buffer size on the learning algorithm in this case.

(iii) Modify the DQN implementation to train from each transition as it is received, rather than via the replay buffer. Compare this to the previous results.

(iv) Starting from an implementation of the C51 agent, implement the signed distributional algorithm from Section 9.6. Train both C51 and the signed

78. https://github.com/google/dopamine

79. https://github.com/deepmind/dqn_zoo

algorithm on Acrobot. Visualize the return-distribution function predictions made for the same initial state, over time. Plot the average undiscounted return obtain by either algorithm over time. Explain your findings.

(v) Train both QR-DQN and IQN agents on Acrobot, and visualize the distributional predictions made for the same initial state. For IQN, this will require querying the deep neural network for multiple values of τ. Visualize those predictions at different points during an episode; do they behave as expected? △

Exercise 10.2 (*). In this exercise, you will apply deep RL methods to the MinAtar domain (Young and Tian 2019), which is a set of simplified versions of Atari 2600 games (ASTERIX, BREAKOUT, FREEWAY, SEAQUEST, SPACE INVADERS).

(i) Using MinAtar's built-in human-play example, play each game for ten episodes and log your scores. How much variability do you see between episodes? Plot the scores versus games played; does your performance improve over time?

(ii) Implement a random agent that takes actions uniformly at random. Evaluate the random agent on each MinAtar game for at least ten episodes and log the agent's scores for each game and episode.

(iii) Train a C51 agent to play BREAKOUT in MinAtar for 2 million frames (approximately 2.5 hours on a GPU). Use the default hyperparameters for the range of predicted returns ($\theta_1 = -10$ and $\theta_m = 10$). Use your recorded scores from above (for human and random players) to compute C51's human-normalized score. How do these compare with those reported for C51 on the corresponding games in Atari 2600?

(iv) Based upon your performance and that of the random agent, compute a reasonable estimate of the maximum achievable discounted return in each game. Train C51, again for 2 million frames, using this maximum return to set the particle locations. Compare this agent's performance in terms of human-normalized score with that of the default C51 agent above. How do they compare?

(v) Train a QR-DQN agent on the MinAtari BREAKOUT game, evaluating in terms of human-normalized score. Inspecting the learned return distributions, how do these distributions compare with those learned by C51? Do any of the quantile estimates exceed your estimated maximal discounted return? Why might this happen?

(vi) Train DQN, C51, and QR-DQN agents with your preferred hyperparameters, for 5 million frames with at least three seeds, on each of the five MinAtar games. Compute the human-normalized interquartile mean score

(HNIQM) for each method versus training steps. How do your results compare with those see in Figure 10.4? Note that this exercise will require significantly greater computational resources than others and will in general require running multiple agents in parallel. △

11 Two Applications and a Conclusion

We conclude by highlighting two applications of the core ideas covered in earlier chapters, with the aim to give a sense of the range of domains to which ideas from distributional reinforcement learning have been and may eventually be applied.

11.1 Multiagent Reinforcement Learning

The core setting studied in this book is the interaction between an agent and its environment. The model of the environment as an unchanging, static Markov decision process is a good fit for many problems of interest. However, a notable exception is the case in which the agent finds itself interacting with other learning agents. Such settings arise in games, both competitive and cooperative, as well as real-world interactions such as in autonomous driving.

Interactions between distinct agents lead to an incredibly rich space of learning problems. What is possible is governed by considerations such as how many agents there are, whether their interests are aligned or competing, whether they have the same information about the environment, whether they must act concurrently or sequentially, and whether they can directly communicate with each other. We choose to focus here on just one of many models for cooperative multiagent interactions.

Definition 11.1 (Boutilier 1996). A *multiagent Markov decision process* (MMDP) is a Markov decision process $(\mathcal{X}, \mathcal{A}, \xi_0, P_\mathcal{X}, P_\mathcal{R})$ in which the action set $\mathcal{A}$ has a factorized structure $\mathcal{A} = \prod_{i=1}^{N} \mathcal{A}_i$, for some integer $N \in \mathbb{N}^+$ and finite nonempty sets $\mathcal{A}_i$. We refer to N as the *number of players* in the MMDP. △

An N-player MMDP describes N agents interacting with an environment. At each stage, agent i selects an action $a_i \in \mathcal{A}_i$ ($i = 1, \ldots, N$), knowing the current state $x \in \mathcal{X}$ of the MMDP, but without knowledge of the actions of the other agents. All agents observe the reward resulting from the joint action $(a_1, \ldots, a_N)$

and share the joint goal of maximizing the discounted sum of rewards arising in the MMDP; their interests are perfectly aligned.

To compute a joint optimal policy for the agents, one approach is to treat the problem as an MDP and use either dynamic programming or temporal-difference learning methods to compute an optimal policy. These methods assume *centralized* computation of the policy, which is then communicated to the agents to execute.

By contrast, the *decentralized control problem* is for the agents to arrive at a joint optimal policy through direct interaction with the environment and without any centralized or interagent communication; this is pertinent when communication between agents is impossible or costly, and a model of the environment is not known. Thus, the agents jointly interact with the environment, producing transitions of the form $(x, (a_1, \ldots, a_N), r, x')$; agent i observes only (x, a_i, r, x') and must learn from transitions of this form, without observing the actions of other agents that influenced the transition.

Example 11.2. The *partially stochastic climbing game* (Kapetanakis and Kudenko 2002; Claus and Boutilier 1998) is an MMDP with a single non-terminal state (also known as a matrix game), two players, and three actions per player. The reward distributions for each combination of the players' actions are shown on the left-hand side of Figure 11.1; the first player's actions index the rows of this matrix, and the second player's actions index the columns. All rewards are deterministic, except for the case of the central element, where the distribution is uniform over the set $\{0, 14\}$. This environment represents a coordination challenge for the two agents: the optimal strategy is for both to take the first action, but if either agent deviates from this strategy (by exploring the value of other actions, for example), negative rewards of large magnitude are incurred. △

A concrete example of an approach to the decentralized control problem is for each agent to independently implement Q-learning with these transitions (Tan 1993). The center panels of Figure 11.1 show the result of the agents using Q-learning to learn in the partially stochastic climbing game. Both agents act using an ε-greedy policy, with ε decaying linearly during the interaction (beginning at 1 and ending at 0), and use a step size of $\alpha = 0.001$ to update their action values. Due to the exploration the agents are undertaking, the first action is judged as worse than the third action by both agents, and both quickly move to using the third action, hence not discovering the optimal behavior for this environment.

The failure of the Q-learning agents to reach the optimal behavior stems from the fact that from the point of view of an individual agent, the environment it is

$$\begin{pmatrix} 11 & -30 & 0 \\ -30 & \mathcal{U}(\{0,14\}) & 6 \\ 0 & 0 & 5 \end{pmatrix}$$

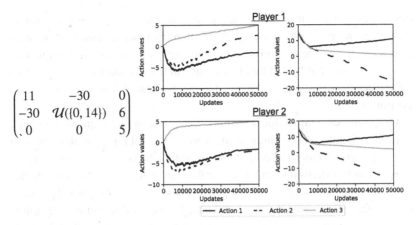

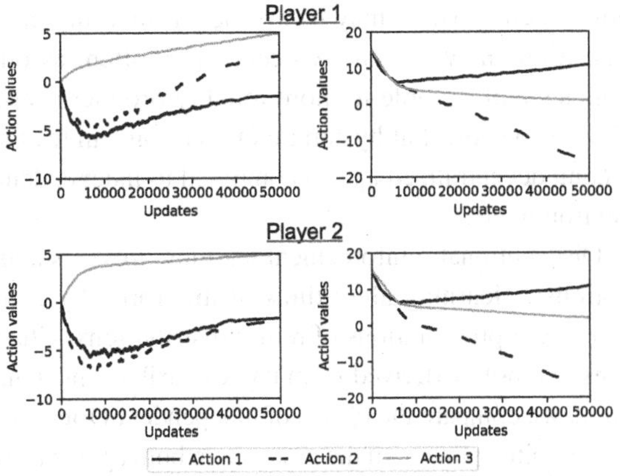

Figure 11.1
Left: Table specifying reward distributions for the partially stochastic climbing game. **Right**: Learned action values for each player–action combination under Q-learning (first column) and a distributional algorithm (second column).

interacting with is no longer Markov; it contains other learning agents, which may adapt their behavior as time progresses, and in particular in response to changes in the behavior of the individual agent itself. Redesigning learning rules such as Q-learning to take into account the changing behavior of other agents in the environment is a core means of encouraging better cooperation between agents in such settings in multiagent reinforcement learning.

Hysteretic Q-learning (Matignon et al. 2007; HQL) is a modification of Q-learning that swaps the usual risk-neutral value update for a rule that instead tends to learn an *optimistic* estimate of the value associated with an action. Specifically, given an observed transition (x, a, r, x'), HQL performs the update

$$Q(x,a) \leftarrow Q(x,a) + \big(\alpha \mathbb{1}\{\Delta > 0\} + \beta \mathbb{1}\{\Delta < 0\}\big)\Delta\,,$$

where $\Delta = r + \gamma \max_{a' \in \mathcal{X}} Q(x', a) - Q(x, a)$ is the TD error associated with the transition. Here, $0 < \beta < \alpha$ are asymmetric step size parameters associated with negative and positive TD errors. By making larger updates in response to positive TD errors, the learnt Q-values end up placing more weight on high-reward outcomes. In fact, this update can be shown to be equivalent to following the negative gradient of the expectile loss encountered in Section 8.6:

$$Q(x,a) \leftarrow Q(x,a) + (\alpha + \beta)|\mathbb{1}_{\{\Delta<0\}} - \tau|\Delta\,,$$

with $\tau = {}^{\alpha}/_{\alpha+\beta}$. The values learnt by HQL are therefore a kind of optimistic summary of the agent's observations. The motivation for learning values in this way is that low-reward outcomes may be due to the exploratory behavior from other agents, which may be avoided as learning progresses, while rewarding transitions may eventually occur more often, as other agents improve their policies and are able to more reliably produce these outcomes. Matignon et al. (2007) show that hysteretic Q-learning can lead to improved coordination among decentralized agents compared to independent Q-learning in a range of environments.

Distributional reinforcement learning provides a natural framework to build optimistic learning algorithms of this form, by combining an algorithm for learning representations of return distributions (Chapters 5 and 6) with a risk-sensitive policy derived from these distributions (Chapter 7). To illustrate this point, we compare the results of independent Q-learning on the partially stochastic climbing game with the case where both agents use a distributional algorithm in which distributions are updated using categorical TD updates. We take distributions supported on $\{-30, -29, \ldots, 30\}$ and define greedy actions defined in a risk-sensitive manner; in particular, the greedy action is the one with the greatest expectile at level τ, calculated from the categorical distribution estimates (see Chapter 7), with τ linearly decaying from 0.9 to 0.7 throughout the course of learning.

Figure 11.1 shows the learnt action values by both distributional agents in this setting; the exploration schedule and step sizes are the same as for the independent Q-learning agents. This level of optimism means that action values are not overly influenced by the exploration of other agents and is also not too high so as to be distracted by the (stochastic) outcome of fourteen available when both agents play the second action, and indeed the agents converge to the optimal joint policy in this case. We remark, however, that the optimism level chosen here is tuned to illustrate the beneficial effects that are possible with distributional approaches to decentralized cooperative learning, and in general, other choices of risk-sensitive policies will not lead to optimal behavior in this environment. This is illustrative of a broader tension: while we would like to be optimistic about the behavior of other learning agents, the approach inevitably leads to optimism in aleatoric environment randomness (in this example, the randomness in the outcome when both players select the second action). With both distributional and nondistributional approaches to decentralized multiagent learning, it is difficult to treat these sources of randomness differently from one another.

The majority of work in distributional multiagent reinforcement learning has focused on the case of large-scale environments, using deep reinforcement learning approaches such as those described in Chapter 10. Lyu and Amato (2020) introduce Likelihood Hysteretic IQN, which uses return distribution learnt by an IQN architecture to adapt the level of optimism used in value function estimates throughout training. Da Silva et al. (2019) also found benefits from using risk-sensitive policies based on learnt return distributions. In the centralized training, decentralized execution regime (Oliehoek et al. 2008), Sun et al. (2021) and Qiu et al. (2021) empirically explore the combination of distributional reinforcement learning with previously established value function factorization methods (Sunehag et al. 2017; Rashid et al. 2018; Rashid et al. 2020). Deep distributional reinforcement learning agents have also been successfully employed in cooperative multiagent environments without making any use of learnt return distributions beyond expected values. The Rainbow agent (Hessel et al. 2018), which makes use of the C51 algorithm described in Chapter 10, forms a baseline for the Hanabi challenge (Bard et al. 2020). Combinations of deep reinforcement learning with distributional reinforcement learning have found application in a variety of multiagent problems to date; we expect there to be further experimentally driven research in this area of application and also remark that the theoretical understanding of how such algorithms perform is largely open.

11.2 Computational Neuroscience

Machine learning and reinforcement learning often take inspiration from psychology, neuroscience, and animal behavior. Examples include convolutional neural networks (LeCun and Bengio 1995), experience replay (Lin 1992), episodic control (Pritzel et al. 2017), and navigation by grid cells (Banino et al. 2018). Conversely, algorithms developed for artificial agents have proven useful as computational models for building theories regarding the mechanisms of learning in humans and animals; some authors have argued, for example, on the plausibility of backpropagation in the brain (Lillicrap et al. 2016a). As we will see in this section, distributional reinforcement learning is also useful in this regard and serves to explain some of the fine-grained behavior of dopaminergic neurons in the brain.

Dopamine (DA) is a neurotransmitter associated with learning, motivation, motor control, and attention. Dopaminergic neurons, especially those concentrated in the ventral tegmental area (VTA) and substantia nigra pars compacta (SNc) regions of the midbrain, release dopamine along several pathways projecting throughout the brain – in particular, to areas known to be involved in reinforcement, motor function, executive functions (such as planning, decision-making, selective attention, and working memory), and associative learning. Furthermore, despite their relatively modest numbers (making up less than 0.001 percent of the neurons in the human brain), they are crucial to the development and functioning of human intelligence. This can be seen especially acutely by dopamine's implication in a range of neurological disorders such as Parkinson's disease, attention-deficit hyperactivity disorder (ADHD), and schizophrenia.

The Rescorla–Wagner model (Rescorla and Wagner 1972) posits that the learning of conditioned behavior in humans and animals is *error-driven*. That is, learning occurs as the consequence of a mismatch between the learner's predictions and the observed outcome. The Rescorla–Wagner equation takes the form of a familiar update rule:[80]

$$V \leftarrow V + \alpha \underbrace{(r - V)}_{\text{error}}, \tag{11.1}$$

where V is the predicted reward, r the observed reward, and α an asymmetric step size parameter. Here, the term α plays the same role as the step size parameter introduced in Chapter 3 but describes the modeled rate at which the animal learns rather than a parameter proper.[81]

Rescorla and Wagner's model explained, for example, classic experiments in which rabbits learned to blink in response to a light cue predictive of an unpleasant puff of air (an example of Pavlovian conditioning). The model also explained a learning phenomenon called *blocking* (Kamin 1968): having learned that the light cue predicts a puff of air, the rabbits did not become conditioned to a second cue (an audible tone) when that cue was presented concurrently with the light. This gave support to the theory of error-driven learning, as opposed to associative learning purely based on co-occurrence (Pavlov 1927).

Temporal-difference learning is also a type of error-driven learning, one that accounts for the temporally extended nature of prediction. In its simplest form, TD learning is described by the equation

$$V \leftarrow V + \alpha \underbrace{(r + \gamma V' - V)}_{\text{TD error}}, \tag{11.2}$$

which improves on the Rescorla–Wagner model by decomposing the learning target into an immediate reward (observed) and a prediction V' about future rewards (guessed). Just as the Rescorla–Wagner equation explains blocking, temporal-difference learning explains how cues can themselves generate prediction errors (by a process of bootstrapping). This in turn gives rise to the phenomenon of *second-order conditioning*. Second-order conditioning arises when a secondary cue is presented anterior to the main cue, which itself predicts the reward. In this case, the secondary cue elicits a prediction of the future reward, despite only being paired with the main cue and not the reward itself.

80. This notation resembles, but is not quite the same as, that of previous chapters, yet it is common in the field (see, e.g., Ludvig et al. 2011).

81. Admittedly, the difference is subtle.

In one set of experiments, the dopaminergic (DA) neurons of macaque monkeys were recorded as they learned that a light is predictive of the availability of a reward (juice, received by pressing a lever).[82] In the absence of reward, DA neurons exhibit a sustained level of activity, given by the baseline or *tonic* firing rate. Prior to learning, when a reward was delivered, the monkeys' DA neurons showed a sudden, short burst of activity, known as *phasic* firing (Figure 11.2a, top). After learning, the DA neurons' firing rate no longer deviated from the baseline when receiving the reward (Figure 11.2a, middle). However, phasic activity was now observed following the appearance of the cue (CS, for conditional stimulus).

One interpretation for these learning-dependent increases in firing rate is that they encode a positive prediction error. The increase in firing rate at the appearance of the cue, in particular, gives evidence that the cue itself eventually induces a reward-based prediction error (RPE). Even more suggestive of an error-driven learning process, omitting the juice reward following the cue resulted in a *decrease* in firing rate (a negative prediction error) at the time at which a reward was previously received; simultaneously, the cue still resulted in an increased firing rate (Figure 11.2a, bottom).

The RPE interpretation was further extended when Montague et al. (1996) showed that temporal-difference learning predicts the occurrence of a particularly interesting phenomenon found in an early experiment by Schultz et al. (1993). In this experiment, macaque monkeys learned that juice could be obtained by pressing one of two levers in response to a sequence of colored lights. One of two lights (green, the "instruction") first indicated which lever to press. Then, a second light (yellow, the "trigger") indicated when to press the lever and thus receive an apple juice reward – effectively providing a first-order cue.

Figure 11.2b shows recordings from DA neurons after conditioning. When the instruction light was provided at the same time as the trigger light, the DA neurons responded as before: positively in response to the cue. When the instruction occurred consistently one second before the trigger, the DA neurons showed an increase in firing only in response to the earlier of the two cues. However, when the instruction was provided at a random time prior to the trigger, the DA neurons now increased their firing rate in response to both events – encoding a positive error from receiving the unexpected instruction and the necessary error from the unpredictable trigger. In conclusion, varying the time interval between these two lights produced results that could not be

82. For a more complete review of reinforcement learning models of dopaminergic neurons and experimental findings, see Schultz (2002), Glimcher (2011), and Daw and Tobler (2014).

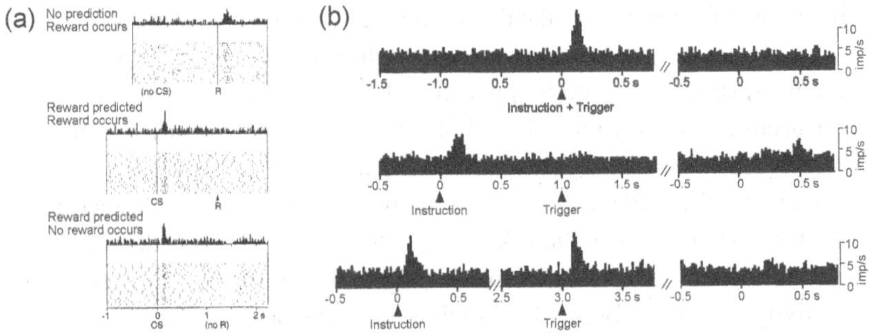

Figure 11.2
(a) DA activity when an unpredicted reward occurs, when a cue predicts a reward and it occurs, and when a cue predicts a reward but it is omitted. The data are presented both in raster plots showing firing of a single dopaminergic neuron and as peri-stimulus time histograms (PSTHs) – histograms capturing neuron firing rate over time. Conditioned stimulus (CS) marks the onset of the cue, with delivery or omission of reward indicated by (R) or (no R). From Schultz et al. (1997). Reprinted with permission from AAAS. **(b)** PSTHs averaged over a population of dopamine neurons for three conditions examining temporal credit assignment. From Schultz et al. (1993), copyright 1993 Society for Neuroscience.

completely explained by the Rescorla–Wagner model but were consistent with TD learning.

The temporal-difference learning model of dopaminergic neurons suggests that, in aggregate, these neurons modulate their firing rate in response to unexpected rewards or in response to an anticipated reward failing to appear. In particular, the model makes two predictions: first, that deviations from the tonic firing rate should be proportional to the magnitude of the prediction error (because the TD error in Equation 11.2 is linear in r), and second, that the tonic firing rate in a trained animal should correspond to the situation in which the received reward matches the expected value (that is, $r + \gamma V' = V$, in which case there is no prediction error).

For a given DA neuron, let us call *reversal point* the amount of reward r_0 for which, if a reward $r < r_0$ is received, the neuron expresses a negative error, and if a reward $r > r_0$ is received, it expresses a positive error.[83] Under the TD learning model, individual neurons should show approximately identical reversal points (up to an estimation error) and should weigh positive and negative errors equally (Figure 11.3a). However, experimental evidence suggests

83. Assuming that the return is r (i.e., there is no future value V'). We can more generally define the reversal point with respect to an observed return, but this distinction is not needed here.

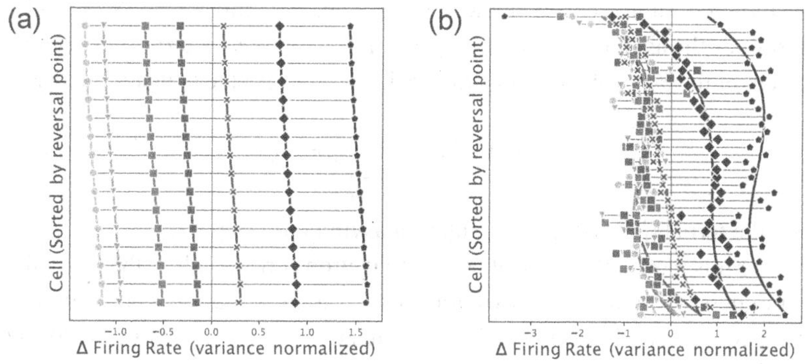

Figure 11.3
(a) The temporal-difference learning model of DA neurons predicts that, while individual neurons may show small variations in reversal point (e.g., due to estimation error), their response should be linear in the TD error and weight positive and negative errors equally. Neurons are sorted from top to bottom in decreasing order of reversal point. **(b)** Measurements of the change in firing rate in response to each of the seven possible reward magnitudes (indicated by marker and shading) for individual dopaminergic neurons in mice (Eshel et al. 2015), sorted in decreasing order of imputed reversal point. These measurements exhibit marked deviation from the linear error-response predicted by the TD learning model.

otherwise – that individual neurons instead respond to the same cue in a manner specific to each neuron and asymmetrically depending on the reward's magnitude (Figure 11.3b).

Eshel et al. (2015) measured the firing rate of individual DA neurons of mice in response to a random reward $0.1, 0.3, 1.2, 2.5, 5, 10$, or $20\ \mu$L of juice, chosen uniformly at random for each trial. Figure 11.4a shows the change in firing rate in response to each reward, after conditioning, as a function of each neuron's imputed reversal point (see Dabney et al. 2018 for details). The analysis illustrates a marked asymmetry in the response of individual neurons to reward; the neurons with the lowest reversal points, in particular, increase their firing rate for almost all rewards.

We may explain this phenomenon by considering a per-neuron update rule that incorporates an asymmetric step size, known as the *distributional TD model*. Because the neurons' change in firing rate does in general vary monotonically with the magnitude of the reward, it is natural to consider an incremental algorithm derived from expectile dynamic programming (Section 8.6). As before, let $(\tau_i)_{1=1}^m$ be values in the interval $(0, 1)$, and $(\theta_i)_{i=1}^m$ a set of adjustable locations. Here, i corresponds to an individual DA neuron, such that θ_i denotes the predicted future reward for which this neuron computes an error, and τ_i determines the asymmetry in its step size. For a sample reward r, the negative gradient of the expectile loss (Equation 8.13) with respect to θ_i yields the update rule

$$\theta_i \leftarrow \theta_i + \alpha \underbrace{|\mathbb{1}_{\{r<\theta_i\}} - \tau_i|(r - \theta_i)}_{\text{expectile error}}, \tag{11.3}$$

Here, the term $|\mathbb{1}_{\{r<\theta_i\}} - \tau_i|$ constitutes an asymmetric step size.

Under this model, the reversal point of a neuron corresponds to the prediction θ_i, and therefore a neuron's deviation from its tonic firing rate corresponds to the expectile error. In turn, the *slope* or rate at which the firing rate is reduced or increased as a function of the error reflects in some sense the neuron's "step size" $\alpha|\mathbb{1}_{\{g<\theta_i\}} - \tau_i|$. By measuring the slope of a neuron's change in firing rate for rewards smaller and larger than the imputed reversal point, one finds that different neurons indeed exhibit asymmetric slopes around their reversal point (Figure 11.4a).

Given the slopes α^+ and α^- above and below the reversal point, respectively, for an individual neuron, we can recover an estimate of the asymmetry parameter τ_i according to

$$\tau_i = \frac{\alpha^+}{\alpha^+ + \alpha^-} .$$

With this change of variables, one finds a strong correlation between individual neurons' reversal points (θ_i) and their inferred asymmetries (τ_i); see Figure 11.4b. This gives evidence that the diversity in responses to rewards of different magnitudes is structured consistent with an expectile representation of the distribution learned through asymmetric scaling of prediction errors, that is, evidence supporting the distributional TD model of dopamine.

As a whole, these results suggest that the behavior of dopaminergic neurons is best modeled not with a single global update rule, such as in TD learning, but rather a collection of update rules that together describe a richer prediction about future rewards – a distributional prediction. While the downstream uses of such a prediction remain to be identified, one can naturally imagine that there should be behavioral correlates involving risk and uncertainty. Other open questions around distributional RL in the brain include: What are the biological mechanisms that give rise to the diverse asymmetric responses in DA neurons? How, and to what degree, are DA neurons and those that encode reversal points coupled, as required by the distributional TD model? Does distributional RL confer representation learning benefits in biological agents as it does in artificial agents?

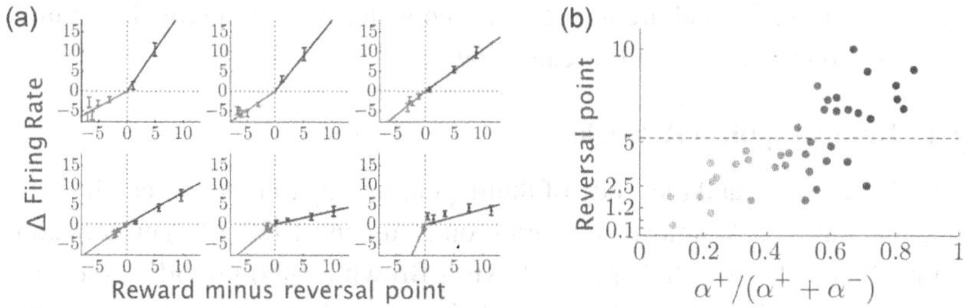

Figure 11.4
(a) Examples of the change in firing rate in response to various reward magnitudes for individual dopaminergic (DA) neurons showing asymmetry about the reversal point. Each plot corresponds to an individual DA neuron, and each point within, with error bars showing standard deviation over trials, shows that neuron's change in firing rate upon receiving one of the seven reward magnitudes. Solid lines correspond to the piecewise linear best fit around the reversal point. **(b)** Estimated asymmetries strongly correlate with reversal points as predicted by distributional TD. For all measured DA neurons ($n = 40$), we show the estimated reversal point versus the cell's asymmetry. We observe a strong positive correlation between the two, as predicted by the distributional TD model.

11.3 Conclusion

The act of learning is fundamentally an anticipatory activity. It allows us to deduce that eating certain kinds of foods might be hazardous to our health and consequently avoid them. It helps the footballer decide how to kick the ball into the opposite team's net and the goalkeeper to prepare for the save before the kick is even made. It informs us that studying leads to better grades; experience teaches us to avoid the highway at rush hour. In a rich, complex world, many phenomena carry a part of unpredictability, which in reinforcement learning we model as randomness. In that respect, learning to predict the full range of possible outcomes – the return distribution – is only natural: it improves our understanding of the environment "for free," in the sense that it can be done in parallel with the usual learning of expected returns.

For the authors of this book, the roots of the distributional perspective lie in deep reinforcement learning, as a technique for obtaining more accurate representations of the world. By now, it is clear that this is but one potential application. Distributional reinforcement learning has proven useful in settings far beyond what was expected, including to model the behaviors of coevolving agents and the dynamics of dopaminergic neurons. We expect this trend to continue and look forward to seeing its greater application in mathematical finance, engineering, and life sciences. We hope this book will provide a sturdy foundation on which these ideas can be built.

11.4 Bibliographical Remarks

11.1. Game theory and the study of multiagent interactions is a research discipline that dates back almost a century (von Neumann 1928; Morgenstern and von Neumann 1944). Shoham and Leyton-Brown (2009) provide a modern summary of a wide range of topics relating to multiagent interactions, and Oliehoek and Amato (2016) provide a recent overview from a reinforcement learning perspective. The MMDP model described here was introduced by Boutilier (1996), and forms a special case of the general class of Markov games (Shapley 1953; van der Wal 1981; Littman 1994). A commonly encountered generalization of the MMDP is the Dec-POMDP (Bernstein et al. 2002), which also allows for partial observations of the state. Lauer and Riedmiller (2000) propose an optimistic algorithm with convergence guarantees in deterministic MMDPs, and many other (nondistributional) approaches to decentralized control in MMDPs have since been considered in the literature (see, e.g., Bowling and Veloso 2002; Panait et al. 2003; Panait et al. 2006; Matignon et al. 2007, 2012; Wei and Luke 2016), including in combination with deep reinforcement learning (Tampuu et al. 2017; Omidshafiei et al. 2017; Palmer et al. 2018; Palmer et al. 2019). There is some overlap between certain classes of these techniques and distribution reinforcement learning in stateless environments, as noted by Rowland et al. (2021), on which the distributional example in this section is based.

11.2. A thorough review of the research surrounding computational models of DA neurons is beyond the scope of this book. For the machine learning researcher, Niv (2009) and Sutton and Barto (2018) provide a broad discussion and historical account of the connections between neuroscience and reinforcement learning; see also the primer by Ludvig et al. (2011) for a concise introduction to the topic and the work by Daw (2003) for a neuroscientific perspective on computational models. Other recent, neuroscience-focused overviews are provided by Shah (2012), Daw and Tobler (2014), and Lowet et al. (2020). Here we highlight a few key works due to their historical relevance, as well as those that provide context into both compatible and competing hypotheses surrounding dopamine-based learning in the brain.

As discussed in Section 11.2, Montague et al. (1996) and Schultz et al. (1997) provided the early experimental findings that led to the formulation of the temporal-difference model of dopamine. These results followed mounting evidence of limitations in the Rescorla–Wagner model (Schultz 1986; Schultz and Romo 1990; Ljungberg et al. 1992; Miller et al. 1995).

Dopamine's role in learning (White and Viaud 1991), motivation (Mogenson et al. 1980; Cagniard et al. 2006), motor control (Barbeau 1974), and attention (Nieoullon 2002) has been extensively studied and we recommend the interested reader consult Wise (2004) for a thorough review.

We arrived at our claim of less than 0.001 percent of the brain's neurons being dopaminergic based upon the following two results. First, there are approximately 86 ± 8 billion neurons in the adult human brain (Azevedo et al. 2009), with only 400,000 to 600,000 dopaminergic neurons in the midbrain, which itself contains approximately 75 percent of all DA neurons in the human brain (Hegarty et al. 2013).

Much of the work untangling the role of DA in the brain was borne out of studying associated neurological disorders. The loss of midbrain DA neurons is seen as the neurological hallmark of Parkinson's disease (Hornykiewicz 1966; German et al. 1989), while ADHD is associated with reduced DA activity (Olsen et al. 2021), and the connections between dysregulation of the dopamine system and schizophrenia have continued to be studied and refined for many years (Braver et al. 1999; Howes and Kapur 2009).

Recently, Muller et al. (2021) used distributional RL to model reward-related responses in the prefrontal cortex (PFC). This may suggest a more ubiquitous role for distributional RL in the brain.

While the distributional TD model posits that DA neurons differ in their sensitivity to positive versus negative prediction errors, several alternative models have been proposed to explain the observed diversity in dopaminergic response. Kurth-Nelson and Redish (2009) propose that the brain encodes value with a distributed representation over temporal discounts, with a multitude of value prediction channels differing in their discount factor. Such a model can readily explain observations of purported *hyperbolic discounting* in humans and animals. We also note that these neuroscientific models have themselves inspired recent work in deep RL that combines multiple discount factors and distributional predictions (Fedus et al. 2019).

Another line of research proposes to generalize temporal-difference learning to prediction errors over reward-predictive features (Schultz 2016; Gardner et al. 2018). These are motivated by findings in neuroscience, which have shown that DA neurons may increase their firing in response to unexpected changes in sensory features, independent of anticipated reward (Takahashi et al. 2017; Stalnaker et al. 2019). This generalization of the TD model is grounded in the concept of successor representations (Dayan 1993), but is perhaps more precisely characterized as *successor features* (Barreto et al. 2017), where the features are themselves predictive of reward.

Tano et al. (2020) propose a temporal-difference learning algorithm for distributional reinforcement learning which uses a variety of discount factors, reward sensitivities, and multistep updates, allowing the population to make distributional predictions with a linear operator. The advantage of such a model is that it is local, in the sense that there need not be any communication between the various value prediction channels, whereas distributional TD assumes significant communication among the DA neurons. Relatedly, Chapman and Kaelbling (1991) consider estimating the value function by decomposing it into the total discounted probability of individual reward outcomes.

Notation

Notation	Description
$\mathbb{R}$	The set of real numbers
$\mathbb{N}, \mathbb{N}^+$	The set of natural numbers including (excluding) zero: $\{0, 1, 2, \dots\}$
$\mathbb{P}$	The probability of one or many random variables producing the given outcomes
$\mathscr{P}(Y)$	The space of probability distributions over a set Y
F_ν, F_ν^{-1}	The cumulative distribution function (CDF) and inverse CDF, respectively, for distribution ν
δ_θ	Dirac delta distribution at $\theta \in \mathbb{R}$, a probability distribution that assigns probability 1 to outcome θ
$\mathcal{N}(\mu, \sigma^2)$	Normal distribution with mean μ and variance σ^2
$\mathcal{U}([a, b])$	Uniform distribution over $[a, b]$, with $a, b \in \mathbb{R}$
$\mathcal{U}(\{a, b, \dots\})$	Uniform distribution over the set $\{a, b, \dots\}$
$Z \sim \nu$	The random variable Z, with probability distribution ν
z, Z	Capital letters generally denote random variables and lowercase letters their realizations or expectations. Notable exceptions are V, Q, and P
$x \in \mathcal{X}$	A state x in the state space $\mathcal{X}$
$a \in \mathcal{A}$	An action a in the action space $\mathcal{A}$
$r \in \mathcal{R}$	A reward r from the set $\mathcal{R}$
γ	Discount factor
$R_t, X_{t+1} \sim P(\cdot, \cdot \mid X_t, A_t)$	Joint probability distribution of reward and next states in terms of the current state and action
$P_{\mathcal{X}}$	Transition kernel

Notation	Description
$P_{\mathcal{R}}$	Reward distribution function
ξ_0	Initial state distribution
$x_{\varnothing}$	A terminal state
$N_{\mathcal{X}}, N_{\mathcal{A}}, N_{\mathcal{R}}$	Size of the state, action, and reward spaces (when finite)
π	A policy; usually stationary and Markov, mapping states to distributions over actions
π^*	An optimal policy
k	Iteration number or index of a sample trajectory
t	Time step or time index
$(X_t, A_t, R_t)_{t\geq 0}$	A trajectory of random variables for state, action, and reward produced through interaction with a Markov decision process
$X_{0:t-1}$	A sequence of random variables
T	Length of an episode
$\mathbb{P}_\pi$	The distribution over trajectories induced by a Markov decision process and a policy π
$\mathbb{E}_\pi$	The expectation operator for the distribution over trajectories induced by $\mathbb{P}_\pi$
G	A random-variable function or random return
Var, Var_π	Variance of a distribution generally and variance under the distribution $\mathbb{P}_\pi$
$V^\pi(x)$	The value function for policy π at state $x \in \mathcal{X}$
$Q^\pi(x, a)$	The state-action value function for policy π at state $x \in \mathcal{X}$ and taking action $a \in \mathcal{A}$
$Z \overset{\mathcal{D}}{=} Z'$	Equality in distribution of two random variables Z, Z'
$\mathcal{D}(Z \mid Y)$	The conditional probability distribution of a random variable Z given Y
G^π	The random-variable function for policy π
η	A return-distribution function
$\eta^\pi(x)$	The return-distribution function for policy π at state $x \in \mathcal{X}$
$f_\# \nu$	Pushforward distribution passing distribution ν through the function f
$\mathrm{b}_{r,\gamma}$	Bootstrap function with reward r and discount γ
$R_{\text{MIN}}, R_{\text{MAX}}, V_{\text{MIN}}, V_{\text{MAX}}$	Minimum and maximum possible reward and return within an MDP

Notation	Description
$N_k(x)$	Number of visits to state $x \in \mathcal{X}$ up to but excluding iteration k
m	Number of particles or parameters of the distribution representation
$\{\theta_1, \ldots, \theta_m\}$	Support of a categorical distribution representation, with $\theta_i < \theta_j$ for $i < j$
ς_m	The gap between consecutive locations for the support of a categorical representation with m locations
$\hat{V}^\pi(x), \hat{\eta}^\pi(x)$	An estimate of the value function or return distribution function at state x under policy π
α, α_k	The step size in an update expression and the step size used for iteration k
$A \leftarrow B$	Denotes updating the variable A with the contents of variable B
Π_c	The categorical projection (Sections 3.5 and 5.6)
Π_Q	The quantile projection (Section 5.6)
T^π, T	The policy-evaluation Bellman operator and Bellman optimality operator, respectively
$\mathcal{T}^\pi, \mathcal{T}$	The policy-evaluation distributional Bellman operator and distributional optimality operator, respectively
OU	An operator O applied to a point $U \in M$, where (M, d) is a metric space.
$\|\cdot\|_\infty$	Supremum norm on a vector space
w_p	p-Wasserstein distance
ℓ_p	ℓ_p distance between probability distributions
ℓ_2	Cramér distance
$\bar{d}$	The supremum extension of a probability metric d to return-distribution functions, where the supremum is taken over states
$\Gamma(\nu, \nu')$	The set of couplings, joint probability distributions, of $\nu, \nu' \in \mathscr{P}(\mathbb{R})$
$\mathscr{P}_p(\mathbb{R})$	The set of distributions with finite pth moments
$\mathscr{P}_d(\mathbb{R})$	The set of distributions with finite d-distance to the distribution δ_0 and finite first moment. Also referred to as the finite domain of d
CVaR	Conditional value at risk

Notation	Description
$\mathscr{F}$	A probability distribution representation
$\mathscr{F}_{\mathrm{E}}$	Empirical probability distribution representation
$\mathscr{F}_{\mathrm{N}}$	Normal probability distribution representation, parameterized by mean and variance
$\mathscr{F}_{\mathrm{C},m}$	m-categorical probability distribution representation
$\mathscr{F}_{\mathrm{Q},m}$	m-quantile probability distribution representation
$\Pi_{\mathscr{F}}$	A projection onto the probability distribution representation $\mathscr{F}$
$\lfloor z \rfloor, \lceil z \rceil$	Floor and ceiling operations, mapping $z \in \mathbb{R}$ to the nearest integer that is less or equal (floor), or greater or equal (ceiling), than z
d	A probability metric, typically used for the purposes of contraction analysis
$\mathcal{L}_{\tau}(\theta)$	Quantile regression loss function for target threshold $\tau \in (0, 1)$ and location estimate $\theta \in \mathbb{R}$
$\mathbb{1}_{\{u\}}$	An indicator function that takes the value 1 when u is true and 0 otherwise; also $\mathbb{1}\{\cdot\}$
$J(\pi)$	Objective function for a control problem
$\mathcal{G}$	Greedy policy operator, produces a policy that is greedy with respect to a given action-value function
$T^{\mathcal{G}}, \mathcal{T}^{\mathcal{G}}$	The Bellman and distributional Bellman optimality operators derived from greedy selection rule $\mathcal{G}$
$J_{\rho}(\pi)$	A risk-sensitive control objective function, with risk measure ρ
ψ	A statistical functional or sketch
ξ_{π}	Steady-state distribution under policy π
$\phi(x)$	State representation for state x, a mapping $\phi : \mathcal{X} \to \mathbb{R}^n$
$\Pi_{\phi,\xi}$	Projection onto the linear subspace generated by ϕ with state weighting ξ
$\mathscr{M}(\mathbb{R})$	Space of signed probability measures over the reals
$\ell_{\xi,2}$	Weighted Cramér distance over return-distribution functions, with state weighting given by ξ
Π_{ϕ,ξ,ℓ_2}	Projection onto the linear subspace generated by ϕ, minimizing the $\ell_{\xi,2}$ distance
$\mathcal{L}$	Loss function
$\mathcal{H}_{\kappa}$	The Huber loss with threshold $\kappa \geq 0$

References

Achab, Mastane. 2020. Ranking and risk-aware reinforcement learning. PhD diss., Institut Polytechnique de Paris.

Agarwal, Rishabh, Dale Schuurmans, and Mohammad Norouzi. 2020. An optimistic perspective on offline reinforcement learning. In *Proceedings of the International Conference on Machine Learning.*

Agarwal, Rishabh, Max Schwarzer, Pablo Samuel Castro, Aaron Courville, and Marc G. Bellemare. 2021. Deep reinforcement learning at the edge of the statistical precipice. In *Advances in Neural Information Processing Systems.*

Aigner, D. J., Takeshi Amemiya, and Dale J. Poirier. 1976. On the estimation of production frontiers: Maximum likelihood estimation of the parameters of a discontinuous density function. *International Economic Review* 17 (2): 377–396.

Aldous, David J., and Antar Bandyopadhyay. 2005. A survey of max-type recursive distributional equations. *The Annals of Applied Probability* 15 (2): 1047–1110.

Alsmeyer, Gerold. 2012. Random recursive equations and their distributional fixed points. *Unpublished manuscript.*

Altman, Eitan. 1999. *Constrained Markov decision processes.* Vol. 7. CRC Press.

Ambrosio, Luigi, Nicola Gigli, and Giuseppe Savaré. 2005. *Gradient flows: In metric spaces and in the space of probability measures.* Springer Science & Business Media.

Amortila, Philip, Marc G. Bellemare, Prakash Panangaden, and Doina Precup. 2019. Temporally extended metrics for Markov decision processes. In *SafeAI: AAAI Workshop on Artificial Intelligence Safety.*

Amortila, Philip, Doina Precup, Prakash Panangaden, and Marc G. Bellemare. 2020. A distributional analysis of sampling-based reinforcement learning algorithms. In *Proceedings of the International Conference on Artificial Intelligence and Statistics.*

Arjovsky, Martin, Soumith Chintala, and Léon Bottou. 2017. Wasserstein GAN. In *Proceedings of the International Conference on Machine Learning.*

Artzner, Philippe, Freddy Delbaen, Jean-Marc Eber, and David Heath. 1999. Coherent measures of risk. *Mathematical Finance* 9 (3): 203–228.

Artzner, Philippe, Freddy Delbaen, Jean-Marc Eber, David Heath, and Hyejin Ku. 2007. Coherent multiperiod risk adjusted values and Bellman's principle. *Annals of Operations Research* 152 (1): 5–22.

Arulkumaran, Kai, Marc Peter Deisenroth, Miles Brundage, and Anil Anthony Bharath. 2017. A brief survey of deep reinforcement learning. *IEEE Signal Processing Magazine, Special Issue on Deep Learning for Image Understanding.*

Auer, Peter, Mark Herbster, and Manfred K. Warmuth. 1995. Exponentially many local minima for single neurons. In *Advances in Neural Information Processing Systems.*

Azar, Mohammad Gheshlaghi, Rémi Munos, Mohammad Ghavamzadeh, and Hilbert J. Kappen. 2011. Speedy Q-learning. In *Advances in Neural Information Processing Systems.*

Azar, Mohammad Gheshlaghi, Rémi Munos, and Hilbert J. Kappen. 2012. On the sample complexity of reinforcement learning with a generative model. In *Proceedings of the International Conference on Machine Learning.*

Azar, Mohammad Gheshlaghi, Rémi Munos, and Hilbert J. Kappen. 2013. Minimax PAC bounds on the sample complexity of reinforcement learning with a generative model. *Machine Learning* 91 (3): 325–349.

Azar, Mohammad Gheshlaghi, Ian Osband, and Rémi Munos. 2017. Minimax regret bounds for reinforcement learning. In *Proceedings of the International Conference on Machine Learning.*

Azevedo, Frederico A. C., Ludmila R. B. Carvalho, Lea T. Grinberg, José Marcelo Farfel, Renata E. L. Ferretti, Renata E. P. Leite, Wilson Jacob Filho, Roberto Lent, and Suzana Herculano-Houzel. 2009. Equal numbers of neuronal and nonneuronal cells make the human brain an isometrically scaled-up primate brain. *Journal of Comparative Neurology* 513 (5): 532–541.

Bahdanau, Dzmitry, Kyunghyun Cho, and Yoshua Bengio. 2015. Neural machine translation by jointly learning to align and translate. In *Proceedings of the International Conference on Learning Representations.*

Baird, Leemon C. 1995. Residual algorithms: Reinforcement learning with function approximation. In *Proceedings of the International Conference on Machine Learning.*

Baird, Leemon C. 1999. Reinforcement learning through gradient descent. PhD diss., Carnegie Mellon University.

Banach, Stefan. 1922. Sur les opérations dans les ensembles abstraits et leur application aux équations intégrales. *Fundamenta Mathematicae* 3 (1): 133–181.

Banino, Andrea, Caswell Barry, Benigno Uria, Charles Blundell, Timothy Lillicrap, Piotr Mirowski, Alexander Pritzel, Martin J. Chadwick, Thomas Degris, Joseph Modayil, Greg Wayne, Hubert Soyer, Fabio Viola, Brian Zhang, Ross Goroshin, Neil Rabinowitz, Razvan Pascanu, Charlie Beattie, Stig Petersen, Amir Sadik, Stephen Gaffney, Helen King, Koray Kavukcuoglu, Demis Hassabis, Raia Hadsell, and Dharshan Kumaran. 2018. Vector-based navigation using grid-like representations in artificial agents. *Nature* 557 (7705): 429–433.

Barbeau, André. 1974. Drugs affecting movement disorders. *Annual Review of Pharmacology* 14 (1): 91–113.

Bard, Nolan, Jakob N. Foerster, Sarath Chandar, Neil Burch, Marc Lanctot, H. Francis Song, Emilio Parisotto, Vincent Dumoulin, Subhodeep Moitra, Edward Hughes, Iain Dunning, Shibl Mourad, Hugo Larochelle, Marc G. Bellemare, and Michael Bowling. 2020. The Hanabi challenge: A new frontier for AI research. *Artificial Intelligence* 280:103216.

Barnard, Etienne. 1993. Temporal-difference methods and Markov models. *IEEE Transactions on Systems, Man, and Cybernetics* 23 (2): 357–365.

Barreto, André, Will Dabney, Rémi Munos, Jonathan J. Hunt, Tom Schaul, Hado van Hasselt, and David Silver. 2017. Successor features for transfer in reinforcement learning. In *Advances in Neural Information Processing Systems.*

Barth-Maron, Gabriel, Matthew W. Hoffman, David Budden, Will Dabney, Dan Horgan, Dhruva TB, Alistair Muldal, Nicolas Heess, and Timothy Lillicrap. 2018. Distributed distributional deterministic policy gradients. In *Proceedings of the International Conference on Learning Representations.*

Barto, Andrew G., Steven J. Bradtke, and Satinder P. Singh. 1995. Learning to act using real-time dynamic programming. *Artificial Intelligence* 72 (1): 81–138.

Barto, Andrew G., Richard S. Sutton, and Charles W. Anderson. 1983. Neuronlike adaptive elements that can solve difficult learning control problems. *IEEE Transactions on Systems, Man, and Cybernetics* 13 (5): 834–846.

Bäuerle, Nicole, and Jonathan Ott. 2011. Markov decision processes with average-value-at-risk criteria. *Mathematical Methods of Operations Research* 74 (3): 361–379.

Beattie, Charles, Joel Z. Leibo, Denis Teplyashin, Tom Ward, Marcus Wainwright, Heinrich Küttler, Andrew Lefrancq, Simon Green, Víctor Valdés, Amir Sadik, Julian Schrittwieser, Keith Anderson, Sarah York, Max Cant, Adam Cain, Adrian Bolton, Stephen Gaffney, Helen King, Demis Hassabis, Shane Legg, and Stig Petersen. 2016. DeepMind Lab. *arXiv preprint arXiv:1612.03801.*

Bellemare, Marc G., Salvatore Candido, Pablo Samuel Castro, Jun Gong, Marlos C. Machado, Subhodeep Moitra, Sameera S. Ponda, and Ziyu Wang. 2020. Autonomous navigation of stratospheric balloons using reinforcement learning. *Nature* 588 (7836): 77–82.

Bellemare, Marc G., Will Dabney, Robert Dadashi, Adrien Ali Taiga, Pablo Samuel Castro, Nicolas Le Roux, Dale Schuurmans, Tor Lattimore, and Clare Lyle. 2019a. A geometric perspective on optimal representations for reinforcement learning. In *Advances in Neural Information Processing Systems.*

Bellemare, Marc G., Will Dabney, and Rémi Munos. 2017a. A distributional perspective on reinforcement learning. In *Proceedings of the International Conference on Machine Learning.*

Bellemare, Marc G., Ivo Danihelka, Will Dabney, Shakir Mohamed, Balaji Lakshminarayanan, Stephan Hoyer, and Rémi Munos. 2017b. The Cramer distance as a solution to biased Wasserstein gradients. *arXiv preprint arXiv:1705.10743.*

Bellemare, Marc G., Yavar Naddaf, Joel Veness, and Michael Bowling. 2013a. The Arcade Learning Environment: An evaluation platform for general agents. *Journal of Artificial Intelligence Research* 47 (June): 253–279.

Bellemare, Marc G., Yavar Naddaf, Joel Veness, and Michael Bowling. 2015. The Arcade Learning Environment: An evaluation platform for general agents, extended abstract. In *European Workshop on Reinforcement Learning.*

Bellemare, Marc G., Georg Ostrovski, Arthur Guez, Philip S. Thomas, and Rémi Munos. 2016. Increasing the action gap: New operators for reinforcement learning. In *Proceedings of the AAAI Conference on Artificial Intelligence.*

Bellemare, Marc G., Nicolas Le Roux, Pablo Samuel Castro, and Subhodeep Moitra. 2019b. Distributional reinforcement learning with linear function approximation. In *Proceedings of the International Conference on Artificial Intelligence and Statistics.*

Bellemare, Marc G., Joel Veness, and Michael Bowling. 2012a. Investigating contingency awareness using Atari 2600 games. In *Proceedings of the AAAI Conference on Artificial Intelligence.*

Bellemare, Marc G., Joel Veness, and Michael Bowling. 2012b. Sketch-based linear value function approximation. In *Advances in Neural Information Processing Systems.*

Bellemare, Marc G., Joel Veness, and Michael Bowling. 2013b. Bayesian learning of recursively factored environments. In *Proceedings of the International Conference on Machine Learning.*

Bellini, Fabio, and Elena Di Bernardino. 2017. Risk management with expectiles. *The European Journal of Finance* 23 (6): 487–506.

Bellini, Fabio, Bernhard Klar, Alfred Müller, and Emanuela Rosazza Gianin. 2014. Generalized quantiles as risk measures. *Insurance: Mathematics and Economics* 54:41–48.

Bellman, Richard E. 1957a. A Markovian decision process. *Journal of Mathematics and Mechanics* 6 (5): 679–684.

Bellman, Richard E. 1957b. *Dynamic programming.* Dover Publications.

Benveniste, Albert, Michel Métivier, and Pierre Priouret. 2012. *Adaptive algorithms and stochastic approximations.* Springer Science & Business Media.

Bernstein, Daniel S., Robert Givan, Neil Immerman, and Shlomo Zilberstein. 2002. The complexity of decentralized control of Markov decision processes. *Mathematics of Operations Research* 27 (4): 819–840.

Bertsekas, Dimitri P. 1994. *Generic rank-one corrections for value iteration in Markovian decision problems.* Technical report. Massachusetts Institute of Technology.

Bertsekas, Dimitri P. 1995. A counterexample to temporal differences learning. *Neural Computation* 7 (2): 270–279.

Bertsekas, Dimitri P. 2011. Approximate policy iteration: A survey and some new methods. *Journal of Control Theory and Applications* 9 (3): 310–335.

Bertsekas, Dimitri P. 2012. *Dynamic programming and optimal control.* 4th ed. Vol. 2. Athena Scientific.

Bertsekas, Dimitri P., and Sergey Ioffe. 1996. *Temporal differences-based policy iteration and applications in neuro-dynamic programming.* Technical report. Massachusetts Institute of Technology.

Bertsekas, Dimitri P., and John N. Tsitsiklis. 1996. *Neuro-dynamic programming.* Athena Scientific.

Bhandari, Jalaj, and Daniel Russo. 2021. On the linear convergence of policy gradient methods for finite MDPs. In *Proceedings of the International Conference on Artificial Intelligence and Statistics.*

Bhonker, Nadav, Shai Rozenberg, and Itay Hubara. 2017. Playing SNES in the Retro Learning Environment. In *Proceedings of the International Conference on Learning Representations.*

Bickel, Peter J., and David A. Freedman. 1981. Some asymptotic theory for the bootstrap. *The Annals of Statistics* 9 (6): 1196–1217.

Billingsley, Patrick. 2012. *Probability and measure.* 4th ed. John Wiley & Sons.

Bishop, Christopher M. 2006. *Pattern recognition and machine learning.* Springer.

Bobkov, Sergey, and Michel Ledoux. 2019. *One-dimensional empirical measures, order statistics, and Kantorovich transport distances.* American Mathematical Society.

Bodnar, Cristian, Adrian Li, Karol Hausman, Peter Pastor, and Mrinal Kalakrishnan. 2020. Quantile QT-OPT for risk-aware vision-based robotic grasping. In *Proceedings of Robotics: Science and Systems.*

Borkar, Vivek S. 1997. Stochastic approximation with two time scales. *Systems & Control Letters* 29 (5): 291–294.

Borkar, Vivek S. 2008. *Stochastic approximation: A dynamical systems viewpoint.* Cambridge University Press.

Borkar, Vivek S., and Sean P. Meyn. 2000. The ODE method for convergence of stochastic approximation and reinforcement learning. *SIAM Journal on Control and Optimization* 38 (2): 447–469.

Bottou, Léon. 1998. Online learning and stochastic approximations. *On-line Learning in Neural Networks* 17 (9): 142.

Boutilier, Craig. 1996. Planning, learning and coordination in multiagent decision processes. In *Proceedings of the Conference on Theoretical Aspects of Rationality and Knowledge.*

Bowling, Michael, and Manuela Veloso. 2002. Multiagent learning using a variable learning rate. *Artificial Intelligence* 136 (2): 215–250.

Boyan, Justin, and Andrew W. Moore. 1995. Generalization in reinforcement learning: Safely approximating the value function. In *Advances in Neural Information Processing Systems.*

Boyd, Stephen, and Lieven Vandenberghe. 2004. *Convex optimization.* Cambridge University Press.

Bradtke, Steven J., and Andrew G. Barto. 1996. Linear least-squares algorithms for temporal difference learning. *Machine Learning* 22 (1): 33–57.

Braver, Todd S., Deanna M. Barch, and Jonathan D. Cohen. 1999. Cognition and control in schizophrenia: A computational model of dopamine and prefrontal function. *Biological Psychiatry* 46 (3): 312–328.

Brooks, Steve, Andrew Gelman, Galin Jones, and Xiao-Li Meng. 2011. *Handbook of Markov chain Monte Carlo.* CRC Press.

Brown, Daniel, Scott Niekum, and Marek Petrik. 2020. Bayesian robust optimization for imitation learning. In *Advances in Neural Information Processing Systems.*

Browne, Cameron B., Edward Powley, Daniel Whitehouse, Simon M. Lucas, Peter I. Cowling, Philipp Rohlfshagen, Stephen Tavener, Diego Perez, Spyridon Samothrakis, and Simon Colton. 2012. A survey of Monte Carlo tree search methods. *IEEE Transactions on Computational Intelligence and AI in Games* 4 (1): 1–43.

Cabi, Serkan, Sergio Gómez Colmenarejo, Alexander Novikov, Ksenia Konyushkova, Scott Reed, Rae Jeong, Konrad Zolna, Yusuf Aytar, David Budden, Mel Vecerik, Oleg Sushkov, David Barker, Jonathan Scholz, Misha Denil, Nando de Freitas, and Ziyu Wang. 2020. Scaling data-driven robotics with reward sketching and batch reinforcement learning. In *Proceedings of Robotics: Science and Systems.*

Cagniard, Barbara, Peter D. Balsam, Daniela Brunner, and Xiaoxi Zhuang. 2006. Mice with chronically elevated dopamine exhibit enhanced motivation, but not learning, for a food reward. *Neuropsychopharmacology* 31 (7): 1362–1370.

Carpin, Stefano, Yinlam Chow, and Marco Pavone. 2016. Risk aversion in finite Markov decision processes using total cost criteria and average value at risk. In *Proceedings of the IEEE International Conference on Robotics and Automation.*

Castro, Pablo S., Subhodeep Moitra, Carles Gelada, Saurabh Kumar, and Marc G. Bellemare. 2018. Dopamine: A research framework for deep reinforcement learning. *arXiv preprint arXiv:1812.06110.*

Ceron, Johan Samir Obando, and Pablo Samuel Castro. 2021. Revisiting Rainbow: Promoting more insightful and inclusive deep reinforcement learning research. In *Proceedings of the International Conference on Machine Learning.*

Chandak, Yash, Scott Niekum, Bruno Castro da Silva, Erik Learned-Miller, Emma Brunskill, and Philip S. Thomas. 2021. Universal off-policy evaluation. In *Advances in Neural Information Processing Systems.*

Chapman, David, and Leslie Pack Kaelbling. 1991. Input generalization in delayed reinforcement learning: An algorithm and performance comparisons. In *Proceedings of the International Joint Conference on Artificial Intelligence.*

Chen, Jinglin, and Nan Jiang. 2019. Information-theoretic considerations in batch reinforcement learning. In *Proceedings of the International Conference on Machine Learning.*

Chopin, Nicolas, and Omiros Papaspiliopoulos. 2020. *An introduction to sequential Monte Carlo.* Springer.

Chow, Yinlam. 2017. Risk-sensitive and data-driven sequential decision making. PhD diss., Stanford University.

Chow, Yinlam, and Mohammad Ghavamzadeh. 2014. Algorithms for CVaR optimization in MDPs. In *Advances in Neural Information Processing Systems.*

Chow, Yinlam, Mohammad Ghavamzadeh, Lucas Janson, and Marco Pavone. 2018. Risk-constrained reinforcement learning with percentile risk criteria. *Journal of Machine Learning Research* 18 (1): 6070–6120.

Chow, Yinlam, Aviv Tamar, Shie Mannor, and Marco Pavone. 2015. Risk-sensitive and robust decision-making: A CVaR optimization approach. In *Advances in Neural Information Processing Systems.*

Chung, Kun-Jen, and Matthew J. Sobel. 1987. Discounted MDPs: Distribution functions and exponential utility maximization. *SIAM Journal on Control and Optimization* 25 (1): 49–62.

Chung, Wesley, Somjit Nath, Ajin Joseph, and Martha White. 2018. Two-timescale networks for nonlinear value function approximation. In *Proceedings of the International Conference on Learning Representations.*

Claus, Caroline, and Craig Boutilier. 1998. The dynamics of reinforcement learning in cooperative multiagent systems. In *Proceedings of the AAAI Conference on Artificial Intelligence.*

Clements, William R., Benoit-Marie Robaglia, Bastien Van Delft, Reda Bahi Slaoui, and Sebastien Toth. 2020. Estimating risk and uncertainty in deep reinforcement learning. In *Workshop on Uncertainty and Robustness in Deep Learning at the International Conference on Machine Learning.*

Cobbe, Karl, Chris Hesse, Jacob Hilton, and John Schulman. 2020. Leveraging procedural generation to benchmark reinforcement learning. In *Proceedings of the International Conference on Machine Learning.*

Cormen, Thomas H., Charles E. Leiserson, Ronald L. Rivest, and Clifford Stein. 2001. *Introduction to algorithms.* MIT Press.

Cormode, G., and S. Muthukrishnan. 2005. An improved data stream summary: The count-min sketch and its applications. *Journal of Algorithms* 55 (1): 58–75.

Cuturi, Marco. 2013. Sinkhorn distances: Lightspeed computation of optimal transport. In *Advances in Neural Information Processing Systems.*

Da Silva, Felipe Leno, Anna Helena Reali Costa, and Peter Stone. 2019. Distributional reinforcement learning applied to robot soccer simulation. In *Adaptive and Learning Agents Workshop at the International Conference on Autonomous Agents and Multiagent Systems.*

Dabney, Will, André Barreto, Mark Rowland, Robert Dadashi, John Quan, Marc G. Bellemare, and David Silver. 2020a. The value-improvement path: Towards better representations for reinforcement learning. In *Proceedings of the AAAI Conference on Artificial Intelligence.*

Dabney, Will, Zeb Kurth-Nelson, Naoshige Uchida, Clara Kwon Starkweather, Demis Hassabis, Rémi Munos, and Matthew Botvinick. 2020b. A distributional code for value in dopamine-based reinforcement learning. *Nature* 577 (7792): 671–675.

Dabney, Will, Georg Ostrovski, David Silver, and Rémi Munos. 2018a. Implicit quantile networks for distributional reinforcement learning. In *Proceedings of the International Conference on Machine Learning.*

Dabney, Will, Mark Rowland, Marc G. Bellemare, and Rémi Munos. 2018b. Distributional reinforcement learning with quantile regression. In *AAAI Conference on Artificial Intelligence.*

Dai, Bo, Albert Shaw, Lihong Li, Lin Xiao, Niao He, Zhen Liu, Jianshu Chen, and Le Song. 2018. SBEED: Convergent reinforcement learning with nonlinear function approximation. In *Proceedings of the International Conference on Machine Learning.*

Daw, Nathaniel D. 2003. *Reinforcement learning models of the dopamine system and their behavioral implications.* Carnegie Mellon University.

Daw, Nathaniel D., and Philippe N. Tobler. 2014. Value learning through reinforcement: The basics of dopamine and reinforcement learning. In *Neuroeconomics,* edited by Paul W. Glimcher and Ernst Fehr, 283–298. Academic Press.

Dayan, Peter. 1992. The convergence of TD(λ) for general λ. *Machine Learning* 8 (3–4): 341–362.

Dayan, Peter. 1993. Improving generalization for temporal difference learning: The successor representation. *Neural Computation* 5 (4): 613–624.

Dayan, Peter, and Terrence J. Sejnowski. 1994. TD(λ) converges with probability 1. *Machine Learning* 14 (3): 295–301.

Dearden, Richard, Nir Friedman, and Stuart Russell. 1998. Bayesian Q-learning. In *Proceedings of the AAAI Conference on Artificial Intelligence.*

Degrave, Jonas, Federico Felici, Jonas Buchli, Michael Neunert, Brendan Tracey, Francesco Carpanese, Timo Ewalds, Roland Hafner, Abbas Abdolmaleki, Diego de las Casas, Craig Donner, Leslie Fritz, Cristian Galperti, Andrea Huber, James Keeling, Maria Tsimpoukelli, Jackie Kay, Antoine Merle, Jean-Marc Moret, Seb Noury, Federico Pesamosca, David Pfau, Olivier Sauter, Cristian Sommariva, Stefano Coda, Basil Duval, Ambrogio Fasoli, Pushmeet Kohli, Koray Kavukcuoglu, Demis Hassabis, and Martin Riedmiller. 2022. Magnetic control of tokamak plasmas through deep reinforcement learning. *Nature* 602:414–419.

Delage, Erick, and Shie Mannor. 2010. Percentile optimization for Markov decision processes with parameter uncertainty. *Operations Research* 58 (1): 203–213.

Denardo, Eric V., and Uriel G. Rothblum. 1979. Optimal stopping, exponential utility, and linear programming. *Mathematical Programming* 16 (1): 228–244.

Derman, Cyrus. 1970. *Finite state Markovian decision processes.* Academic Press.

Diaconis, Persi, and David Freedman. 1999. Iterated random functions. *SIAM Review* 41 (1): 45–76.

Doan, Thang, Bogdan Mazoure, and Clare Lyle. 2018. GAN Q-learning. *arXiv preprint arXiv:1805.04874.*

Doob, J. L. 1994. *Measure theory.* Springer.

Doucet, Arnaud, Nando De Freitas, and Neil Gordon. 2001. *Sequential Monte Carlo methods in practice.* Springer.

Doucet, Arnaud, and Adam M. Johansen. 2011. A tutorial on particle filtering and smoothing: Fifteen years later. In *The Oxford handbook of nonlinear filtering,* edited by Dan Crisan and Boris Rozovskii. Oxford University Press.

Duan, Jingliang, Yang Guan, Shengbo Eben Li, Yangang Ren, Qi Sun, and Bo Cheng. 2021. Distributional soft actor-critic: Off-policy reinforcement learning for addressing value estimation errors. *IEEE Transactions on Neural Networks and Learning Systems.*

Dvoretzky, Aryeh. 1956. On stochastic approximation. In *Proceedings of the Berkeley Symposium on Mathematical Statistics and Probability,* 39–55.

Dvoretzky, Aryeh, Jack Kiefer, and Jacob Wolfowitz. 1956. Asymptotic minimax character of the sample distribution function and of the classical multinomial estimator. *The Annals of Mathematical Statistics* 27 (3): 642–669.

Engel, Yaakov, Shie Mannor, and Ron Meir. 2003. Bayes meets Bellman: The Gaussian process approach to temporal difference learning. In *Proceedings of the International Conference on Machine Learning.*

Engel, Yaakov, Shie Mannor, and Ron Meir. 2007. Bayesian reinforcement learning with Gaussian process temporal difference methods. *Unpublished manuscript.*

Engert, Martin. 1970. Finite dimensional translation invariant subspaces. *Pacific Journal of Mathematics* 32 (2): 333–343.

Ernst, Damien, Pierre Geurts, and Louis Wehenkel. 2005. Tree-based batch mode reinforcement learning. *Journal of Machine Learning Research* 6:503–556.

Eshel, Neir, Michael Bukwich, Vinod Rao, Vivian Hemmelder, Ju Tian, and Naoshige Uchida. 2015. Arithmetic and local circuitry underlying dopamine prediction errors. *Nature* 525 (7568): 243–246.

Even-Dar, Eyal, and Yishay Mansour. 2003. Learning rates for Q-learning. *Journal of Machine Learning Research* 5 (1): 1–25.

Farahmand, Amir-massoud. 2011. Action-gap phenomenon in reinforcement learning. In *Advances in Neural Information Processing Systems.*

Farahmand, Amir-massoud. 2019. Value function in frequency domain and the characteristic value iteration algorithm. In *Advances in Neural Information Processing Systems.*

Fedus, William, Carles Gelada, Yoshua Bengio, Marc G. Bellemare, and Hugo Larochelle. 2019. Hyperbolic discounting and learning over multiple horizons. In *Multi-Disciplinary Conference on Reinforcement Learning and Decision-Making.*

Feinberg, Eugene A. 2000. Constrained discounted Markov decision processes and Hamiltonian cycles. *Mathematics of Operations Research* 25 (1): 130–140.

Ferns, Norm, Prakash Panangaden, and Doina Precup. 2004. Metrics for finite Markov decision processes. In *Proceedings of the Conference on Uncertainty in Artificial Intelligence.*

Ferns, Norman, and Doina Precup. 2014. Bisimulation metrics are optimal value functions. In *Proceedings of the Conference on Uncertainty in Artificial Intelligence.*

Filar, Jerzy A., Dmitry Krass, and Keith W. Ross. 1995. Percentile performance criteria for limiting average Markov decision processes. *IEEE Transactions on Automatic Control* 40 (1): 2–10.

Fortunato, Meire, Mohammad Gheshlaghi Azar, Bilal Piot, Jacob Menick, Ian Osband, Alex Graves, Vlad Mnih, Rémi Munos, Demis Hassabis, Olivier Pietquin, Charles Blundell, and Shane Legg. 2018. Noisy networks for exploration. In *Proceedings of the International Conference on Learning Representations.*

François-Lavet, Vincent, Peter Henderson, Riashat Islam, Marc G. Bellemare, and Joelle Pineau. 2018. An introduction to deep reinforcement learning. *Foundations and Trends® in Machine Learning* 11 (3–4): 219–354.

Freirich, Dror, Tzahi Shimkin, Ron Meir, and Aviv Tamar. 2019. Distributional multivariate policy evaluation and exploration with the Bellman GAN. In *Proceedings of the International Conference on Machine Learning.*

Gardner, Matthew P. H., Geoffrey Schoenbaum, and Samuel J. Gershman. 2018. Rethinking dopamine as generalized prediction error. *Proceedings of the Royal Society B* 285 (1891): 20181645.

German, Dwight C., Kebreten Manaye, Wade K. Smith, Donald J. Woodward, and Clifford B. Saper. 1989. Midbrain dopaminergic cell loss in Parkinson's disease: Computer visualization. *Annals of Neurology* 26 (4): 507–514.

Ghavamzadeh, Mohammad, Shie Mannor, Joelle Pineau, and Aviv Tamar. 2015. Bayesian reinforcement learning: A survey. *Foundations and Trends® in Machine Learning* 8 (5–6): 359–483.

Ghosh, Dibya, and Marc G. Bellemare. 2020. Representations for stable off-policy reinforcement learning. In *Proceedings of the International Conference on Machine Learning.*

Ghosh, Dibya, Marlos C. Machado, and Nicolas Le Roux. 2020. An operator view of policy gradient methods. In *Advances in Neural Information Processing Systems.*

Gilbert, Hugo, Paul Weng, and Yan Xu. 2017. Optimizing quantiles in preference-based Markov decision processes. In *Proceedings of the AAAI Conference on Artificial Intelligence.*

Glimcher, Paul W. 2011. Understanding dopamine and reinforcement learning: The dopamine reward prediction error hypothesis. *Proceedings of the National Academy of Sciences* 108 (Suppl. 3): 15647–15654.

Goodfellow, Ian, Aaron Courville, and Yoshua Bengio. 2016. *Deep learning.* MIT Press.

Goodfellow, Ian, Jean Pouget-Abadie, Mehdi Mirza, Bing Xu, David Warde-Farley, Sherjil Ozair, Aaron Courville, and Yoshua Bengio. 2014. Generative adversarial nets. In *Advances in Neural Information Processing Systems.*

Gordon, Geoffrey. 1995. Stable function approximation in dynamic programming. In *Proceedings of the International Conference on Machine Learning.*

Gordon, Neil J., David J. Salmond, and Adrian F. M. Smith. 1993. Novel approach to nonlinear/non-Gaussian Bayesian state estimation. *IEE Proceedings F (Radar and Signal Processing)* 140 (2): 107–113.

Graesser, Laura, and Wah Loon Keng. 2019. *Foundations of deep reinforcement learning: Theory and practice in Python.* Addison-Wesley Professional.

Gretton, Arthur, Karsten M. Borgwardt, Malte J. Rasch, Bernhard Schölkopf, and Alexander Smola. 2012. A kernel two-sample test. *Journal of Machine Learning Research* 13 (1): 723–773.

Grünewälder, Steffen, and Klaus Obermayer. 2011. The optimal unbiased value estimator and its relation to LSTD, TD and MC. *Machine Learning* 83 (3): 289–330.

Gruslys, Audrunas, Will Dabney, Mohammad Gheshlaghi Azar, Bilal Piot, Marc Bellemare, and Rémi Munos. 2018. The Reactor: A fast and sample-efficient actor-critic agent for reinforcement learning. In *Proceedings of the International Conference on Learning Representations.*

Guo, Zhaohan Daniel, Bernardo Avila Pires, Bilal Piot, Jean-Bastien Grill, Florent Altché, Rémi Munos, and Mohammad Gheshlaghi Azar. 2020. Bootstrap latent-predictive representations for multitask reinforcement learning. In *Proceedings of the International Conference on Machine Learning.*

Gurvits, Leonid, Long-Ji Lin, and Stephen José Hanson. 1994. *Incremental learning of evaluation functions for absorbing Markov chains: New methods and theorems.* Technical report. Siemens Corporate Research.

Harmon, Mance E., and Leemon C. Baird. 1996. *A response to Bertsekas' "A counterexample to temporal-differences learning".* Technical report. Wright Laboratory.

Haskell, William B., and Rahul Jain. 2015. A convex analytic approach to risk-aware Markov decision processes. *SIAM Journal on Control and Optimization* 53 (3): 1569–1598.

Hegarty, Shane V., Aideen M. Sullivan, and Gerard W. O'Keeffe. 2013. Midbrain dopaminergic neurons: A review of the molecular circuitry that regulates their development. *Developmental Biology* 379 (2): 123–138.

Heger, Matthias. 1994. Consideration of risk in reinforcement learning. In *Proceedings of the International Conference on Machine Learning.*

Henderson, Peter, Riashat Islam, Philip Bachman, Joelle Pineau, Doina Precup, and David Meger. 2018. Deep reinforcement learning that matters. In *Proceedings of the AAAI Conference on Artificial Intelligence.*

Hessel, Matteo, Joseph Modayil, Hado van Hasselt, Tom Schaul, Georg Ostrovski, Will Dabney, Dan Horgan, Bilal Piot, Mohammad Azar, and David Silver. 2018. Rainbow: Combining improvements in deep reinforcement learning. In *Proceedings of the AAAI Conference on Artificial Intelligence.*

Hochreiter, Sepp, and Jürgen Schmidhuber. 1997. Long short-term memory. *Neural Computation* 9 (8): 1735–1780.

Hornykiewicz, Oleh. 1966. Dopamine (3-hydroxytyramine) and brain function. *Pharmacological Reviews* 18 (2): 925–964.

Howard, R. 1960. *Dynamic programming and Markov processes.* MIT Press.

Howard, Ronald A., and James E. Matheson. 1972. Risk-sensitive Markov decision processes. *Management Science* 18 (7): 356–369.

Howes, Oliver D., and Shitij Kapur. 2009. The dopamine hypothesis of schizophrenia: Version III—the final common pathway. *Schizophrenia Bulletin* 35 (3): 549–562.

Hutter, Marcus. 2005. *Universal artificial intelligence: Sequential decisions based on algorithmic probability.* Springer.

Imani, Ehsan, and Martha White. 2018. Improving regression performance with distributional losses. In *Proceedings of the International Conference on Machine Learning.*

Jaakkola, Tommi, Michael I. Jordan, and Satinder P. Singh. 1994. On the convergence of stochastic iterative dynamic programming algorithms. *Neural Computation* 6 (6): 1185–1201.

Jaderberg, Max, Volodymyr Mnih, Wojciech M. Czarnecki, Tom Schaul, Joel Z. Leibo, David Silver, and Koray Kavukcuoglu. 2017. Reinforcement learning with unsupervised auxiliary tasks. In *Proceedings of the International Conference on Learning Representations.*

Janner, Michael, Igor Mordatch, and Sergey Levine. 2020. Generative temporal difference learning for infinite-horizon prediction. In *Advances in Neural Information Processing Systems.*

Jaquette, Stratton C. 1973. Markov decision processes with a new optimality criterion: Discrete time. *The Annals of Statistics* 1 (3): 496–505.

Jaquette, Stratton C. 1976. A utility criterion for Markov decision processes. *Management Science* 23 (1): 43–49.

Jessen, Børge, and Aurel Wintner. 1935. Distribution functions and the Riemann zeta function. *Transactions of the American Mathematical Society* 38 (1): 48–88.

Jiang, Daniel R., and Warren B. Powell. 2018. Risk-averse approximate dynamic programming with quantile-based risk measures. *Mathematics of Operations Research* 43 (2): 554–579.

Jordan, Richard, David Kinderlehrer, and Felix Otto. 1998. The variational formulation of the Fokker–Planck equation. *SIAM Journal on Mathematical Analysis* 29 (1): 1–17.

Kaelbling, Leslie Pack, Michael L. Littman, and Anthony R. Cassandra. 1998. Planning and acting in partially observable stochastic domains. *Artificial Intelligence* 101:99–134.

Kamin, Leon J. 1968. "Attention like" processes in classical conditioning. In *Miami Symposium on the Prediction of Behavior: Aversive Stimulation,* 9–31.

Kantorovich, Leonid V. 1942. On the translocation of masses. *Proceedings of the USSR Academy of Sciences* 37 (7–8): 227–229.

Kapetanakis, Spiros, and Daniel Kudenko. 2002. Reinforcement learning of coordination in cooperative multi-agent systems. In *Proceedings of the AAAI Conference on Artificial Intelligence.*

Kartal, Bilal, Pablo Hernandez-Leal, and Matthew E. Taylor. 2019. Terminal prediction as an auxiliary task for deep reinforcement learning. In *Proceedings of the AAAI Conference on Artificial Intelligence and Interactive Digital Entertainment.*

Kempka, Michał, Marek Wydmuch, Grzegorz Runc, Jakub Toczek, and Wojciech Jaśkowski. 2016. Vizdoom: A Doom-based AI research platform for visual reinforcement learning. In *2016 IEEE Conference on Computational Intelligence and Games,* 1–8.

Keramati, Ramtin, Christoph Dann, Alex Tamkin, and Emma Brunskill. 2020. Being optimistic to be conservative: Quickly learning a CVaR policy. In *Proceedings of the AAAI Conference on Artificial Intelligence.*

Kingma, Diederik, and Jimmy Ba. 2015. Adam: A method for stochastic optimization. In *Proceedings of the International Conference on Learning Representations.*

Koenker, Roger. 2005. *Quantile regression.* Cambridge University Press.

Koenker, Roger, and Gilbert Bassett Jr. 1978. Regression quantiles. *Econometrica* 46 (1): 33–50.

Kolter, J. Zico. 2011. The fixed points of off-policy TD. In *Advances in Neural Information Processing Systems.*

Konidaris, George D., Sarah Osentoski, and Philip S. Thomas. 2011. Value function approximation in reinforcement learning using the Fourier basis. In *Proceedings of the AAAI Conference on Artificial Intelligence.*

Kuan, Chung-Ming, Jin-Huei Yeh, and Yu-Chin Hsu. 2009. Assessing value at risk with care, the conditional autoregressive expectile models. *Journal of Econometrics* 150 (2): 261–270.

Kuhn, Harold W. 1950. A simplified two-person poker. *Contributions to the Theory of Games* 1:97–103.

Kurth-Nelson, Zeb, and A. David Redish. 2009. Temporal-difference reinforcement learning with distributed representations. *PLoS One* 4 (10): e7362.

Kusher, Harold, and Dean Clark. 1978. *Stochastic approximation methods for constrained and unconstrained systems.* Springer.

Kushner, Harold, and G. George Yin. 2003. *Stochastic approximation and recursive algorithms and applications.* Springer Science & Business Media.

Kuznetsov, Arsenii, Pavel Shvechikov, Alexander Grishin, and Dmitry Vetrov. 2020. Controlling overestimation bias with truncated mixture of continuous distributional quantile critics. In *Proceedings of the International Conference on Machine Learning.*

Lagoudakis, Michail G., and Ronald Parr. 2003. Least-squares policy iteration. *Journal of Machine Learning Research* 4:1107–1149.

Lample, Guillaume, and Devendra Singh Chaplot. 2017. Playing FPS games with deep reinforcement learning. In *Proceedings of the AAAI Conference on Artificial Intelligence.*

Laskin, Michael, Aravind Srinivas, and Pieter Abbeel. 2020. CURL: Contrastive unsupervised representations for reinforcement learning. In *Proceedings of the International Conference on Machine Learning.*

Lattimore, Tor, and Marcus Hutter. 2012. PAC bounds for discounted MDPs. In *Proceedings of the International Conference on Algorithmic Learning Theory.*

Lattimore, Tor, and Csaba Szepesvári. 2020. *Bandit algorithms.* Cambridge University Press.

Lauer, Martin, and Martin Riedmiller. 2000. An algorithm for distributed reinforcement learning in cooperative multi-agent systems. In *Proceedings of the International Conference on Machine Learning.*

Le Lan, Charline, Stephen Tu, Adam Oberman, Rishabh Agarwal, and Marc G. Bellemare. 2022. On the generalization of representations in reinforcement learning. In *Proceedings of the International Conference on Artificial Intelligence and Statistics.*

LeCun, Yann, and Yoshua Bengio. 1995. Convolutional networks for images, speech, and time series. In *The handbook of brain theory and neural networks,* edited by Michael A. Arbib. MIT Press.

Lee, Daewoo, Boris Defourny, and Warren B. Powell. 2013. Bias-corrected Q-learning to control max-operator bias in Q-learning. In *Symposium on Adaptive Dynamic Programming And Reinforcement Learning.*

Levine, Sergey. 2018. Reinforcement learning and control as probabilistic inference: Tutorial and review. *arXiv preprint arXiv:1805.00909.*

Levine, Sergey, Chelsea Finn, Trevor Darrell, and Pieter Abbeel. 2016. End-to-end training of deep visuomotor policies. *Journal of Machine Learning Research* 17 (1): 1334–1373.

Li, Xiaocheng, Huaiyang Zhong, and Margaret L. Brandeau. 2022. Quantile Markov decision processes. *Operations Research* 70 (3): 1428–1447.

Lillicrap, Timothy P., Daniel Cownden, Douglas B. Tweed, and Colin J. Akerman. 2016a. Random synaptic feedback weights support error backpropagation for deep learning. *Nature Communications* 7 (1): 1–10.

Lillicrap, Timothy P., Jonathan J. Hunt, Alexander Pritzel, Nicolas Heess, Tom Erez, Yuval Tassa, David Silver, and Daan Wierstra. 2016b. Continuous control with deep reinforcement learning. In *Proceedings of the International Conference on Learning Representations.*

Lin, Gwo Dong. 2017. Recent developments on the moment problem. *Journal of Statistical Distributions and Applications* 4 (1): 1–17.

Lin, L. J. 1992. Self-improving reactive agents based on reinforcement learning, planning and teaching. *Machine Learning* 8 (3): 293–321.

Lin, Zichuan, Li Zhao, Derek Yang, Tao Qin, Tie-Yan Liu, and Guangwen Yang. 2019. Distributional reward decomposition for reinforcement learning. In *Advances in Neural Information Processing Systems.*

Lipovetzky, Nir, Miquel Ramirez, and Hector Geffner. 2015. Classical planning with simulators: Results on the Atari video games. In *Proceedings of International Joint Conference on Artificial Intelligence.*

Littman, Michael L. 1994. Markov games as a framework for multi-agent reinforcement learning. In *Proceedings of the International Conference on Machine Learning.*

Littman, Michael L., and Csaba Szepesvári. 1996. A generalized reinforcement-learning model: Convergence and applications. In *Proceedings of the International Conference on Machine Learning.*

Liu, Jun S. 2001. *Monte Carlo strategies in scientific computing.* Springer.

Liu, Qiang, and Dilin Wang. 2016. Stein variational gradient descent: A general purpose Bayesian inference algorithm. In *Advances in Neural Information Processing Systems.*

Liu, Quansheng. 1998. Fixed points of a generalized smoothing transformation and applications to the branching random walk. *Advances in Applied Probability* 30 (1): 85–112.

Ljung, Lennart. 1977. Analysis of recursive stochastic algorithms. *IEEE Transactions on Automatic Control* 22 (4): 551–575.

Ljungberg, Tomas, Paul Apicella, and Wolfram Schultz. 1992. Responses of monkey dopamine neurons during learning of behavioral reactions. *Journal of Neurophysiology* 67 (1): 145–163.

Lowet, Adam S., Qiao Zheng, Sara Matias, Jan Drugowitsch, and Naoshige Uchida. 2020. Distributional reinforcement learning in the brain. *Trends in Neurosciences* 43 (12): 980–997.

Ludvig, Elliot A., Marc G. Bellemare, and Keir G. Pearson. 2011. A primer on reinforcement learning in the brain: Psychological, computational, and neural perspectives. In *Computational neuroscience for advancing artificial intelligence: Models, methods and applications,* edited by Eduardo Alonso and Esther Mondragón. IGI Global.

Luo, Yudong, Guiliang Liu, Haonan Duan, Oliver Schulte, and Pascal Poupart. 2021. Distributional reinforcement learning with monotonic splines. In *Proceedings of the International Conference on Learning Representations.*

Lyle, Clare, Pablo Samuel Castro, and Marc G. Bellemare. 2019. A comparative analysis of expected and distributional reinforcement learning. In *Proceedings of the AAAI Conference on Artificial Intelligence.*

Lyle, Clare, Mark Rowland, Georg Ostrovski, and Will Dabney. 2021. On the effect of auxiliary tasks on representation dynamics. In *Proceedings of the International Conference on Artificial Intelligence and Statistics.*

Lyu, Xueguang, and Christopher Amato. 2020. Likelihood quantile networks for coordinating multi-agent reinforcement learning. In *Proceedings of the International Conference on Autonomous Agents and Multiagent Systems.*

Machado, Marlos C., Marc G. Bellemare, Erik Talvitie, Joel Veness, Matthew Hausknecht, and Michael Bowling. 2018. Revisiting the Arcade Learning Environment: Evaluation protocols and open problems for general agents. *Journal of Artificial Intelligence Research* 61:523–562.

MacKay, David J. C. 2003. *Information theory, inference and learning algorithms.* Cambridge University Press.

Maddison, Chris J., Dieterich Lawson, George Tucker, Nicolas Heess, Arnaud Doucet, Andriy Mnih, and Yee Whye Teh. 2017. Particle value functions. In *Proceedings of the International Conference on Learning Representations (Workshop Track).*

Madeira Auraújo, João Guilherme, Johan Samir Obando Ceron, and Pablo Samuel Castro. 2021. Lifting the veil on hyper-parameters for value-based deep reinforcement learning. In *NeurIPS 2021 Workshop: LatinX in AI.*

Maei, Hamid Reza. 2011. Gradient temporal-difference learning algorithms. PhD diss., University of Alberta.

Mandl, Petr. 1971. On the variance in controlled Markov chains. *Kybernetika* 7 (1): 1–12.

Mannor, Shie, Duncan Simester, Peng Sun, and John N. Tsitsiklis. 2007. Bias and variance approximation in value function estimates. *Management Science* 53 (2): 308–322.

Mannor, Shie, and John Tsitsiklis. 2011. Mean-variance optimization in Markov decision processes. In *Proceedings of the International Conference on Machine Learning.*

Markowitz, Harry M. 1952. Portfolio selection. *Journal of Finance* 7:77–91.

Martin, John, Michal Lyskawinski, Xiaohu Li, and Brendan Englot. 2020. Stochastically dominant distributional reinforcement learning. In *Proceedings of the International Conference on Machine Learning.*

Massart, Pascal. 1990. The tight constant in the Dvoretzky-Kiefer-Wolfowitz inequality. *The Annals of Probability* 18 (3): 1269–1283.

Matignon, Laëtitia, Guillaume J. Laurent, and Nadine Le Fort-Piat. 2007. Hysteretic Q-learning: An algorithm for decentralized reinforcement learning in cooperative multi-agent teams. In *IEEE International Conference on Intelligent Robots and Systems.*

Matignon, Laëtitia, Guillaume J. Laurent, and Nadine Le Fort-Piat. 2012. Independent reinforcement learners in cooperative Markov games: A survey regarding coordination problems. *The Knowledge Engineering Review* 27 (1): 1–31.

Mavrin, Borislav, Hengshuai Yao, Linglong Kong, Kaiwen Wu, and Yaoliang Yu. 2019. Distributional reinforcement learning for efficient exploration. In *Proceedings of the International Conference on Machine Learning.*

McCallum, Andrew K. 1995. Reinforcement learning with selective perception and hidden state. PhD diss., University of Rochester.

Meyn, Sean. 2022. *Control systems and reinforcement learning.* Cambridge University Press.

Meyn, Sean P., and Richard L. Tweedie. 2012. *Markov chains and stochastic stability.* Cambridge University Press.

Mihatsch, Oliver, and Ralph Neuneier. 2002. Risk-sensitive reinforcement learning. *Machine Learning* 49 (2): 267–290.

Miller, Ralph R., Robert C. Barnet, and Nicholas J. Grahame. 1995. Assessment of the Rescorla-Wagner model. *Psychological Bulletin* 117 (3): 363.

Mnih, Volodymyr, Koray Kavukcuoglu, David Silver, Andrei A. Rusu, Joel Veness, Marc G. Bellemare, Alex Graves, Martin Riedmiller, Andreas K. Fidjeland, Georg Ostrovski, Stig Petersen, Charles Beattie, Amir Sadik, Ioannis Antonoglou, Helen King, Dharshan Kumaran, Daan Wierstra, Shane Legg, and Demis Hassabis. 2015. Human-level control through deep reinforcement learning. *Nature* 518 (7540): 529–533.

Mogenson, Gordon J., Douglas L. Jones, and Chi Yiu Yim. 1980. From motivation to action: Functional interface between the limbic system and the motor system. *Progress in Neurobiology* 14 (2–3): 69–97.

Monge, Gaspard. 1781. Mémoire sur la théorie des déblais et des remblais. *Histoire de l'Académie Royale des Sciences de Paris:* 666–704.

Montague, P. Read, Peter Dayan, and Terrence J. Sejnowski. 1996. A framework for mesencephalic dopamine systems based on predictive Hebbian learning. *Journal of Neuroscience* 16 (5): 1936–1947.

Montfort, Nick, and Ian Bogost. 2009. *Racing the beam: The Atari video computer system.* MIT Press.

Moore, Andrew W., and Christopher G. Atkeson. 1993. Prioritized sweeping: Reinforcement learning with less data and less time. *Machine Learning* 13 (1): 103–130.

Morgenstern, Oskar, and John von Neumann. 1944. *Theory of games and economic behavior.* Princeton University Press.

Morimura, Tetsuro, Masashi Sugiyama, Hisashi Kashima, Hirotaka Hachiya, and Toshiyuki Tanaka. 2010a. Nonparametric return distribution approximation for reinforcement learning. In *Proceedings of the International Conference on Machine Learning.*

Morimura, Tetsuro, Masashi Sugiyama, Hisashi Kashima, Hirotaka Hachiya, and Toshiyuki Tanaka. 2010b. Parametric return density estimation for reinforcement learning. In *Proceedings of the Conference on Uncertainty in Artificial Intelligence.*

Morton, Thomas E. 1971. On the asymptotic convergence rate of cost differences for Markovian decision processes. *Operations Research* 19 (1): 244–248.

Mott, Bradford W., Stephen Anthony, and the Stella team. 1995–2023. *Stella: A multi-platform Atari 2600 VCS Emulator.* http://stella.sourceforge.net.

Müller, Alfred. 1997. Integral probability metrics and their generating classes of functions. *Advances in Applied Probability* 29 (2): 429–443.

Muller, Timothy H., James L. Butler, Sebastijan Veselic, Bruno Miranda, Timothy E. J. Behrens, Zeb Kurth-Nelson, and Steven W. Kennerley. 2021. Distributional reinforcement learning in prefrontal cortex. *bioRxiv 2021.06.14.448422.*

Munos, Rémi. 2003. Error bounds for approximate policy iteration. In *Proceedings of the International Conference on Machine Learning.*

Munos, Rémi, Tom Stepleton, Anna Harutyunyan, and Marc G. Bellemare. 2016. Safe and efficient off-policy reinforcement learning. In *Advances in Neural Information Processing Systems.*

Murphy, Kevin P. 2012. *Machine learning: A probabilistic perspective.* MIT Press.

Naddaf, Yavar. 2010. Game-independent AI agents for playing Atari 2600 console games. Master's thesis, University of Alberta.

Naesseth, Christian A., Fredrik Lindsten, and Thomas B. Schön. 2019. Elements of sequential Monte Carlo. *Foundations and Trends® in Machine Learning* 12 (3): 307–392.

Nair, Vinod, and Geoffrey E. Hinton. 2010. Rectified linear units improve restricted Boltzmann machines. In *Proceedings of the International Conference on Machine Learning.*

Nam, Daniel W., Younghoon Kim, and Chan Y. Park. 2021. GMAC: A distributional perspective on actor-critic framework. In *Proceedings of the International Conference on Machine Learning.*

Neininger, Ralph. 1999. Limit laws for random recursive structures and algorithms. PhD diss., University of Freiburg.

Neininger, Ralph. 2001. On a multivariate contraction method for random recursive structures with applications to Quicksort. *Random Structures & Algorithms* 19 (3–4): 498–524.

Neininger, Ralph, and Ludger Rüschendorf. 2004. A general limit theorem for recursive algorithms and combinatorial structures. *The Annals of Applied Probability* 14 (1): 378–418.

Newey, Whitney K., and James L. Powell. 1987. Asymmetric least squares estimation and testing. *Econometrica* 55 (4): 819–847.

Nguyen, Thanh Tang, Sunil Gupta, and Svetha Venkatesh. 2021. Distributional reinforcement learning via moment matching. In *Proceedings of the AAAI Conference on Artificial Intelligence.*

Nieoullon, André. 2002. Dopamine and the regulation of cognition and attention. *Progress in Neurobiology* 67 (1): 53–83.

Nikolov, Nikolay, Johannes Kirschner, Felix Berkenkamp, and Andreas Krause. 2019. Information-directed exploration for deep reinforcement learning. In *Proceedings of the International Conference on Learning Representations.*

Niv, Yael. 2009. Reinforcement learning in the brain. *Journal of Mathematical Psychology* 53 (3): 139–154.

Olah, Chris, Arvind Satyanarayan, Ian Johnson, Shan Carter, Ludwig Schubert, Katherine Ye, and Alexander Mordvintsev. 2018. The building blocks of interpretability. *Distill.*

Oliehoek, Frans A., and Christopher Amato. 2016. *A concise introduction to decentralized POMDPs.* Springer.

Oliehoek, Frans A., Matthijs T. J. Spaan, and Nikos Vlassis. 2008. Optimal and approximate Q-value functions for decentralized POMDPs. *Journal of Artificial Intelligence Research* 32 (1): 289–353.

Olsen, Ditte, Niels Wellner, Mathias Kaas, Inge E. M. de Jong, Florence Sotty, Michael Didriksen, Simon Glerup, and Anders Nykjaer. 2021. Altered dopaminergic firing pattern and novelty response underlie ADHD-like behavior of SorCS2-deficient mice. *Translational Psychiatry* 11 (1): 1–14.

Omidshafiei, Shayegan, Jason Pazis, Christopher Amato, Jonathan P. How, and John Vian. 2017. Deep decentralized multi-task multi-agent reinforcement learning under partial observability. In *Proceedings of the International Conference on Machine Learning.*

Owen, Art B. 2013. *Monte Carlo theory, methods and examples.*

Palmer, Gregory, Rahul Savani, and Karl Tuyls. 2019. Negative update intervals in deep multi-agent reinforcement learning. In *Proceedings of the International Conference on Autonomous Agents and Multiagent Systems.*

Palmer, Gregory, Karl Tuyls, Daan Bloembergen, and Rahul Savani. 2018. Lenient multi-agent deep reinforcement learning. In *Proceedings of the International Conference on Autonomous Agents and Multiagent Systems.*

Panait, Liviu, Keith Sullivan, and Sean Luke. 2006. Lenient learners in cooperative multiagent systems. In *Proceedings of the International Conference on Autonomous Agents and Multiagent Systems.*

Panait, Liviu, R. Paul Wiegand, and Sean Luke. 2003. Improving coevolutionary search for optimal multiagent behaviors. In *Proceedings of the International Joint Conference on Artificial Intelligence.*

Panaretos, Victor M., and Yoav Zemel. 2020. *An invitation to statistics in Wasserstein space.* Springer Nature.

Parr, Ronald, Lihong Li, Gavin Taylor, Christopher Painter-Wakefield, and Michael L. Littman. 2008. An analysis of linear models, linear value-function approximation, and feature selection for reinforcement learning. In *Proceedings of the International Conference on Machine Learning.*

Parr, Ronald, Christopher Painter-Wakefield, Lihong Li, and Michael Littman. 2007. Analyzing feature generation for value-function approximation. In *Proceedings of the International Conference on Machine Learning.*

Pavlov, Ivan P. 1927. *Conditioned reflexes: An investigation of the physiological activity of the cerebral cortex.* Oxford University Press.

Peres, Yuval, Wilhelm Schlag, and Boris Solomyak. 2000. Sixty years of Bernoulli convolutions. In *Fractal geometry and stochastics II,* edited by Christoph Bandt, Siegfried Graf, and Martina Zähle. Springer.

Peyré, Gabriel, and Marco Cuturi. 2019. Computational optimal transport: With applications to data science. *Foundations and Trends® in Machine Learning* 11 (5–6): 355–607.

Prashanth, L. A., and Michael Fu. 2021. Risk-sensitive reinforcement learning. *arXiv preprint arXiv:1810.09126.*

Prashanth, L. A., and Mohammad Ghavamzadeh. 2013. Actor-critic algorithms for risk-sensitive MDPs. In *Advances in Neural Information Processing Systems.*

Precup, Doina, Richard S. Sutton, and Satinder P. Singh. 2000. Eligibility traces for off-policy policy evaluation. In *Proceedings of the International Conference on Machine Learning.*

Pritzel, Alexander, Benigno Uria, Sriram Srinivasan, Adria Puigdomenech Badia, Oriol Vinyals, Demis Hassabis, Daan Wierstra, and Charles Blundell. 2017. Neural episodic control. In *Proceedings of the International Conference on Machine Learning.*

Puterman, Martin L. 2014. *Markov decision processes: Discrete stochastic dynamic programming.* John Wiley & Sons.

Puterman, Martin L., and Moon Chirl Shin. 1978. Modified policy iteration algorithms for discounted Markov decision problems. *Management Science* 24 (11): 1127–1137.

Qiu, Wei, Xinrun Wang, Runsheng Yu, Xu He, Rundong Wang, Bo An, Svetlana Obraztsova, and Zinovi Rabinovich. 2021. RMIX: Learning risk-sensitive policies for cooperative reinforcement learning agents. In *Advances in Neural Information Processing Systems.*

Qu, Chao, Shie Mannor, and Huan Xu. 2019. Nonlinear distributional gradient temporal-difference learning. In *Proceedings of the International Conference on Machine Learning.*

Quan, John, and Georg Ostrovski. 2020. *DQN Zoo: Reference implementations of DQN-based agents.* Version 1.0.0. http://github.com/deepmind/dqn_zoo.

Rachev, Svetlozar T., Lev Klebanov, Stoyan V. Stoyanov, and Frank Fabozzi. 2013. *The methods of distances in the theory of probability and statistics.* Springer Science & Business Media.

Rachev, Svetlozar T., and Ludger Rüschendorf. 1995. Probability metrics and recursive algorithms. *Advances in Applied Probability* 27 (3): 770–799.

Rashid, Tabish, Mikayel Samvelyan, Christian Schroeder de Witt, Gregory Farquhar, Jakob N. Foerster, and Shimon Whiteson. 2020. Monotonic value function factorisation for deep multi-agent reinforcement learning. *Journal of Machine Learning Research* 21 (1): 7234–7284.

Rashid, Tabish, Mikayel Samvelyan, Christian Schroeder de Witt, Gregory Farquhar, Jakob Foerster, and Shimon Whiteson. 2018. QMIX: Monotonic value function factorisation for deep multi-agent reinforcement learning. In *Proceedings of the International Conference on Machine Learning.*

Rescorla, Robert A., and Allan R. Wagner. 1972. A theory of Pavlovian conditioning: Variations in the effectiveness of reinforcement and nonreinforcement. In *Classical conditioning II: Current Research and Theory,* edited by Abraham J. Black and William F. Prosaky, 64–99. Appleton-Century-Crofts.

Riedmiller, M. 2005. Neural fitted Q iteration – first experiences with a data efficient neural reinforcement learning method. In *Proceedings of the European Conference on Machine Learning.*

Riedmiller, Martin, Thomas Gabel, Roland Hafner, and Sascha Lange. 2009. Reinforcement learning for robot soccer. *Autonomous Robots* 27 (1): 55–73.

Rizzo, Maria L., and Gábor J. Székely. 2016. Energy distance. *Wiley Interdisciplinary Reviews: Computational Statistics* 8 (1): 27–38.

Robbins, Herbert, and Sutton Monro. 1951. A stochastic approximation method. *The Annals of Mathematical Statistics* 22 (3): 400–407.

Robbins, Herbert, and David Siegmund. 1971. A convergence theorem for non negative almost supermartingales and some applications. In *Optimizing methods in statistics,* edited by Jagdish S. Rustagi, 233–257. Academic Press.

Robert, Christian, and George Casella. 2004. *Monte Carlo statistical methods.* Springer Science & Business Media.

Rockafellar, R. Tyrrell, and Stanislav Uryasev. 2000. Optimization of conditional value-at-risk. *Journal of Risk* 2:21–42.

Rockafellar, R. Tyrrell, and Stanislav Uryasev. 2002. Conditional value-at-risk for general loss distributions. *Journal of Banking & Finance* 26 (7): 1443–1471.

Rösler, Uwe. 1991. A limit theorem for "Quicksort." *RAIRO-Theoretical Informatics and Applications* 25 (1): 85–100.

Rösler, Uwe. 1992. A fixed point theorem for distributions. *Stochastic Processes and Their Applications* 42 (2): 195–214.

Rösler, Uwe. 2001. On the analysis of stochastic divide and conquer algorithms. *Algorithmica* 29 (1): 238–261.

Rösler, Uwe, and Ludger Rüschendorf. 2001. The contraction method for recursive algorithms. *Algorithmica* 29 (1–2): 3–33.

Rowland, Mark, Marc G. Bellemare, Will Dabney, Rémi Munos, and Yee Whye Teh. 2018. An analysis of categorical distributional reinforcement learning. In *Proceedings of the International Conference on Artificial Intelligence and Statistics.*

Rowland, Mark, Robert Dadashi, Saurabh Kumar, Rémi Munos, Marc G. Bellemare, and Will Dabney. 2019. Statistics and samples in distributional reinforcement learning. In *Proceedings of the International Conference on Machine Learning.*

Rowland, Mark, Shayegan Omidshafiei, Daniel Hennes, Will Dabney, Andrew Jaegle, Paul Muller, Julien Pérolat, and Karl Tuyls. 2021. Temporal difference and return optimism in cooperative multi-agent reinforcement learning. In *Adaptive and Learning Agents Workshop at the International Conference on Autonomous Agents and Multiagent Systems.*

Rubner, Yossi, Carlo Tomasi, and Leonidas J. Guibas. 1998. A metric for distributions with applications to image databases. In *Sixth International Conference on Computer Vision.*

Rudin, Walter. 1976. *Principles of mathematical analysis.* McGraw-Hill.

Rumelhart, David E., Geoffrey E. Hinton, and Ronald J. Williams. 1986. Learning representations by back-propagating errors. *Nature* 323 (6088): 533–536.

Rummery, Gavin A., and Mahesan Niranjan. 1994. *On-line Q-learning using connectionist systems.* Technical report. Cambridge University Engineering Department.

Rüschendorf, Ludger. 2006. On stochastic recursive equations of sum and max type. *Journal of Applied Probability* 43 (3): 687–703.

Rüschendorf, Ludger, and Ralph Neininger. 2006. A survey of multivariate aspects of the contraction method. *Discrete Mathematics & Theoretical Computer Science* 8:31–56.

Ruszczyński, Andrzej. 2010. Risk-averse dynamic programming for Markov decision processes. *Mathematical Programming* 125 (2): 235–261.

Samuel, Arthur L. 1959. Some studies in machine learning using the game of checkers. *IBM Journal of Research and Development* 11 (6): 601–617.

Santambrogio, Filippo. 2015. *Optimal transport for applied mathematicians: Calculus of variations, PDEs and modeling.* Birkhäuser.

Särkkä, Simo. 2013. *Bayesian filtering and smoothing.* Cambridge University Press.

Sato, Makoto, Hajime Kimura, and Shibenobu Kobayashi. 2001. TD algorithm for the variance of return and mean-variance reinforcement learning. *Transactions of the Japanese Society for Artificial Intelligence* 16 (3): 353–362.

Schaul, Tom, John Quan, Ioannis Antonoglou, and David Silver. 2016. Prioritized experience replay. In *Proceedings of the International Conference on Learning Representations.*

Scherrer, Bruno. 2010. Should one compute the temporal difference fix point or minimize the Bellman residual? The unified oblique projection view. In *Proceedings of the International Conference on Machine Learning.*

Scherrer, Bruno. 2014. Approximate policy iteration schemes: A comparison. In *Proceedings of the International Conference on Machine Learning.*

Scherrer, Bruno, and Boris Lesner. 2012. On the use of non-stationary policies for stationary infinite-horizon Markov decision processes. In *Advances in Neural Information Processing Systems.*

Schlegel, Matthew, Andrew Jacobsen, Zaheer Abbas, Andrew Patterson, Adam White, and Martha White. 2021. General value function networks. *Journal of Artificial Intelligence Research (JAIR)* 70:497–543.

Schultz, Wolfram. 1986. Responses of midbrain dopamine neurons to behavioral trigger stimuli in the monkey. *Journal of Neurophysiology* 56 (5): 1439–1461.

Schultz, Wolfram. 2002. Getting formal with dopamine and reward. *Neuron* 36 (2): 241–263.

Schultz, Wolfram. 2016. Dopamine reward prediction-error signalling: A two-component response. *Nature Reviews Neuroscience* 17 (3): 183–195.

Schultz, Wolfram, Paul Apicella, and Tomas Ljungberg. 1993. Responses of monkey dopamine neurons to reward and conditioned stimuli during successive steps of learning a delayed response task. *Journal of Neuroscience* 13 (3): 900–913.

Schultz, Wolfram, Peter Dayan, and P. Read Montague. 1997. A neural substrate of prediction and reward. *Science* 275 (5306): 1593–1599.

Schultz, Wolfram, and Ranulfo Romo. 1990. Dopamine neurons of the monkey midbrain: Contingencies of responses to stimuli eliciting immediate behavioral reactions. *Journal of Neurophysiology* 63 (3): 607–624.

Shah, Ashvin. 2012. Psychological and neuroscientific connections with reinforcement learning. In *Reinforcement learning,* edited by Marco Wiering and Martijn Otterlo, 507–537. Springer.

Shapiro, Alexander, Darinka Dentcheva, and Andrzej Ruszczynski. 2009. *Lectures on stochastic programming: Modeling and theory.* SIAM.

Shapley, Lloyd S. 1953. Stochastic games. *Proceedings of the National Academy of Sciences* 39 (10): 1095–1100.

Shen, Yun, Wilhelm Stannat, and Klaus Obermayer. 2013. Risk-sensitive Markov control processes. *SIAM Journal on Control and Optimization* 51 (5): 3652–3672.

Shoham, Yoav, and Kevin Leyton-Brown. 2009. *Multiagent systems.* Cambridge University Press.

Silver, David, Aja Huang, Chris J. Maddison, Arthur Guez, Laurent Sifre, George van den Driessche, Julian Schrittwieser, Ioannis Antonoglou, Veda Panneershelvam, Marc Lanctot, Sander Dieleman, Dominik Grewe, John Nham, Nal Kalchbrenner, Ilya Sutskever, Timothy Lillicrap, Madeleine Leach, Koray Kavukcuoglu, Thore Graepel, and Demis Hassabis. 2016. Mastering the game of Go with deep neural networks and tree search. *Nature* 529 (7587): 484–489.

Singh, Satinder P., and Richard S. Sutton. 1996. Reinforcement learning with replacing eligibility traces. *Machine Learning* 22:123–158.

Sobel, Matthew J. 1982. The variance of discounted Markov decision processes. *Journal of Applied Probability* 19 (4): 794–802.

Solomyak, Boris. 1995. On the random series $\Sigma \pm \lambda^n$ (an Erdős problem). *Annals of Mathematics* 142 (3): 611–625.

Stalnaker, Thomas A., James D. Howard, Yuji K. Takahashi, Samuel J. Gershman, Thorsten Kahnt, and Geoffrey Schoenbaum. 2019. Dopamine neuron ensembles signal the content of sensory prediction errors. *eLife* 8:e49315.

Steinbach, Marc C. 2001. Markowitz revisited: Mean-variance models in financial portfolio analysis. *SIAM Review* 43 (1): 31–85.

Strang, Gilbert. 1993. *Introduction to linear algebra.* Wellesley-Cambridge Press.

Such, Felipe Petroski, Vashisht Madhavan, Rosanne Liu, Rui Wang, Pablo Samuel Castro, Yulun Li, Ludwig Schubert, Marc G. Bellemare, Jeff Clune, and Joel Lehman. 2019. An Atari model zoo for analyzing, visualizing, and comparing deep reinforcement learning agents. In *Proceedings of the International Joint Conference on Artificial Intelligence.*

Sun, Wei-Fang, Cheng-Kuang Lee, and Chun-Yi Lee. 2021. DFAC framework: Factorizing the value function via quantile mixture for multi-agent distributional Q-learning. In *Proceedings of the International Conference on Machine Learning.*

Sunehag, Peter, Guy Lever, Audrunas Gruslys, Wojciech Marian Czarnecki, Vinicius Zambaldi, Max Jaderberg, Marc Lanctot, Nicolas Sonnerat, Joel Z. Leibo, Karl Tuyls, and Thore Graepel. 2017. Value-decomposition networks for cooperative multi-agent learning. *arXiv preprint arXiv:1706.05296.*

Sutton, Richard S. 1984. Temporal credit assignment in reinforcement learning. PhD diss., University of Massachusetts, Amherst.

Sutton, Richard S. 1988. Learning to predict by the methods of temporal differences. *Machine Learning* 3 (1): 9–44.

Sutton, Richard S. 1995. TD models: Modeling the world at a mixture of time scales. In *Proceedings of the International Conference on Machine Learning.*

Sutton, Richard S. 1996. Generalization in reinforcement learning: Successful examples using sparse coarse coding. In *Advances in Neural Information Processing Systems.*

Sutton, Richard S. 1999. Open theoretical questions in reinforcement learning. In *European Conference on Computational Learning Theory.*

Sutton, Richard S., and Andrew G. Barto. 2018. *Reinforcement learning: An introduction.* MIT Press.

Sutton, Richard S., Hamid Reza Maei, Doina Precup, Shalabh Bhatnagar, David Silver, Csaba Szepesvári, and Eric Wiewiora. 2009. Fast gradient-descent methods for temporal-difference learning with linear function approximation. In *Proceedings of the International Conference on Machine Learning.*

Sutton, Richard S., David A. McAllester, Satinder P. Singh, and Yishay Mansour. 2000. Policy gradient methods for reinforcement learning with function approximation. In *Advances in Neural Information Processing Systems.*

Sutton, Richard S., Joseph Modayil, Michael Delp, Thomas Degris, Patrick M. Pilarski, Adam White, and Doina Precup. 2011. Horde: A scalable real-time architecture for learning knowledge from unsupervised sensorimotor interaction. In *Proceedings of the International Conference on Autonomous Agents and Multiagents Systems.*

Sutton, Richard S., Doina Precup, and Satinder Singh. 1999. Between MDPs and semi-MDPs: A framework for temporal abstraction in reinforcement learning. *Artificial Intelligence* 112 (1–2): 181–211.

Sutton, Richard S., Csaba Szepesvári, and Hamid Reza Maei. 2008a. A convergent $O(n)$ temporal-difference algorithm for off-policy learning with linear function approximation. In *Advances in Neural Information Processing Systems.*

Sutton, Richard S., Csaba Szespesvári, Alborz Geramifard, and Michael Bowling. 2008b. Dyna-style planning with linear function approximation and prioritized sweeping. In *Proceedings of the Conference on Uncertainty in Artificial Intelligence.*

Székely, Gabor J. 2002. *E-statistics: The energy of statistical samples.* Technical report 02-16. Bowling Green State University, Department of Mathematics and Statistics.

Székely, Gábor J., and Maria L. Rizzo. 2013. Energy statistics: A class of statistics based on distances. *Journal of Statistical Planning and Inference* 143 (8): 1249–1272.

Szepesvári, Csaba. 1998. The asymptotic convergence-rate of Q-learning. In *Advances in Neural Information Processing Systems.*

Szepesvári, Csaba. 2010. *Algorithms for reinforcement learning.* Morgan & Claypool Publishers.

Szepesvári, Csaba. 2020. *Constrained MDPs and the reward hypothesis.* https://readingsml.blogspot.com/2020/03/constrained-mdps-and-reward-hypothesis.html. Accessed June 25, 2021.

Takahashi, Yuji K., Hannah M. Batchelor, Bing Liu, Akash Khanna, Marisela Morales, and Geoffrey Schoenbaum. 2017. Dopamine neurons respond to errors in the prediction of sensory features of expected rewards. *Neuron* 95 (6): 1395–1405.

Tamar, Aviv, Dotan Di Castro, and Shie Mannor. 2012. Policy gradients with variance related risk criteria. In *Proceedings of the International Conference on Machine Learning.*

Tamar, Aviv, Dotan Di Castro, and Shie Mannor. 2013. Temporal difference methods for the variance of the reward to go. In *Proceedings of the International Conference on Machine Learning.*

Tamar, Aviv, Dotan Di Castro, and Shie Mannor. 2016. Learning the variance of the reward-to-go. *Journal of Machine Learning Research* 17 (1): 361–396.

Tamar, Aviv, Yonatan Glassner, and Shie Mannor. 2015. Optimizing the CVaR via sampling. In *Proceedings of the AAAI Conference on Artificial Intelligence.*

Tampuu, Ardi, Tambet Matiisen, Dorian Kodelja, Ilya Kuzovkin, Kristjan Korjus, Juhan Aru, Jaan Aru, and Raul Vicente. 2017. Multiagent cooperation and competition with deep reinforcement learning. *PloS One* 12 (4): e0172395.

Tan, Ming. 1993. Multi-agent reinforcement learning: Independent vs. cooperative agents. In *Proceedings of the International Conference on Machine Learning.*

Tano, Pablo, Peter Dayan, and Alexandre Pouget. 2020. A local temporal difference code for distributional reinforcement learning. In *Advances in Neural Information Processing Systems.*

Taylor, James W. 2008. Estimating value at risk and expected shortfall using expectiles. *Journal of Financial Econometrics* 6 (2): 231–252.

Tesauro, Gerald. 1995. Temporal difference learning and TD-Gammon. *Communications of the ACM* 38 (3): 58–68.

Tessler, Chen, Guy Tennenholtz, and Shie Mannor. 2019. Distributional policy optimization: An alternative approach for continuous control. In *Advances in Neural Information Processing Systems.*

Tieleman, T., and G. Hinton. 2012. *rmsprop: Divide the gradient by a running average of its recent magnitude.* COURSERA: Neural Networks for Machine Learning.

Toussaint, Marc. 2009. Robot trajectory optimization using approximate inference. In *Proceedings of the International Conference on Machine Learning.*

Toussaint, Marc, and Amos Storkey. 2006. Probabilistic inference for solving discrete and continuous state Markov decision processes. In *Proceedings of the International Conference on Machine Learning.*

Tsitsiklis, John N. 1994. Asynchronous stochastic approximation and Q-learning. *Machine Learning* 16 (3): 185–202.

Tsitsiklis, John N. 2002. On the convergence of optimistic policy iteration. *Journal of Machine Learning Research* 3:59–72.

Tsitsiklis, John N., and Benjamin Van Roy. 1997. An analysis of temporal-difference learning with function approximation. *IEEE Transactions on Automatic Control* 42 (5): 674–690.

Tulcea, Cassius T. Ionescu. 1949. Mesures dans les espaces produits. *Atti Accademia Nazionale Lincei Rend* 8 (7): 208–211.

van den Oord, Aäron, Nal Kalchbrenner, and Koray Kavukcuoglu. 2016. Pixel recurrent neural networks. In *Proceedings of the International Conference on Machine Learning.*

van der Vaart, Aad W. 2000. *Asymptotic statistics.* Cambridge University Press.

van der Wal, Johannes. 1981. *Stochastic dynamic programming: Successive approximations and nearly optimal strategies for Markov decision processes and Markov games.* Stichting Mathematisch Centrum.

van Hasselt, Hado, Arthur Guez, Matteo Hessel, Volodymyr Mnih, and David Silver. 2016a. Learning values across many orders of magnitude. In *Advances in Neural Information Processing Systems.*

van Hasselt, Hado, Arthur Guez, and David Silver. 2016b. Deep reinforcement learning with double Q-learning. In *Proceedings of the AAAI Conference on Artificial Intelligence.*

Vaswani, Ashish, Noam Shazeer, Niki Parmar, Jakob Uszkoreit, Llion Jones, Aidan N. Gomez, Łukasz Kaiser, and Illia Polosukhin. 2017. Attention is all you need. In *Advances in Neural Information Processing Systems.*

Vecerik, Mel, Oleg Sushkov, David Barker, Thomas Rothörl, Todd Hester, and Jon Scholz. 2019. A practical approach to insertion with variable socket position using deep reinforcement learning. In *IEEE International Conference on Robotics and Automation.*

Veness, Joel, Marc G. Bellemare, Marcus Hutter, Alvin Chua, and Guillaume Desjardins. 2015. Compress and control. In *Proceedings of the AAAI Conference on Artificial Intelligence.*

Veness, Joel, Kee Siong Ng, Marcus Hutter, William T. B. Uther, and David Silver. 2011. A Monte-Carlo AIXI approximation. *Journal of Artificial Intelligence Resesearch* 40:95–142.

Vershik, A. M. 2013. Long history of the Monge-Kantorovich transportation problem. *The Mathematical Intelligencer* 35 (4): 1–9.

Vieillard, Nino, Olivier Pietquin, and Matthieu Geist. 2020. Munchausen reinforcement learning. In *Advances in Neural Information Processing Systems.*

Villani, Cédric. 2003. *Topics in optimal transportation.* Graduate Studies in Mathematics. American Mathematical Society.

Villani, Cédric. 2008. *Optimal transport: Old and new.* Springer Science & Business Media.

von Neumann, John. 1928. Zur Theorie der Gesellschaftsspiele. *Mathematische Annalen* 100 (1): 295–320.

Wainwright, Martin J., and Michael I. Jordan. 2008. Graphical models, exponential families, and variational inference. *Foundations and Trends® in Machine Learning* 1 (1–2): 1–305.

Walton, Neil. 2021. *Lecture notes on stochastic control.* Unpublished manuscript.

Wang, Ziyu, Tom Schaul, Matteo Hessel, Hado van Hasselt, Marc Lanctot, and Nando Freitas. 2016. Dueling network architectures for deep reinforcement learning. In *Proceedings of the International Conference on Machine Learning.*

Watkins, Christopher J. C. H. 1989. Learning from delayed rewards. PhD diss., King's College, Cambridge.

Watkins, Christopher J. C. H., and Peter Dayan. 1992. Q-learning. *Machine Learning* 8 (3–4): 279–292.

Weed, Jonathan, and Francis Bach. 2019. Sharp asymptotic and finite-sample rates of convergence of empirical measures in Wasserstein distance. *Bernoulli* 25 (4A): 2620–2648.

Wei, Ermo, and Sean Luke. 2016. Lenient learning in independent-learner stochastic cooperative games. *Journal of Machine Learning Research* 17 (1): 2914–2955.

Werbos, Paul J. 1982. Applications of advances in nonlinear sensitivity analysis. In *System modeling and optimization,* edited by Rudolph F. Drenick and Frank Kozin, 762–770. Springer.

White, D. J. 1988. Mean, variance, and probabilistic criteria in finite Markov decision processes: A review. *Journal of Optimization Theory and Applications* 56 (1): 1–29.

White, Martha. 2017. Unifying task specification in reinforcement learning. In *Proceedings of the International Conference on Machine Learning.*

White, Martha, and Adam White. 2016. A greedy approach to adapting the trace parameter for temporal difference learning. In *Proceedings of the International Conference on Autonomous Agents and Multiagent Systems.*

White, Norman M., and Marc Viaud. 1991. Localized intracaudate dopamine D2 receptor activation during the post-training period improves memory for visual or olfactory conditioned emotional responses in rats. *Behavioral and Neural Biology* 55 (3): 255–269.

Widrow, Bernard, and Marcian E. Hoff. 1960. Adaptive switching circuits. In *WESCON Convention Record Part IV.*

Williams, David. 1991. *Probability with martingales.* Cambridge University Press.

Wise, Roy A. 2004. Dopamine, learning and motivation. *Nature Reviews Neuroscience* 5 (6): 483–494.

Wurman, Peter R., Samuel Barrett, Kenta Kawamoto, James MacGlashan, Kaushik Subramanian, Thomas J. Walsh, Roberto Capobianco, Alisa Devlic, Franziska Eckert, Florian Fuchs, Leilani Gilpin, Piyush Khandelwal, Varun Kompella, HaoChih Lin, Patrick MacAlpine, Declan Oller, Takuma Seno, Craig Sherstan, Michael D. Thomure,

Houmehr Aghabozorgi, Leon Barrett, Rory Douglas, Dion Whitehead, Peter Dürr, Peter Stone, Michael Spranger, and Hiroaki Kitano. 2022. Outracing champion Gran Turismo drivers with deep reinforcement learning. *Nature* 602 (7896): 223–228.

Yang, Derek, Li Zhao, Zichuan Lin, Tao Qin, Jiang Bian, and Tie-Yan Liu. 2019. Fully parameterized quantile function for distributional reinforcement learning. In *Advances in Neural Information Processing Systems.*

Young, Kenny, and Tian Tian. 2019. MinAtar: An Atari-inspired testbed for thorough and reproducible reinforcement learning experiments. *arXiv preprint arXiv:1903.03176.*

Yue, Yuguang, Zhendong Wang, and Mingyuan Zhou. 2020. Implicit distributional reinforcement learning. In *Advances in Neural Information Processing Systems.*

Zhang, Shangtong, and Hengshuai Yao. 2019. QUOTA: The quantile option architecture for reinforcement learning. In *Proceedings of the AAAI Conference on Artificial Intelligence.*

Zhou, Fan, Zhoufan Zhu, Qi Kuang, and Liwen Zhang. 2021. Non-decreasing quantile function network with efficient exploration for distributional reinforcement learning. In *Proceedings of the International Joint Conference on Artificial Intelligence.*

Ziegel, Johanna F. 2016. Coherence and elicitability. *Mathematical Finance* 26 (4): 901–918.

Zolotarev, Vladimir M. 1976. Metric distances in spaces of random variables and their distributions. *Sbornik: Mathematics* 30 (3): 373–401.

Index

Adaptive Computation and Machine Learning
Francis Bach, editor

Bioinformatics: The Machine Learning Approach, Pierre Baldi and Søren Brunak, 1998

Reinforcement Learning: An Introduction, Richard S. Sutton and Andrew G. Barto, 1998

Graphical Models for Machine Learning and Digital Communication, Brendan J. Frey, 1998

Learning in Graphical Models, edited by Michael I. Jordan, 1999

Causation, Prediction, and Search, second edition, Peter Spirtes, Clark Glymour, and Richard Scheines, 2000

Principles of Data Mining, David J. Hand, Heikki Mannila, and Padhraic Smyth, 2000

Bioinformatics: The Machine Learning Approach, second edition, Pierre Baldi and Søren Brunak, 2001

Learning Kernel Classifiers: Theory and Algorithms, Ralf Herbrich, 2002

Learning with Kernels: Support Vector Machines, Regularization, Optimization, and Beyond, Bernhard Schölkopf and Alexander J. Smola, 2002

Introduction to Machine Learning, Ethem Alpaydın, 2004

Gaussian Processes for Machine Learning, Carl Edward Rasmussen and Christopher K. I. Williams, 2006

Semi-Supervised Learning, edited by Olivier Chapelle, Bernhard Schölkopf, and Alexander Zien, 2006

The Minimum Description Length Principle, Peter D. Grünwald, 2007

Introduction to Statistical Relational Learning, edited by Lise Getoor and Ben Taskar, 2007

Probabilistic Graphical Models: Principles and Techniques, Daphne Koller and Nir Friedman, 2009

Introduction to Machine Learning, second edition, Ethem Alpaydın, 2010

Machine Learning in Non-Stationary Environments: Introduction to Covariate Shift Adaptation, Masashi Sugiyama and Motoaki Kawanabe, 2012

Boosting: Foundations and Algorithms, Robert E. Schapire and Yoav Freund, 2012

Foundations of Machine Learning, Mehryar Mohri, Afshin Rostamizadeh, and Ameet Talwalker, 2012

Machine Learning: A Probabilistic Perspective, Kevin P. Murphy, 2012

Introduction to Machine Learning, third edition, Ethem Alpaydın, 2014

Deep Learning, Ian Goodfellow, Yoshua Bengio, and Aaron Courville, 2017

Elements of Causal Inference: Foundations and Learning Algorithms, Jonas Peters, Dominik Janzing, and Bernhard Schölkopf, 2017

Machine Learning for Data Streams, with Practical Examples in MOA, Albert Bifet, Ricard Gavaldà, Geoffrey Holmes, Bernhard Pfahringer, 2018

Reinforcement Learning: An Introduction, second edition, Richard S. Sutton and Andrew G. Barto, 2018

Foundations of Machine Learning, second edition, Mehryar Mohri, Afshin Rostamizadeh, and Ameet Talwalker, 2019

Introduction to Natural Language Processing, Jacob Eisenstein, 2019

Introduction to Machine Learning, fourth edition, Ethem Alpaydın, 2020

Knowledge Graphs: Fundamentals, Techniques, and Applications, Mayank Kejriwal, Craig A. Knoblock, and Pedro Szekely, 2021

Probabilistic Machine Learning: An Introduction, Kevin P. Murphy, 2022

Machine Learning from Weak Supervision: An Empirical Risk Minimization Approach, Masashi Sugiyama, Han Bao, Takashi Ishida, Nan Lu, Tomoya Sakai, and Gang Niu, 2022

Introduction to Online Convex Optimization, second edition, Elad Hazan, 2022

Distributional Reinforcement Learning, Marc G. Bellemare, Will Dabney, and Mark Rowland, 2023